175
Structure and Bonding

Aims and Scope

Structure and Bonding is a publication which uniquely bridges the journal and book format. Organized into topical volumes, the series publishes in depth and critical reviews on all topics concerning structure and bonding. With over 50 years of history, the series has developed from covering theoretical methods for simple molecules to more complex systems.

Topics addressed in the series now include the design and engineering of molecular solids such as molecular machines, surfaces, two dimensional materials, metal clusters and supramolecular species based either on complementary hydrogen bonding networks or metal coordination centers in metal-organic framework materials (MOFs). Also of interest is the study of reaction coordinates of organometallic transformations and catalytic processes, and the electronic properties of metal ions involved in important biochemical enzymatic reactions.

Volumes on physical and spectroscopic techniques used to provide insights into structural and bonding problems, as well as experimental studies associated with the development of bonding models, reactivity pathways and rates of chemical processes are also relevant for the series.

Structure and Bonding is able to contribute to the challenges of communicating the enormous amount of data now produced in contemporary research by producing volumes which summarize important developments in selected areas of current interest and provide the conceptual framework necessary to use and interpret mega-databases.

We welcome proposals for volumes in the series within the scope mentioned above. Structure and Bonding offers our authors and readers:

- OnlineFirst publication. Each chapter is published online as it is finished, ahead of the print volume
- Wide dissemination. The chapters and the volume will be available on our platform SpringerLink, one of the largest collections of scholarly content in the world. SpringerLink attracts more than 50 million users at 15.000 institutions worldwide.
- Easy manuscript preparation. Authors do not have to spend their valuable time on the layout of their contribution. Springer will take care of all the layout related issues and will provide support throughout the complete process.

More information about this series at http://www.springer.com/series/430

Luis Gómez-Hortigüela

Editor

Insights into the Chemistry of Organic Structure-Directing Agents in the Synthesis of Zeolitic Materials

With contributions by

B. Bernardo-Maestro · M.Á. Camblor · P.A. Cox ·
L. Gómez-Hortigüela · L.B. McCusker · M. Moliner ·
C. Paris · F. Rey · J. Simancas · S. Smeets · A. Turrina ·
A.E. Watts · P.A. Wright

 Springer

Editor
Luis Gómez-Hortigüela
Instituto de Catálisis y Petroleoquímica - CSIC
Madrid, Spain

ISSN 0081-5993   ISSN 1616-8550   (electronic)
Structure and Bonding
ISBN 978-3-319-74288-5   ISBN 978-3-319-74289-2   (eBook)
https://doi.org/10.1007/978-3-319-74289-2

Library of Congress Control Number: 2018935595

Printed on acid-free paper

This Springer imprint is published by the registered company Springer International Publishing AG part of Springer Nature.
The registered company address is: Gewerbestrasse 11, 6330 Cham, Switzerland

Preface

Since their first use, the addition of organic species (usually organic cations) to the synthesis gels of zeolite materials has triggered the most important revolution in the preparation and discovery of new zeolitic materials with unprecedented properties. These organic species drive the crystallization pathway toward a particular zeolite framework type with specific chemical properties, and hence they are referred to as organic structure-directing agents. During the crystallization process, these organic agents organize the inorganic units into a particular geometry from which viable nuclei of the host-guest system will be produced, which will promote the crystallization of a certain zeolitic material. The use of organic structure-directing agents has enabled the discovery of a large number of new zeolite topologies, with very different pore shapes and sizes; in addition, they have made accessible particular chemical compositions of zeolite networks. Both of these factors have greatly amplified the number of applications of this important class of materials in different sectors of the chemical industry.

Traditionally, the main properties whereby researchers designed new organic structure-directing agents focused on their size, shape, and hydrophobicity (chapter "Introduction to the Zeolite Structure-Directing Phenomenon by Organic Species: General Aspects"). However, and due to the complexity of the crystallization process of zeolite materials, a deep knowledge of the chemical role played by these organic species was not available. As a consequence, research on the use of organic structure-directing agents was mainly based on a trial-and-error basis, carrying out systematic explorations of the structure-directing ability of different organic species and synthesizing new materials, but basing the research efforts only on empirical evidence and chemical intuition. Nevertheless, one of the major aims in this research field is the rational design of organic structure-directing agents for the synthesis of a targeted zeolite material. This, which was not possible just a few years ago, is starting to become accessible thanks to the great advancement of mainly two physico-chemical techniques that have greatly enhanced our knowledge of the organic structure-directing phenomenon: (1) diffraction techniques (chapter "Location of Organic Structure-Directing Agents in Zeolites Using Diffraction

Techniques"), which now allow to find the location of the organic species within zeolite frameworks, enabling the establishment of structural relationships between zeolite topologies and the geometrical properties of the organic structure-directing agents that drive their crystallization, and (2) molecular modeling techniques (chapter "Molecular Modelling of Structure Direction Phenomena"), which have allowed to understand, from a molecular level, the structure-directing mode of action of these organic species. In addition, and based in part on the knowledge gained through these techniques, in recent years new chemical concepts are emerging in structure-direction of zeolite frameworks by organic species. In this context, the spectrum of organic compounds, which was originally almost exclusively based on quaternary ammonium cations (based on nitrogen), has been notably amplified by the use of alternative organic cations based on phosphonium or sulfonium ions (chapter "Beyond Nitrogen OSDAs"), or by organometallic compounds (chapter "Metal Complexes as Structure-Directing Agents for Zeolites and Related Micro porous Materials"), providing new structure-directing agents with novel chemical properties that can be transferred into zeolite materials. Not only the use of new types of organic compounds but also the appropriate use of intermolecular interactions that trigger a remarkable supramolecular chemistry during the structure-direction phenomenon has provided with new zeolitic materials (chapter "Role of Supramolecular Chemistry During Templating Phenomenon in Zeolite Synthesis"). The geometrical correspondence between organic structure-directing agents and zeolite frameworks has additionally enabled a strategy to tackle one of the greatest challenges in zeolite (and materials) science, the search for enantiopure chiral solids able to perform enantioselective operations (chapter "Chiral Organic Structure-Directing Agents"). All these aspects that will be covered in the different chapters of this book have notably enhanced our knowledge on the organic structure-directing phenomenon during the crystallization of zeolites, and have led to new complex materials with novel and sometimes unprecedented properties that have greatly expanded the range of applications of zeolites in different fields.

Madrid, Spain   Luis Gómez-Hortigüela

Contents

Struct Bond (2018) 175: 1–42
DOI: 10.1007/430_2017_8
© Springer International Publishing AG 2017
Published online: 10 October 2017

Introduction to the Zeolite Structure-Directing Phenomenon by Organic Species: General Aspects

Luis Gómez-Hortigüela and Miguel Á. Camblor

Abstract During the last years, a tremendous progress has been achieved in the application of new zeolite materials in many different sectors through different pioneering innovations in the field of zeolite synthesis. At the very core of the production of these new zeolite materials lies the use of organic species as structure-directing agents (SDA), which has been recognized as the most important factor to determine the zeolite product rendered after the crystallization process. These organic species organize the inorganic zeolitic units and drive the crystallization pathway towards the production of particular zeolite framework types. This structure-direction phenomenon frequently works in combination with several other factors related to the chemical composition of the synthesis gels, mainly use of fluoride, concentration (H_2O/T ratio), and presence of different heteroatoms, which are also relevant for the crystallization of particular zeolite materials. Several properties determine the structure-directing effect of these organic species, especially their molecular size and shape, hydrophobicity, rigidity vs flexibility, and hydrothermal stability. The properties of the zeolitic materials synthesized can be tuned up to a certain point through the use of rationally selected organic species with particular physico-chemical features as SDA. In this introductory chapter, we briefly review the history of the use of organic cations as SDAs, and give the fundaments of the different aspects related to this structure-direction phenomenon and factors affecting it, explaining the main properties of SDAs, providing some examples of recent uses and trends of organic SDAs, as well as the host–guest chemistry involved. In addition, we pay particular attention to the use of

L. Gómez-Hortigüela (✉)
Instituto de Catálisis y Petroleoquímica (ICP-CSIC), C/Marie Curie 2, 28049 Madrid, Spain
e-mail: lhortiguela@icp.csic.es

M.Á. Camblor (✉)
Instituto de Ciencia de Materiales de Madrid (ICMM-CSIC), C/Sor Juana Inés de la Cruz 3, 28049 Madrid, Spain
e-mail: macamblor@icmm.csic.es

imidazolium-based organic cations as SDAs because of their current relevance in the synthesis of new zeolite materials.

Keywords Host–guest • Imidazolium • Structure-directing agents • Templates • Zeolites

Contents

1   Introduction: Use of Organic Cations in the Synthesis of Zeolites

Zeolites are an important class of inorganic microporous crystalline materials whose oxide-based network is composed of corner-sharing TO_4 atoms, where T refers to a tetrahedral atom, most commonly Si and Al. Since their discovery, zeolites have found widespread applications in many different areas, especially in catalysis, adsorption/separation, and ion-exchange processes [1]. Apart from these traditional uses of zeolites that have met a tremendous economic impact on different sectors of the chemical industry, remarkably on the petrochemical sector, new uses of zeolitic materials are emerging in the last years in different applications such as in luminescence, electricity, magnetism, medicine, and microelectronics [2]. Such a wide range of applications in very different processes is directly associated with the particular characteristics of these materials: (1) Their unique microporous structure, with porous systems of channels and/or cavities of molecular dimensions which allow them to behave as molecular sieves or as shape-selective catalysts, and (2) their versatile range of compositions of the oxide networks that can build up the different frameworks, most commonly Si and Al, but also extended to a large variety of other tetrahedral atoms such as B, Ge, P, and

Ga. The oxide composition determines in a large extent important properties of the zeolite, like concentration, nature and strength of catalytic sites, and polar (hydrophilic/hydrophobic) character of the channel surface. Therefore, a careful control of these two characteristics of zeolite materials, their porous network and their chemical composition, is fundamental in order to optimize their application in many different processes.

Zeolites are found as minerals in nature. Barrer pioneered in the late 1940s the study of the synthesis of zeolite materials in the laboratory by transformation of mineral phases in the presence of different salts at very high temperature [3]. However, the fundamental breakthrough in the field was achieved by Milton, who departed from Barrer's mineralogical approach by using more reactive gels under softer conditions [4], which afforded in the late 1940s and early 1950s the discovery of over a dozen of new zeolites, including zeolites A and X [5], which rapidly met an amazing commercial success [6]. A subsequent fundamental milestone for the production of synthetic zeolites was provided in 1961 by Barrer and Denny [7], and by Kerr and Kokotailo [8], who first reported the use of organic ammonium cations in the synthesis of zeolites. The former reported the synthesis of several already known zeolites, including A and X, while the latter reported zeolite ZK-4, a silica-rich version of zeolite A [9]. So far, synthetic zeolites prepared in the presence of inorganic cations were rich in Al because of the necessity of charge-balancing the abundant presence of inorganic cations. However, due to their smaller charge-to-volume ratio (because of their larger size), the introduction of organic cations enabled the synthesis of zeolites with lower amounts of Al; this represented a significant discovery since high Si/Al ratios (higher than 10) in zeolites provide a high hydrothermal stability and strong Brønsted acidity. The successful introduction of tetramethylammonium in the synthesis of zeolites soon triggered the use of other larger tetraalkylammonium cations which were soluble in the aqueous hydrothermal synthesis media. The use of tetraethylammonium by Mobil researchers led to the discovery in 1967 of the first high-silica zeolite, the key zeolite beta (BEA), by Wadlinger, Kerr, and Rosinsky, with Si/Al ratios between 5 and 100 [10]. Soon after that, ZSM-5, another decisive zeolite for the chemical industry, was discovered in 1972 by using tetrapropylammonium [11], while a further increase in the number of carbons of the alkyl substituent to four (tetrabutylammonium) led to ZSM-11 [12]; both were multidimensional zeolites that reflected the branched structure of the organic cations used. ZSM-5 and Beta zeolites prepared by Mobil researchers represent two of the most successful acid zeolites used nowadays as industrial catalysts. On the other hand, Kerr opened up the possibility of using cyclic and more complex organic ammonium compounds (instead of tetraalkyl-ammonium cations) in the synthesis of zeolites, with his synthesis of ZK-5 by using the dimethylammonium derivative of 1,4-diazabicyclo [2]-octane, which then became a typical building block for organic ammonium cations [13]. Years later, the addition of organic amines (rather than quaternary ammonium cations) to aluminophosphate gels led to the discovery of another important class of zeolitic materials, the aluminophosphate (AlPO$_4$) family, by Wilson et al. [14, 15]. This involved not only the discovery of new framework topologies (some with no counterpart among zeolites), but also a major departure in the framework composition from the typical aluminosilicate network to a more hydrophilic aluminophosphate network.

All these new zeolite materials occluded the organic cations within their porous networks, which led to the idea that the organic cations templated in some way the synthesis of those frameworks, with the organic species directing the crystallization towards specific zeolites [16, 17]; as quoted from [16], "templating is the phenomenon occurring during either the gelation or the nucleation process whereby the organic molecule organizes oxide tetrahedra into a particular geometric topology around itself and thus provides the initial building block for a particular structure type." In this context, the authors argued that the incorporation of an organic cation resulted in a specific chemical environment which could affect and determine the nucleation process, and in turn, the crystallization of a particular zeolite framework. Currently, organic species used in the synthesis of zeolites tend to be more loosely termed as structure-directing agents (SDA).

Since then, the addition of organic molecules to the synthesis gels of zeolite materials has led to a great number of discoveries of new zeolite materials, both in terms of new (usually more hydrophobic silica-based) compositions and, especially, of new zeolite topologies [18, 19]. A wide variety of typical organic chemical reactions has been applied to produce new and increasingly more complex organic SDAs [20], such as Diels–Alder reactions [21], Michael additions [22], and enamine chemistry [23], many of which have led to new frameworks. In this introductory chapter, we will deal with general aspects of the structure-directing phenomenon, including the properties that organic species should fulfill to efficiently promote the crystallization of particular zeolite frameworks (Sect. 2), factors (not intrinsically related to the organic species) that influence or even determine the structure-directing phenomenon (Sect. 3), and the host–guest chemistry that governs the interaction between the nascent zeolite frameworks and the organic species encapsulated (Sect. 4). Finally, due to their current relevance, a special section will be devoted to the use of imidazolium-based SDAs in the synthesis of new zeolite materials (Sect. 5). This chapter will explain the fundaments of structure-direction together with examples of the most recent research trends on each particular aspect of this phenomenon. Subsequent chapters in this Volume will then deal with specific types of species and tools to study the structure-directing phenomenon.

2 Properties of Organic Structure-Directing Agents

Throughout the years, the use of organic species to control the zeolite crystallization has permitted the synthesis of a large number of new zeolite structures with unique framework topologies and chemical compositions that have produced a tremendous impact on the chemical industry as acid catalysts [24, 25], and as adsorbents in separation and purification processes [6]. As previously mentioned, the role of these organic cations during the synthesis of zeolites has been proposed as (1) *template effect*, to account for the fact that a certain structural relationship between the molecular size and shape of the SDA and that of the zeolite framework that crystallizes is observed, with such relationship occurring during the formation of the zeolite nuclei [16], (2) *structure-directing effect*, term related to the fact that the addition of these organic species leads to the crystallization of a particular framework type that would not be produced in its absence, and (3) *pore-filling effect*, in the sense that the organic SDAs end up encapsulated within the porous zeolite structure and provide stability to the open zeolite framework through the development of non-bonded host–guest interactions, as will be explained more in detail in Sect. 4. In any case, all these three effects occur at the same time with more or less definition during the structure-directing phenomenon by organic species [26], although sometimes these terms have been distinctly used to emphasize particular roles of the organic species during the structure-directing phenomenon [27]. Several properties of the organic SDAs strongly affect its structure-direction performance and the chemical and structural properties of the zeolite framework crystallized, including their size and shape, their hydrophobic/hydrophilic nature, their hydrothermal stability, and their rigidity/flexibility with the consequent possibility of a large conformational space. All these properties will in combination determine the ability of an organic species to direct more or less specifically the synthesis of a particular framework type with a characteristic porous topology.

2.1 Size and Shape

The original use of tetramethylammonium cations as SDAs led to small cavities (like the sodalite spherical cavity) with a size related to that of the guest species hosted within, which represented a first indication about the relationship between the size and shape of the organic SDA and that of the void volume of the zeolite that hosts it. However, there is not always a straightforward relationship between both. For instance, this is evidenced by the fact that tetraethylammonium (TEA) led to the discovery of zeolite beta, with a high void volume and 12-membered-rings (12MR) 3-dimensional channels, while larger tetrapropylammonium (TPA) and tetrabutylammonium (TBA) cations led to zeolites based on 10-MR channels, ZSM-5 and ZSM-11, respectively. The reason for this unexpected behavior of the relatively small TEA cations directing the formation of the very open-framework of zeolite beta was later discovered to be related to a clustering mode of action of TEA cations during structure-direction of zeolite beta: Up to six TEA cations are clustered together in the channel intersections of the BEA framework [28]. This observation indicates that packing effects, i.e., how several organic cations pack and fit among each other when occluded in porous systems, can also affect the nature of the framework that crystallizes; this is particularly important in frameworks based on channels (rather than cages) microporous systems.

Small quaternary ammonium cations tend to give clathrasil structures, with cavities linked by small windows through which the guest species cannot diffuse (hence they are considered as 0-dimensional pore systems) [18]. As soon as the size of the organic species is enlarged (at least in one direction), zeolite frameworks based on channel systems start to crystallize. Zones and coworkers found that the number of clathrasil structures crystallized with organic cations reduced dramatically while simultaneously increasing the number of open zeolite frameworks produced when the C/N ratio of the SDAs increased beyond 9 (see Fig. 3 in [29]). This was of course due to the enlargement of the organic cations that could not fit any more in the small cavities of the clathrasil structures and required larger cavities or channels to accommodate. During the structure-direction of cage-based zeolite frameworks, where packing effects are generally not important since usually only one organic cation sites within the cavities, the influence of an increase in size of the organic SDA is clearly apparent, and results in the formation of larger cavities. This can be clearly appreciated in Fig. 1, where cage-based zeolites with cavities of increasing dimensions are produced upon an increase in size of the cyclic ammonium compound. It can be observed that not only the size of the molecules and that of the void space of the resulting hosting cavities are related, but also the shape of the organic SDA clearly determines the shape of the resulting cavity. Indeed, the preferred SDA cations for the synthesis of small-pore zeolites with large cavities, materials that are important as catalysts for the methanol-to-olefin process, involve cyclic or polycyclic ammonium cations that can be hosted in such cavities and stabilize them.

This relationship between size and shape is also apparent when considering zeolite frameworks based on channel (rather than cage) systems; however, as previously mentioned, in this case packing effects can sometimes lead to unexpected behaviors, like the crystallization of zeolite beta with a small cation like TEA. There is extensive evidence in the literature that an increase in size and rigidity of the organic SDAs has resulted in the crystallization of new zeolite frameworks with larger pores [30–33]. On the other hand, not only the size but also the shape of the organic cation has a profound impact on the channel system of the zeolite that crystallizes: linear cations tend to give zeolite frameworks based on one-dimensional channels. An example of this is provided in Fig. 2 (middle), where a large rigid and linear cation gives MTW, a zeolite system based on 12-MR one-dimensional channels. In contrast, branched organic cations usually lead to multipore channel systems, with each molecular branch occupying different channels. The classical example of this is the structure-direction of tetrapropylammonium (TPA) cations towards the formation of ZSM-5 (MFI), where each of the four propyl chains site on a channel, two on the sinusoidal ones and two on the straight ones, clearly showing the relationship between the SDA molecular shape and that of the hosting zeolite framework (see Fig. 2, top). In this context, recent studies have shown that a combination of a rational design of both the molecular size (adjusted to the channel size, with alkyl linear chains being optimum for 10-MR channels and cyclic ammonium cations for 12-MR channels) and the molecular shape (with four large substituents that site in the channel intersections with

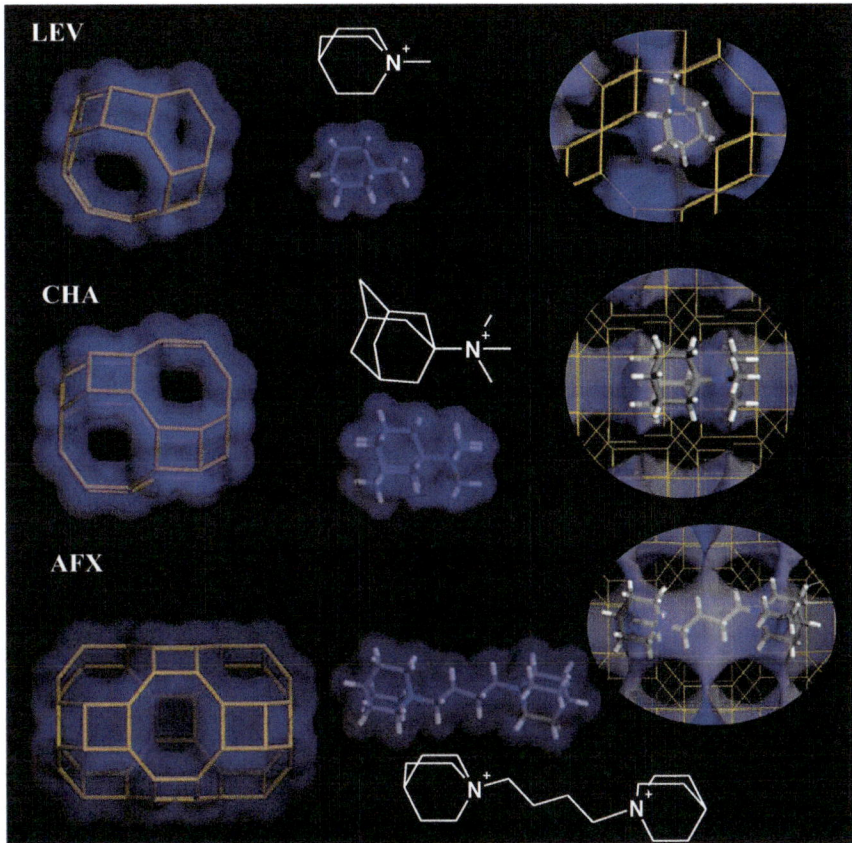

Fig. 1 Cage-based zeolite frameworks with different cage sizes synthesized with related organic cations of increasing size. Adapted from [25]

each molecular branch located on a channel) have allowed the discovery of new multipore zeolites with 12-MR and 10-MR channel systems [34–36]. This concept is illustrated in Fig. 2: The SDA used for the discovery of ITQ-39 (ITN) was based on a polycyclic rigid unit which typically directs the formation of one-dimensional 12-MR channels (like MTW, see Fig. 2, middle) and on quaternary N atoms with propyl substituents which typically site on channel intersections (like TPA in MFI zeolite, see Fig. 2, top) with propyl branches directing 10-MR channels, thus resulting in the formation of a new multipore zeolite system based on 10- and 12-MR channels. This multipore zeolites based on channels of different size can have important applications since they can enable a kind of *molecular traffic control* of reactants and products through the different channel systems [37]. This and many other examples available in the literature clearly evidence the geometrical relationship between the size and shape of the SDA species and those of the zeolite framework that crystallizes around [29, 38–43].

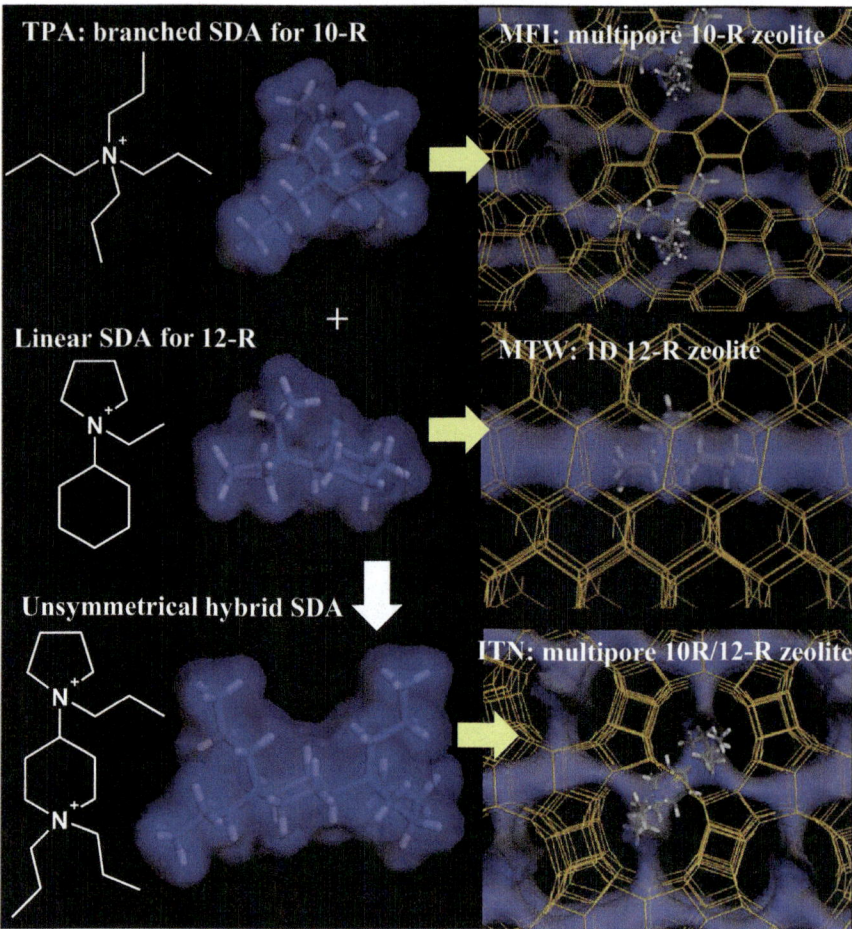

Fig. 2 Rational design of SDAs for multipore zeolites based on 10-ring and 12-ring multi-dimensional channel systems [34]. Adapted with permission from [31]; Copyright 2015 The Royal Society of Chemistry

The relevance of the molecular size and shape of the organic SDAs is nicely illustrated by the different structure-directing effect of distinct diastereoisomers (*cis/trans*, *endo/exo*, *syn/anti*, etc.). For instance, Tsuji and coworkers showed that among the different diastereoisomers of 1-benzyl-1-methyl-4-(trimethylene)-piperidinium (*trans–trans*, *cis–cis* or *trans–cis* piperidinium), only the *trans–trans* isomer was able to direct the formation of zeolite beta (BEA), while the *cis–cis* isomer led to ZSM-12 (MTW) in pure silica preparations [44] (see Fig. 3, top). In their extensive work on SDAs, Zones and coworkers have found several examples of diastereoselective structure-directing effects: For instance, the *cis*-isomer of *N*, *N*-diethyldecahydroquinolinium was selective for SSZ-48 (SFE) in borosilicate composition and SSZ-31 (STO) or SSZ-35 (STF) as aluminosilicates, while the

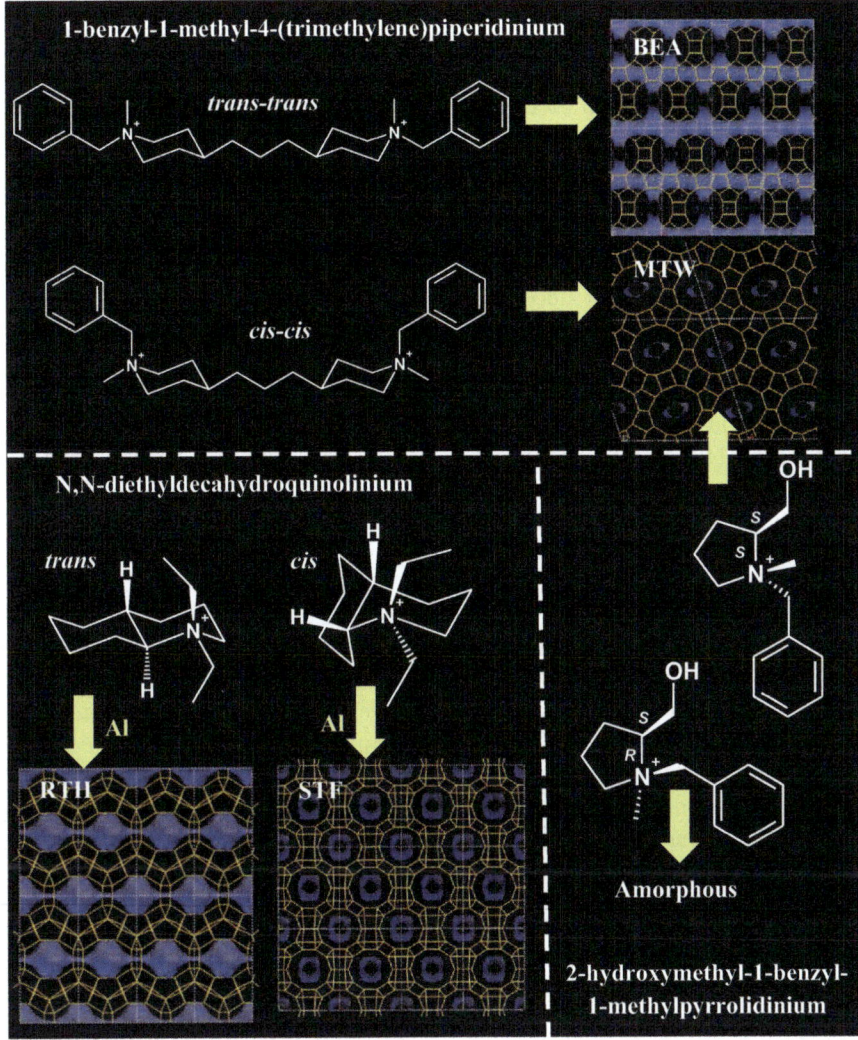

Fig. 3 Diastereoselective structure-directing effects of 1-benzyl-1-methyl-4-(trimethylene) piperidinium (*top*), *N,N*-diethyl-decahydroquinolinium (*bottom-left*) and 2-hydroxymethyl-1-benzyl-1-methylpyrrolidinium (*bottom-right*). Adapted with permission from [44], Copyright 1999 Elsevier Science B.V., and from [45], Copyright 2002 American Chemical Society

pure *trans*-isomer gave a layered silicate in borosilicate composition and SSZ-36 (RTH) or amorphous phase as aluminosilicates [39, 45] (Fig. 3, bottom-left). Another example was provided by García and coworkers, who showed that the (1*S*,2*S*) diastereoisomer of 2-hydroxymethyl-1-benzyl-1-methylpyrrolidinium cation directed the synthesis of the MTW framework, while the (1*S*,2*R*)-isomer was not able to promote the formation of such framework [46] (Fig. 3, bottom-right). Molecular simulations showed a better fit of the molecular shape of the former isomer within the 1-dimensional channels of the MTW framework, which explained the experimental observations.

A recent proposal to take advantage of the possible imprinting of the shape and size of the organic SDAs on zeolite frameworks was reported by Corma and coworkers. If it were possible to use an organic SDA that mimics the transition state of an interesting reaction, it could be presumed that the performance of the zeolite could possibly be maximized [47]. In this strategy, the systems of pores and/or cavities would be produced following the shape and chemical properties of a rationally designed SDA that mimics the transition state of a particular reaction through a molecular-recognition pattern. Because of the structure-directing phenomenon and the relationship between SDAs and zeolite hosts, this synthetic strategy would yield zeolite materials that would stabilize the transition state of such pre-established reactions, hence providing very efficient catalysts for them. Although this proposal has so far not produced any new zeolite, several known materials have been found, in retrospect, to fulfill the SDA/TS relationship and they, indeed, resulted in an improved catalytic performance. Figure 4 gives an example of one of the reactions studied, the disproportionation of toluene to give xylenes, an important reaction in the petrochemical industry. In this reaction, the mechanism goes through a carbocation transition state with a size larger than those of reactants and products (Fig. 4, top). Three different SDAs mimicking the structure of the TS were proposed, with "SDA2" already known to give a zeolite material, ITQ-27 (IWV). When loaded with Ni, this material presented a higher catalytic activity per acid site than the commercial ZSM-5-based catalyst, showing the effectiveness of this new strategy to get improved catalysts for particular reactions through the imprinting of SDA molecular features [47].

2.2   Hydrophobicity/Hydrophilicity

Molecular size and shape of the organic SDA cations is essential in determining the type of microporous system that crystallizes. However, it is generally considered that for an organic cation to effectively direct the synthesis of zeolitic materials, it requires a particular hydrophobic/hydrophilic character. On the one hand, the synthesis of zeolite materials generally takes place in aqueous systems which favor the condensation reactions between the aluminosilicate (or other heteroatoms) species. As a consequence, in order to make contact with such aluminosilicate species through which the structure-direction phenomenon occurs, the organic cations need to be soluble in the aqueous medium, and as such they require certain hydrophilicity. On the other hand, zeolite networks, especially those based on silica-rich compositions, are considered to be rather hydrophobic; consequently, again for structure-direction phenomena to take place, a close contact between the organic and the aluminosilicate species is required. Therefore, organic species need also to

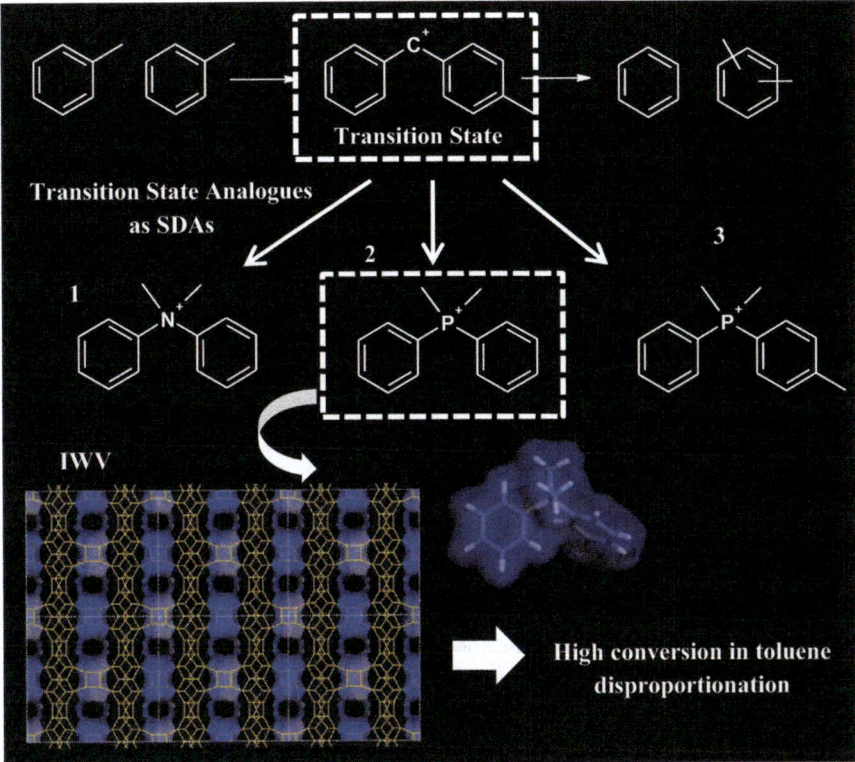

Fig. 4 *Top*: mechanism of the toluene disproportionation to produce xylenes; *middle*: organic SDA cations selected as transition state analogues for this reaction; *bottom*: IWV framework produced by TSA 2 which showed a high conversion in toluene disproportionation. Adapted with permission from [47]; Copyright 2017 American Association for the Advancement of Science

have a certain hydrophobic character in order to promote such contact through development of hydrophobic interactions. Indeed, the mechanism proposed by Burkett and Davis for structure-direction by organic cations [48–50] involved (1) a first formation of hydrophobic hydration spheres of water molecules surrounding the organic cations, on the one hand, and the inorganic oligomeric units, on the other, and (2) a subsequent overlap of the hydrophobic hydration spheres of both organic and inorganic entities, with a consequent release of water molecules, which would be driven by the development of stabilizing hydrophobic interactions between the organic and inorganic species, and establishing a close contact between them whereby structure-direction will take place through an impression of the molecular size/shape of the SDA on the nascent zeolite nuclei. Hence, a moderate hydrophobic nature is considered best for getting efficient SDA species.

In this respect, a milestone study about the optimal hydrophobicity of organic species required for producing effective SDAs is the work by Zones, Kubota, and coworkers [51]. The authors reported a simple method to estimate the hydrophobicity of an organic quaternary ammonium cation by measuring their partition (as an iodide salt) between water and chloroform. They associated such partition with their C/N ratio as an indication of hydrophobicity and related it with their efficiency in producing zeolite materials. After studying a large series of quaternary ammonium cations with different molecular structure and size, the authors found that cations with C/N ratios greater than 16 showed a large percentage of transfer to chloroform, while this decreased to almost zero as the C/N ratio was reduced below 10. Interestingly, this C/N range between 10 and 16 where a partial transfer to the organic phase is observed was coincident with the window where most of the organic cations which direct the crystallization of zeolite frameworks are found [51]. It was concluded that organic cations have to be moderately hydrophobic in order to act as efficient SDAs: Cations with C/N ratios larger than 16 are too hydrophobic and consequently they will not be soluble enough in the aqueous medium, while cations with C/N ratios smaller than 10 will be too hydrophilic and will establish very strong interactions with water molecules which will be difficult to replace with aluminosilicate species during nucleation. A moderate hydrophobicity (in practice a C/N ratio between 10 and 16) will ensure the SDA to be soluble in water but not interact too strongly with water molecules, while providing a hydrophobic character that generates favorable interactions with silica species [42]. Of course this requirement of moderate hydrophobicity reduces the availability of quaternary ammonium cations to be used as SDAs; nevertheless, recent works show that replacement of N by other heteroatoms such as sulfur may enable the use of cations with larger C/S ratios because of the higher polarity of C–S bonds. This would give large and complex sulfonium compounds with large C/S ratios soluble in water, which would be important for the production of extra-large-pore zeolites [52]. In addition, other species with different polarities have been used as SDAs, as macrocyclic ethers [53], metal complexes [54], or quaternary phosphonium ions [55], which will be discussed in subsequent chapters in this Volume.

Nevertheless, the proposed C/N limiting ratios are not absolutely exclusive. On the larger C/N side, another type of quaternary ammonium cations with different hydrophobic properties has been successfully used as SDA. Amphiphilic cations with a polar head of trimethylammonium and a long hydrophobic alkyl chain $(CH_3(CH_2)_nN(CH_3)_3)^+$ were very successful for the synthesis of mesoporous silica, acting as surfactants and inducing ordered micellar arrangements that were reproduced by the silica [56]. Moteki and coworkers have recently reported the synthesis of all-silica MFI zeolite by using cetyltrimethylammonium (CTA) as SDA (with a long-alkyl chain of 16 C atoms), with a C/N ratio of 19 and a low charge-to-volume ratio that resulted in MFI materials with low charge populations, where the CTA cations were proposed to site aligned with the straight channels (Fig. 5, middle) [57]. The authors found that the chain length of the hydrophobic tail was critical for the crystallization of a particular framework: Cations with alkyl chains of 14 and 16 had similar lengths as that of two TPA cations along the straight channels (Fig. 5, top). On the low C/N side, additionally, the intermediate hydrophobicity concept requiring C/N ratios between 10 and 16 is not hold in the case of

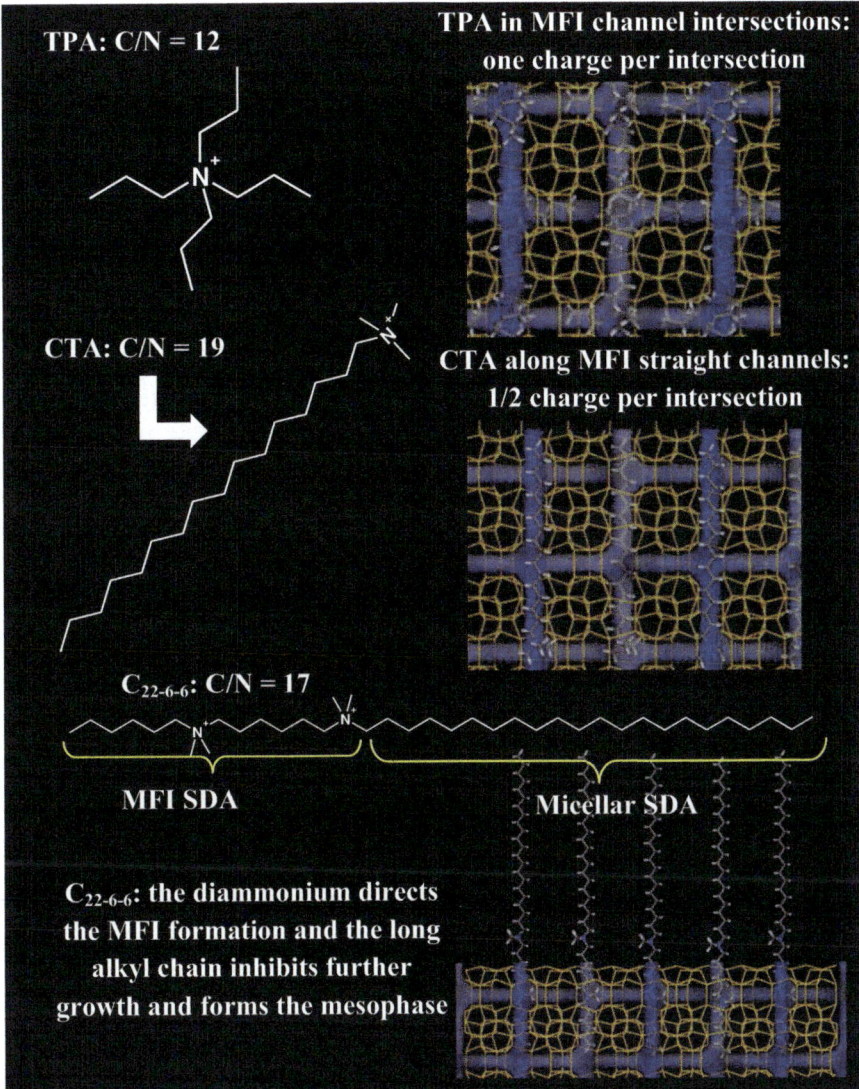

Fig. 5 SDA cations with different hydrophobicity giving place to MFI zeolites with different properties [57, 58]

the synthesis of pure silica zeolites using non-quaternary imidazolium-based cations in fluoride media, where rather hydrophilic cations with C-to-charge ratios as low as 5 work well, as will be discussed in Sect. 5 below.

An important breakthrough in the synthesis of zeolite materials was achieved by Ryoo and coworkers by using appropriately designed SDA cations with particular polarity properties [58]. The authors rationally designed bifunctional surfactants with a long hydrophobic alkyl chain with 22 C atoms on one side and two quaternary ammonium groups linked by a C6 spacer on the other which roughly corresponded to the distance spanning two consecutive MFI unit cells along the *b* axis. Each molecular side played different roles: The diammonium head groups acted as structure-directing species of the MFI framework while the long hydrophobic tails inhibited further crystal growth of the zeolite in the normal direction of the sheet by inducing the formation of a mesoscale micellar structure (Fig. 5, bottom). The rational design of these SDAs with particular hydrophilic/hydrophobic properties allowed the synthesis of ultra thin (2 nm thick) zeolite materials which showed very promising catalytic activity due to a great suppression of the catalyst deactivation through coke deposition during methanol-to-gasoline conversion. By changing the head group, this strategy has been recently extended to other zeolite structures (BEA, MTW, MRE, MWW) [59, 60] as well as to aluminophosphates (AEL, AEI, ATO) [61].

2.3 Hydrothermal Stability

One of the main conditions that SDA species must fulfill is that they display a high hydrothermal stability in order for the organic cations to resist the harsh conditions imposed by the hydrothermal treatments, involving high temperatures and pressures in aqueous systems with usually basic pHs. In this context, quaternary ammonium cations frequently suffer the Hofmann elimination reaction (to give the corresponding alkenes and tertiary amines) when subjected to high pHs and high temperatures in aqueous media, and as a consequence SDA species can be degraded into fragments, thus canceling or at least altering their structure-directing mode of action.

In order to avoid this, the use of phosphonium cations as SDAs for the synthesis of zeolite materials has been recently introduced [55]. These cations do not suffer the Hofmann degradation, and as a consequence they enable the synthesis to be performed under more severe crystallization conditions. In fact, the use of these phosphonium cations has enabled the synthesis of new zeolite frameworks [55, 62]. This will be explained more in detail in [63].

Degradation of the organic SDAs during the crystallization of zeolites may in occasions disturb the observed structure-direction not only as a result of the decreased concentration of the SDA, but also because of the possible formation of new organic species with a structure-directing ability of their own. An extreme case was recently published, where the starting *N,N,N*-trimethyl-tert-butylammonium (TMTBA) reacted with its own Hofmann degradation product (trimethylamine) yielding a smaller tetramethylammonium cation (TMA) [64]. Despite the largely different size and shape of the original and the newly produced cations, both have been shown to structure-direct the crystallization of AST zeolites in fluoride medium [65, 66]. However, the unit cell sizes of the produced AST zeolites were largely different, resulting in their phase segregation. Quite amazingly, the initial crystallization of a TMTBA-containing

AST zeolite was followed by its gradual and finally complete replacement by another AST zeolite with the same framework composition but containing the smaller TMA and displaying a smaller unit cell.

Degradation and rearrangement of organic cations to finally produce a smaller cation may be more frequent than thought. Other examples include: (1) The degradation of certain diquats resulting in the crystallization of TMA-containing phases (AST, RUT, and MTN [67] or Omega [68]); (2) crystallization of TMA-containing AST when using the large N,N,N-trimethylbenzylammonium as SDA (Rojas A and Camblor MA, Unpublished results, 2011); (3) cocrystallization of ITW together with MTW when using 1-benzyl-2,3-dimethylimidazolium, obviously too large to fit into ITW cages, by degradation and rearrangement to produce 1,2,3-trimethylimidazolium [69]; (4) crystallization of CHA phases directed by 1,2-dimethylimidazolium formed by degradation of 1-alkyl-3-methylimidazolium cations during the ionothermal synthesis of aluminophosphates [70]; (5) crystallization of TMA-containing MTN from Hofmann degradation and rearrangement of ($1R,2S$)-dimethylephedrinium (Gómez-Hortigüela L and Gálvez P, Unpublished results); and (6) the use of unstable deep-eutectic mixtures as both solvent and SDA-delivering agent [71].

The stability of the SDA is also relevant because of the need to eliminate it from the zeolite prior to its use. Due to their usually large size, removal of these organics is typically performed by calcination at high temperature in air (or in ozone whose higher oxidizing power enables lower calcination temperatures); such high-temperature treatment and the simultaneous release of water can sometimes lead to a degradation of the zeolite framework. The difficult elimination of the SDAs together with the usually high cost of these organic species, which usually represent the most expensive components of a zeolite synthesis mixture, led Davis and coworkers to devise a new family of SDA species that have the ability to be degraded into simpler organic fragments within the zeolite structures. These fragments can then be easily extracted from the zeolite channels, avoiding the necessity of high-temperature calcination treatments [72–74]. In addition, the extraction of these fragments would enable their recycle and reuse in subsequent synthesis after a reassembly into the original SDA, thus improving the cost-efficiency of zeolite syntheses. One such type of organic species is represented by ketal-containing ammonium compounds, which are stable at the typical high pHs of zeolite synthesis (in alkaline medium), and can then be degraded into ketones and diols at low pHs after the crystallization process. As reported by the authors, this strategy performs best when used in combination with unspecific pore-filling agents easy to be extracted first, thus providing the void space required for H^+ to reach the degradable SDA cations.

2.4   Rigidity vs Flexibility

Another crucial property which determines the mode of action of the organic cations during structure-direction of zeolite frameworks is the molecular rigidity/

flexibility. Again this concept was nicely illustrated in the seminal work by Kubota and coworkers [51], where they studied a large number of organic cations with different rigidities (expressed as the number of tertiary and quaternary connectivities in the molecular structure and the number of rings). The authors found that bulky and rigid cations containing one or more rings with limited conformational flexibility are best candidates for the crystallization of unique new large-pore zeolite frameworks. Figure 6 (top) shows three examples of polycyclic rigid ammonium cations of increasing size that have led to zeolite frameworks of increasing pore dimensions [19]. There is generally a common agreement between zeolite researchers that rigid SDAs are more selective towards the formation of a particular framework type due to its restricted conformational variability. In contrast, flexible cations with a large number of methylene groups enable a wide conformational flexibility, and any of those conformations could in principle be hosted by different pore/cage systems of zeolite structures. Hence, flexible cations are usually considered as less selective SDAs since they can lead to different frameworks depending on the particular conformation; this can be considered a disadvantage since frequently mixture of phases are obtained. In contrast, such conformational variability brings a higher probability of finding appropriate crystallization conditions where they direct the formation of a zeolite material. This is illustrated by the use of hexamethonium cations for the synthesis of zeolite frameworks with very different framework densities and pore systems as a function of the synthesis conditions (mainly the Si/Ge ratio) [19] (Fig. 6, bottom).

Flexibility in the SDA cations enables the occurrence of different conformers, each with a particular ability to promote the crystallization of zeolite frameworks. In this context, widely studied is the case of tetraethylammonium cations which can adopt two different conformations: *tt.tt* (all-trans), where the conformer has a disk-like shape, and *tg.tg* (trans-gauche), where the conformer has a tetrahedral shape (Fig. 7, top). The occurrence of both conformations can be easily monitored by Raman spectroscopy [78]. Indeed, O'Brien and coworkers found a different conformational behavior of tetraethylammonium (TEA) cations in the synthesis of microporous aluminophosphates as a function of the doping level of the gels: High doping levels (of Zn in this case) favored the *tg.tg* conformation and led to the crystallization of ZnAPO-34 (CHA), while the lack of dopants promoted instead the *tt.tt* conformer which led to AlPO-5 (AFI) (Fig. 7, top) [75]. Ikuno and coworkers also studied the crystallization of zeolite beta in the presence of TEA, where the mode of structure-direction of these cations had been shown to occur as molecular clusters of up to six cations sited in the channel intersections. It was observed that the crystallization of zeolite beta required the TEA cation to adopt the *tt.tt* conformation for the formation of viable nuclei, possibly because of a better packing of the clustered cations in the channel intersections [28]. Moreover, very recently Schmidt and coworkers have studied the template–framework interactions of TEA in different zeolite frameworks and related it to the conformation adopted by the SDA in the different structures [79]. They found that the conformation of the TEA cations depends not only on the framework geometry but also on the material composition. The authors found two regimes for the occlusion of the particular

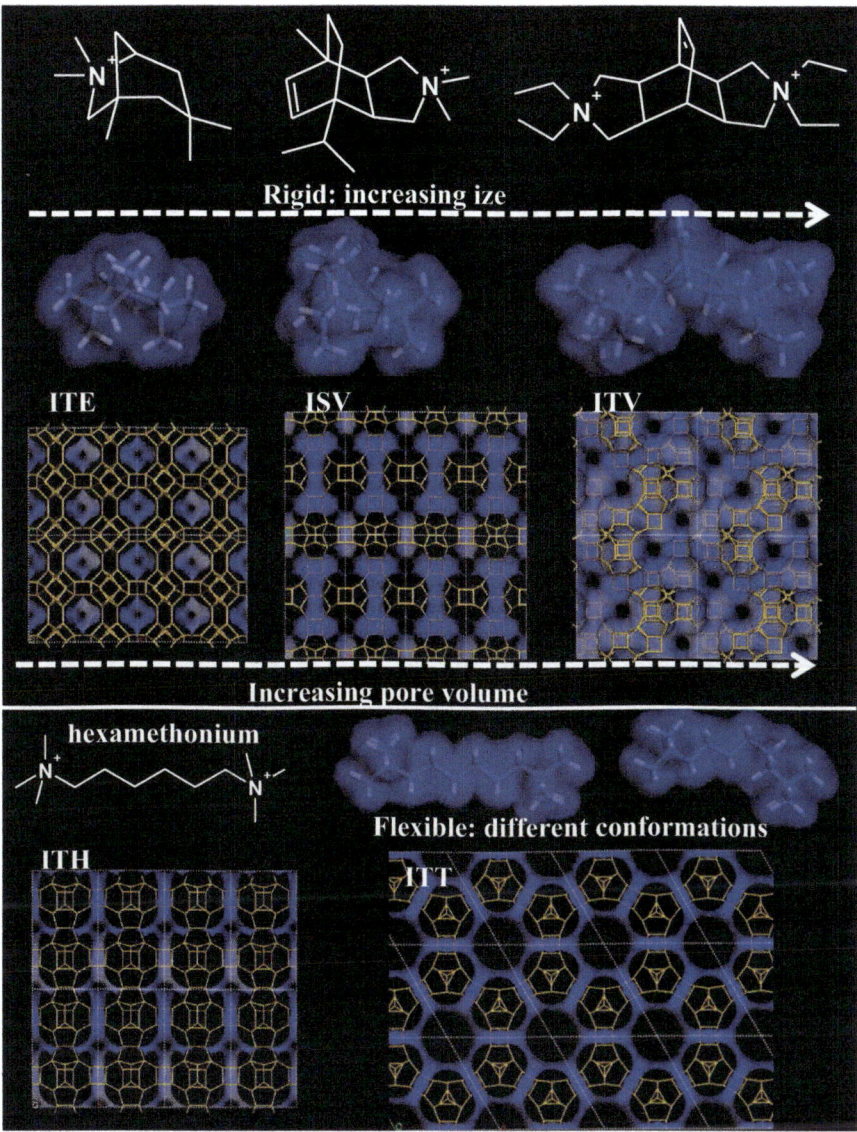

Fig. 6 SDA cations with different rigidity/flexibility properties: *top*: rigid SDAs giving selectively frameworks of larger pore volume; *bottom*: flexible hexamethonium giving different frameworks as a function of Si/Ge ratio. Adapted with permission from [19], Copyright 2013 WILEY-VCH Verlag GmbH & Co. KGaA, Weinheim

conformers: In frameworks where there is a high energy difference upon the confinement of the two conformations, only the most stable conformer is experimentally observed; in contrast, in frameworks where such energy difference is small, the occurrence of the different conformers depends on the composition of the

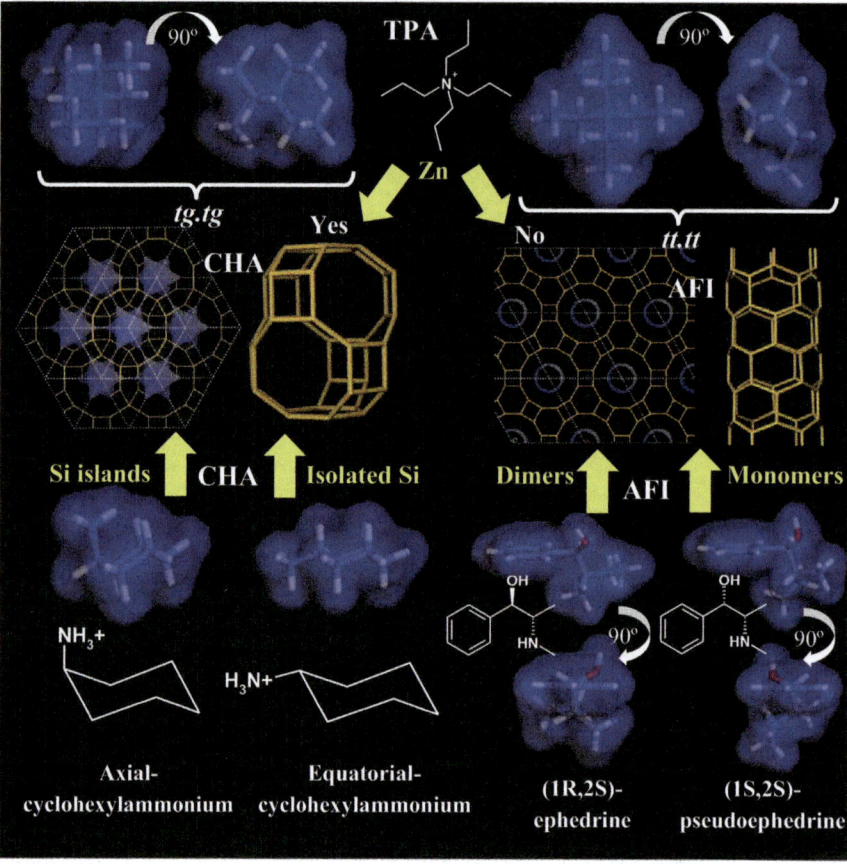

Fig. 7 Effect of conformational space as a consequence of molecular flexibility on SDAs: *tg.tg/tt. tt* conformers of TEA (*top*), axial and equatorial conformers of cyclohexylammonium (*bottom-left*) and conformers of ephedrinium and pseudoephedrinium (*bottom-right*) [75–77]

network. A similar dependence on the presence of heteroatoms had been previously observed by Sánchez and coworkers for the crystallization of ZnAPSO-44 (CHA) in the presence of cyclohexylamine as SDA (Fig. 7, bottom-left) [76]. By preparing materials with varying Zn and Si concentrations, the authors found that cyclohexylammonium in equatorial configuration was the only conformer found in materials where Si^{4+} (and/or Zn^{2+}) was incorporated as isolated species, while the presence of Si islands (at high Si concentrations) involved the occlusion of axial-cyclohexylammonium. This was ascribed to different hydrophilic/hydrophobic interactions as a function of the Si domains.

The high flexibility of certain SDAs has been proposed as a strategy for the discovery of new zeolites, if the conformation of the organics can be controlled, for instance, through the inorganic components of the synthesis mixture. The approach has been followed by Hong's group, after realizing that the pentamethonium diquat (i.e., N,N,N,N',N',N'-hexamethylpentanediammonium) can direct the crystallization of up to five different zeolites (EUO, *MRE, MTW, MWW, and MOR) depending on the Al content and concentration of alkali cations, which results in a different conformation and host–guest interaction in each material [68]. When the flexible 1,4-bis(N-methylpyrrolidinium)butane (1,4-MPB) was used and the Al and Na^+ concentration were systematically varied, up to nine different zeolites were obtained [80]. Interestingly, this included, in a very narrow Al, Na compositional window, the new zeolite TNU-9, a medium pore zeolite with a complex structure [80], as well as TNU-10, a high-silica version of the natural zeolite stilbite [82].

Conformational flexibility has been found also relevant to determine the supramolecular chemistry associated with SDAs during structure-direction. In a series of works, Gómez-Hortigüela and coworkers found a strongly different supramolecular behavior of two closely related diastereoisomers, ($1R,2S$)-ephedrine and ($1S,2S$)-pseudoephedrine, during the crystallization of AFI materials (Fig. 7, bottom-right) [77, 83]. Through a combination of fluorescence and molecular simulations, the authors observed that the ($1R,2S$)-isomer showed a stronger trend to form supramolecular aggregates through π–π stacking interactions in aqueous solution because of its particular most stable conformation with the alkyl chain in an extended configuration. In contrast, the ($1S,2S$)-isomer showed a poorer supramolecular behavior because its different absolute configuration on C1 involved a distinct intramolecular chemistry driven by H-bond interactions, resulting in a stable conformation with the alkyl chain in a folded configuration which provoked a steric hindrance upon the approach of another molecule to form a supramolecular dimer. Again the presence of methylene units and a large conformational space can dramatically affect not only the selectivity of the structure-directing phenomenon but also its mode of action, having important implications in the zeolite materials rendered.

3 Factors Affecting the Structure-Directing Phenomenon

In Sect. 2 we have explained the different physico-chemical properties of the organic SDA cations that determine their structure-direction efficiency and the zeolite product obtained. However, the outcome of a zeolite synthesis is not only affected by the organic cation used as SDA, but many other concomitant factors can also alter the structure-direction phenomenon of a particular organic cation [84]. As a consequence, the use of a SDA in a zeolite synthesis does not necessarily lead to a unique zeolite framework where the host–guest interactions are maximized, but very often one single organic cation can lead to different zeolite materials depending on the synthesis conditions. The most common of these variables affecting the structure-directing phenomenon will be briefly explained in this section.

3.1   Synthesis in Fluoride Media

The synthesis of zeolite materials was traditionally performed at high pH, usually higher than 10, where the hydroxide anion may act as a mineralizer (i.e., a catalysts for the breaking and formation of T–O bonds, T = Si, Al, etc.) favoring the dissolution of the silica and alumina as mobile species into the liquid phase, and allowing the condensation reactions to occur, hence enabling the crystallization process to proceed. However, Flanigen and coworkers first introduced the use of fluoride in the synthesis of zeolites to make pure silica MFI [85], and then systematic work by Guth and coworkers showed that fluoride anions can also play a mineralizing role during the synthesis of zeolites, replacing in this way the hydroxide anions and allowing the synthesis to be carried out at almost-neutral pHs [86–88]. The introduction of the fluoride route enabled a wide range of pHs to operate, affording the use of organic cations which are not stable at high pHs because of the Hofmann elimination reaction. Nonetheless, this synthetic route usually required higher SDA concentrations (generally SDA/SiO$_2$ ratio of 0.5).

Interestingly, this new synthesis method, which is particularly well suited for the synthesis of pure silica zeolites, renders zeolite materials with a very low concentration of connectivity defects since fluoride is usually incorporated within the zeolite materials to charge-balance for the positive charge of the organic cations (instead of siloxy groups in high-silica zeolites prepared by the hydroxide route) [89]; hence, the absence of silanol/siloxy groups makes the resulting material rather hydrophobic [90], modifying its catalytic properties when chemical species of much different polarity are involved [91] and even opening new applications [92]. Additionally, fluoride has been proposed to play a structure-directing role [87], favoring the formation of double 4-rings (D4R), with fluoride usually siting within these cubic units, and hence leading frequently to frameworks rich in such units, as long as the overall zeolite topology can be directed by the organic SDA in question [88, 93]. One argument in favor of such structure-direction by fluoride is that pure or very high-silica zeolites with D4R units have never been synthesized without using this anion during the crystallization. As shown by Zicovich-Wilson and coworkers, the effect is due to an ionization of the Si–O bond induced by occluded fluoride, which causes an increase in the flexibility of the framework that decreases the strain associated with siliceous D4R units and make this kind of zeolites reachable for crystallization [94, 95].

3.2   Effect of Concentration (Amount of Water)

The introduction of fluoride in the synthesis gels allowed for carrying out the zeolite synthesis at much higher concentrations (with very low amounts of water, reaching the reactant rather than solvent level). Synthesis in fluoride media was initially performed with the typical relatively high water contents used in hydroxide media

(H_2O/SiO_2 ratios between 30 and 60) [39]. Under these conditions mostly already known zeolites were synthesized. However, when Camblor and coworkers studied the use of highly concentrated gels for zeolite synthesis by the fluoride route (typically H_2O/SiO_2 ratios below 15 and more typically below 7), a number of new zeolites were discovered. This work revealed an interesting trend in the synthesis of zeolite structures related to the amount of water content in the gel in the presence of fluoride [96]: Phases with lower framework densities are favored at lower water/silica ratios (higher concentration) under the same SDA species [97]. This experimental trend, which was later termed the *Villaescusa rule*, has served as a strategy for the synthesis of new open zeolites [84, 93]. While other groups have also observed and confirmed the same trend [29, 98–100], the Villaescusa rule still lacks a convincing explanation. The rule is clearly illustrated by the structure-directing effect of trimethyladamantammonium, which yielded zeolite structures with very different pore volumes under different H_2O/SiO_2 ratios: *STO at high water amounts ($H_2O/SiO_2 = 15$, with a pore volume of 0.11 cm^3/g and a framework density of 18.7 T/1,000 $Å^3$), STT at intermediate ratios ($H_2O/SiO_2 = 7.5$, with a pore volume of 0.23 cm^3/g and a framework density of 17.0 T/1,000 $Å^3$), and CHA at low water amounts ($H_2O/SiO_2 = 3.0$, with a pore volume of 0.32 cm^3/g and a framework density of 15.1 T/1,000 $Å^3$) (see Fig. 8) [97, 98, 101]. In fact, zeolite materials prepared at high concentrations typically give the highest fluoride and SDA uptakes. Therefore, the amount of water in the synthesis gel can clearly alter the structure-directing mode of action of a particular organic molecule, and has to be taken into account depending on the type of zeolite framework that is targeted.

3.3 Heteroatom Substitution

There exist many reviews on the effect of the introduction of heteroatoms (elements other than Si, Al, or P) in tetrahedral positions of zeolite networks [31, 38], and it is beyond our intention to review this topic here. However, we want to at least mention it since the presence of these heteroatoms, such as Ge, Ga, B, Zn, Be, and Fe, can modify the structure-directing effect of the organic species towards certain zeolite frameworks by stabilizing particular building units. For instance, B, Ga, Be, and Zn may show some tendency to direct the formation of 3-rings, while Ge favors the formation of D4Rs and D3Rs. The isomorphic substitution of divalent or trivalent atoms for Si leads to negatively charged networks which need to be balanced by the positive charge of the organic cations, and this can affect the incorporation of the SDA species and as a consequence their structure-directing mode of action. On the other hand, the optimum T–O bond lengths and T–O–T bond angles can be different for each T atom, and this can have an implication on the type of structural units that are preferred. For example, it is known that Si-O-B and Si-O-Al angles have a narrower range of variation in zeolite networks than Si-O-Si angles, and in fact those Si-O-T angles tend to be smaller than the

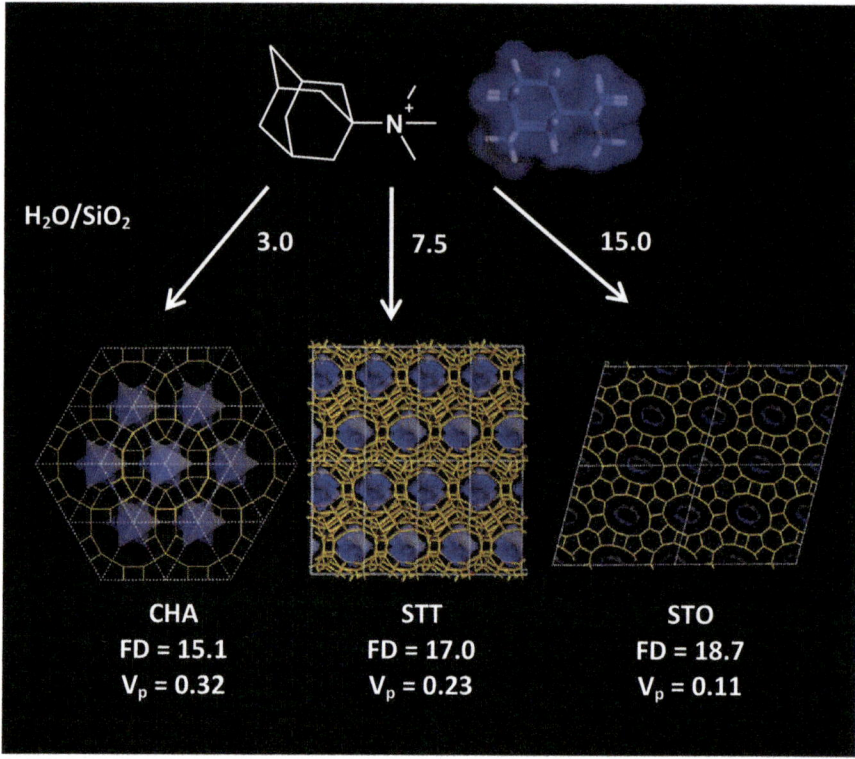

Fig. 8 Effect of H_2O/SiO_2 ratio on the structure-directing effect of trimethyladammantammonium in the synthesis of zeolites. Adapted with permission from [96], Copyright 1999 Kluwer Academic Publishers, and from [98], Copyright 2005 American Chemical Society

corresponding Si-O-Si ones [102]. On the other hand, the incorporation of Ge in the synthesis gels together with the use of novel organic cations has led to a number of new zeolite structures, which very frequently display D4R units characteristic of Ge [38]. Typical Ge-O-Ge angles are smaller than their Si counterparts, and hence its presence can relax the geometric constraints imposed by certain structural units, such as D4Rs and D3Rs [30, 31]. Again the stabilization of these otherwise unstable units can affect the structure-directing role of the organic cations towards particular framework types containing these units, as long as the overall framework can host the SDA species.

4   Host/Guest Interactions During Structure-Direction

The mode of structure-direction of the organic cations to direct the crystallization of particular framework types is driven by the development of host–guest interactions between both species during the nucleation and crystal-growth processes, which results in both kinetic and thermodynamic effects. From the kinetic side, these non-bonded (hydrophobic) interactions are responsible for the silica species to displace the hydrophobic hydration spheres around the organic cations and form the initial organo-inorganic precursors from which crystallization of a particular framework type will occur, as was discussed by Burket and Davis [48–50]. In fact, these interactions are crucial for the structure-direction phenomenon to occur and for the resultant geometric relationship between the size and shape of the organic cations and the void space of the zeolite framework which emerges. In this context, electrostatic interactions between the cationic SDAs and negatively charged inorganic species are vital for keeping these organo-inorganic units assembled long enough for nucleation and crystal growth processes to occur [103]: Electrostatic interactions allow for SDA cations to be adsorbed on the surface of silica fragments and provide in this way a protective shielding of these intrinsically unstable silica units against hydrolysis [104].

The occlusion of SDA species in zeolite frameworks has also a crucial thermodynamic implication for the stabilization of zeolite materials. It is known that the intrinsic stability of silica-based polymorphs increases with the framework density: Quartz is the most stable polymorph of silica, and the stability is reduced when the porosity of the zeolite structure increases, as has been nicely shown by Piccione and coworkers from calorimetric studies [105, 106]. From this point of view, it would be difficult to synthesize zeolite materials from thermodynamic arguments. However, porosity enables the incorporation of guest species, in this case the SDA cations, which may establish strong non-bonded interactions with the nascent zeolite frameworks, providing a relatively strong stabilization to the overall host–guest systems, making the formation of host–guest zeolite frameworks thermodynamically viable [26]. In any case, the energy differences observed in the thermodynamics of different zeolite host–guest systems suggest that the self-assembly process is dependent on a delicate interplay between a large number of weak interactions [30, 105], and sometimes it is difficult to predict which one will dominate the final outcome. In this line, Khan and coworkers have recently shown through Monte-Carlo modeling techniques that the strength of the host–guest interactions is crucial on the type of porous material that is directed: Very strong interactions (>1.2 kcal/mol SDA-oxygen contacts) would lead to 2-dimensional layered materials due to the higher optimization of host–guest attractions in open layered systems. Negligible host–guest interactions would also lead to 2-dimensional materials, while 3-dimensional microporous zeolitic materials are only produced for a relatively narrow range of medium–strength interactions [107]. This might explain the frequent crystallization of low-dimensional AlPO frameworks in the presence of amines because of the abundant presence of very strong H-bond interactions between protonated amines and PO_x^- terminating groups [108].

Two different types of non-bonded forces are established in host–guest SDA-zeolite systems: Van der Waals and electrostatic interactions. The former are short-ranged, and hence very sensitive to the distance between the interacting atoms. Van der Waals interactions relate to the size and shape of the organic species and that of the zeolite structure which crystallizes around, and as a consequence are mainly responsible for the structure-directing effect in terms of the geometric relationship between hosts (their porous system) and guests (SDAs). In this way, van der Waals interactions determine how good the fit between the organic SDAs and the zeolite framework that crystallizes around is. For this reason, calculation of this type of interactions between SDAs and particular zeolite frameworks by molecular modeling techniques has been very useful to predict the ability of the organic species to direct the formation of particular zeolite frameworks: The higher these interactions, the better the host–guest adjustment, and hence the higher the ability of such particular host–guest system to be produced [109]. Originally these modeling techniques allowed for rational explanations of experimentally observed structure-directing trends of organic species, but nowadays these can also be used to a priori predict the best SDA cation to direct the crystallization of a particular framework type [110, 111]. In addition, packing interactions, i.e., interactions between the organic cations when occluded in the confined environments of zeolite frameworks, are also relevant for the structure-directing phenomenon, and determine how the SDA cations pack within zeolite void volumes as a function of their size/shape and electrostatics, driven by steric and repulsive electrostatic interactions. Such packing interactions are particularly important in channel-based zeolite systems where the SDA cations locate close to each other without zeolite walls separating them (as occurs instead in cage-based systems).

Electrostatic interactions are mainly due to the interactions established between the positive charge of the cationic organic SDA species and the negative charge of the zeolite frameworks, which can be connectivity defects (Si-O$^-$ groups), inorganic anions like fluoride incorporated in the frameworks, or low-valent dopants substituting for Si^{4+} (typically Al^{3+} or B^{3+} replacing Si^{4+} or divalent M^{2+} replacing Al^{3+} in AlPO networks). In order to maximize these interactions, it is reasonable to expect that the negative charge associated with the inorganic network will locate as close as possible to the positive charge of the SDA cations. Indeed, recent studies have shown that connectivity defects (SiO$^-$) tend to locate close to the molecular charge center, unless steric shielding avoids it [112, 113]. Such minimization of the $SDA^+ \cdot (zeolite)^-$ distance between charged entities to maximize the electrostatic interactions has been used by us and others in order to direct the spatial incorporation of Al in zeolite networks in particular framework positions by rationally modifying the SDAs used in the synthesis [114–116], or to promote the development of chiral spatial distributions of dopants within achiral zeolite networks [117]. In another work, Lemishko and coworkers have found a competition between the electrostatic interactions of fluoride and P-containing SDA cationic species, which are characterized by a high concentration of positive charge on the phosphorus atoms, and short-ranged van der Waals SDA···zeolite interactions as a function of the molecular structure [118]. The predominance of one or the other type of interaction will determine the mode of action of the structure-directing phenomenon; this is particularly important in phosphonium-based SDAs due to the higher concentration of positive charge on P compared to their N counterparts.

H-bond interactions can also take place if the organic SDA species have H-bond donor groups, as is the case for amine-based SDAs (rather than quaternary ammonium compounds). In our group we have shown that these H-bond interactions determine the supramolecular aggregation of chiral amines [77, 119]. H-bond interactions have been shown to be pivotal for a transfer of chirality from organic chiral species into zeolite frameworks to occur [120], as will be dealt in [121]. Even without amines or alcohol groups, strong C–H···O–Si H-bonds may be stablished and have a role in the structure-direction of zeolites, as shown by Behrens et al. in the case of nonasil synthesized using cobaltocenium and fluoride, where the cation resides fixed in a cage and rotational disorder only occurs at temperatures above the synthesis temperature [122].

Very recently a new type of host–guest interaction has been identified in the synthesis of silicoaluminophosphate molecular sieves by using quaternary ammonium compounds with hydroxyethyl branches [123, 124]. In these SDAs, the terminal hydroxyl moieties display a strong tendency to form complexes with Al through coordination bonds in octahedral geometry, which leads to the formation of Al-O-Al linkages that are retained in the final zeolitic framework and even after calcination. The calcined materials may thus be interpreted as extensively violating the Loewenstein rule (Al-O-Al avoidance in tetrahedral frameworks), although in the as-made materials both paired Al have, in fact, octahedral coordination. The discovery, anyway, leads to a new family of microporous materials with potentially new structural units and frameworks. The authors found that there seems to be an optimal degree of hydrophilicity of the SDA involved to promote the formation of these framework-bound SDA-containing molecular sieves.

Another kind of intermolecular interactions that have been exploited during the structure-directing phenomenon is hydrophobic π–π type interactions between aromatic SDA cations. Such particular interactions have been used to promote the formation of large supramolecular aggregates through π–π stacking of the aromatic rings, leading to large structure-directing entities that will drive the crystallization of zeolite frameworks with large pore volumes [125–127]. This will be analyzed in [128], and hence no further details will be given here.

Hydrophobic interactions are also responsible for the crystallization of extra-large pore zeolites using small and flexible diammonium cations such as hexamethonium, which gives a number of different zeolite structures depending on the synthesis conditions (see Fig. 6 above). These structures have very different porous systems, and hence these SDA dications have to pack in a different fashion within each pore system. These relatively small cations can direct the formation of extra-large pore zeolites like ITQ-33 (ITT) [129]; however, they do not form large micellar arrangements as surfactants do, neither behave like rigid and very large SDA species that direct large pore zeolite frameworks (like those in Fig. 6, top). Therefore, hydrophobic interactions between the aliphatic spacer links of different hexamethonium cations have to be established to efficiently pack these molecules in extra-large pore zeolites and fill up the large void-volume in a cooperative fashion [30].

5   Imidazolium-Based Cations as Structure-Directing Agents

So far in this chapter we have dealt with the properties that organic SDA species have to fulfill in order to efficiently direct the crystallization of zeolite materials, as well as the variables that can alter their mode of action and the host–guest chemistry involved in this phenomenon. Subsequent chapters of this Volume will be devoted to the different types of SDA species and trends in structure-direction, as well as to different tools that are available to understand the structure-directing phenomenon. This last section of the introductory chapter will analyze the use of imidazolium-based cations as SDAs in the synthesis of silica-based zeolite materials. This is, on one hand, because imidazolium cations are attracting an increasing interest in recent years and, on the other hand, because they may well exemplify several of the concepts outlined above.

An imidazolium is an aromatic cation derived from the imidazole ring by sub-stitution reactions in which the N atoms act as the nucleophile. They are not "quaternary" cations, since all the five atoms that build the charged ring have sp^2 hybridization and, thus, cannot have four substituents. The ring itself is consider-ably rigid, small, and planar, while substituents in the ring increase the size and may also increase the flexibility. They are scientifically and technologically relevant because imidazolium salts may behave as ionic liquids [130] and because they can be used to prepare persistent carbenes (Arduengo carbenes) [131]. It is worth noting that aqueous solutions of imidazolium compounds should not be confused with ionic liquids. Several examples of imidazolium cations relevant to this review are represented in Fig. 9, together with their names and acronyms.

The first zeolite synthesis using imidazolium cations was reported by Zones in 1989 [133]. He obtained four already known high-silica zeolites in hydroxide media (ZSM-12, ZSM-22, ZSM-23, and ZSM-48) using 123TMI, 13DMI, 13DiPI, and 13MiPI. However, in recent years several new zeolites have been discovered by combining the structure-directing effect of imidazolium cations (including those used by Zones) with other synthetic strategies for the discovery of new zeolites (Table 1).

5.1   New Zeolites Discovered Using Imidazolium Cations

5.1.1   ITW and CSV: Imidazolium and Fluoride

The first zeolite discovered making use of imidazolium cations was ITQ-12 (ITW) [134], which was obtained with 134TMI (and also its isomer 123TMI) in concen-trated conditions by the fluoride route. This zeolite is a small pore material that contains two types of cages: The D4R, in which fluoride resides, and a larger, obloid, relatively thin and slit-shaped [4^{4}5^{4}6^{4}8^4] cage in where the organic cation is

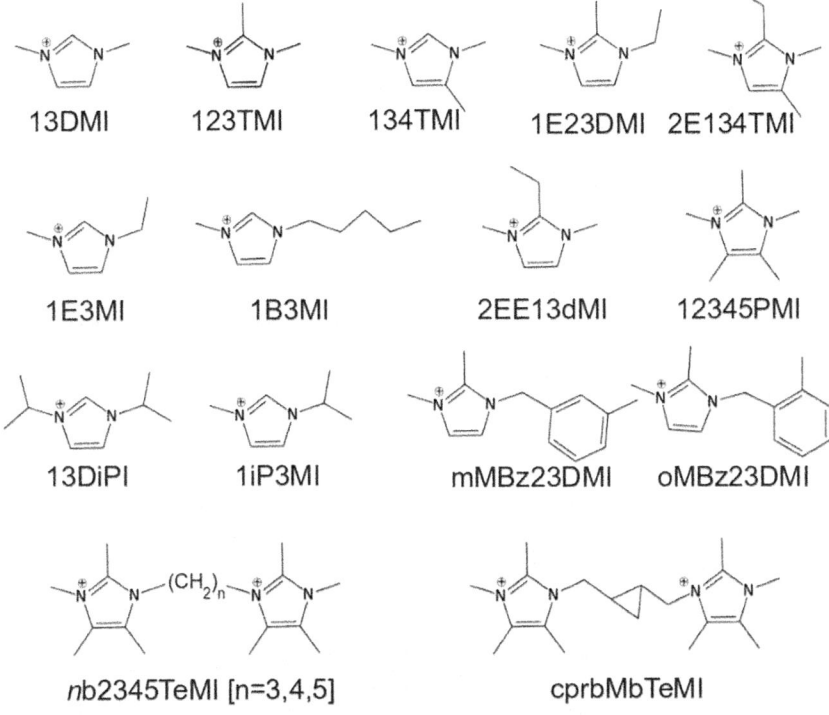

Fig. 9 Several imidazolium cations relevant to this review, with their acronyms, as used here [*13DMI* 1,3-dimethylimidazolium, *123TMI* 1,2,3-trimethylimidazolium, *134TMI* 1,3,4-trimethylimidazolium, *1E23DMI* 1-ethyl-2,3-dimethylimidazolium, *2E134TMI* 2-ethyl-1,3,4trimetylimidazolium, *1E3MI* 1-ethyl-3-methylimidazolium, *1B3MI* 1-butyl-3-methylimidazolium, *2E13DMI* 2-ethyl-1,3-dimethylimidazolium, *12345PMI* 1,2,3,4,5-pentamethylimidazolium, *13DiPI* 1,3-diisopropylimidazolium, *1iP3MI* 1-isopropyl-3-methylimidazolium, *mMBz23DMI* 2,3-dimethyl-1-(*m*-methylbenzyl)-imidazolium, *oMBz23DMI* 2,3-dimethyl-1-(*o*-methylbenzyl)-imidazolium, *nb2345TeMI* '*n*'methylene-bis-(2,3,4,5-tetramethylimidazolium) [where *n* = tri, tetra, penta], *cprbMbTeMI* 3,3´-cyclopropane-1,2-diylbis(methylene)bis(2,3,4,5-tetramethylimidazolium)]. The five cations in the first row produced ITW. The cation at the top right also afforded the chiral pure silica STW. The first cation in the second row produced SIV by the ionothermal route and UOS by the fluoride, germanium route (the same route that afforded UWY with the second cation in the same row and CIT-13 with the benzyl derivatives in the third row). The dication with *n* = 4 in the last row afforded the discovery of CSV. The dication in the bottom right is chiral (not detailed in the figure) and has been used in enantiopure form to produce scalemic conglomerates of germanosilicates and aluminogermanosilicate STW [132]

occluded. This pure SiO_2 material showed a large potential for the separation of propene and propane from its mixtures [140] and was the subject of quite extensive investigation from both the experimental and theoretical perspectives. One interesting result found by density functional theory calculations was that in the as-made material the Si–O bond was significantly more polarized than in the calcined zeolite [94]. This fact provided an important insight into the proposed structure-direction role of fluoride towards D4R silica structures: The increased ionicity of the Si–O

Table 1 New zeolites originally discovered through the use of imidazolium-based SDAs[a]

Zeolite	ZFT[b]	Cation[c]	Framework composition	Route[d]	Ref
ITQ-12	ITW	134TMI	SiO_2	Highly concentrated F^-	[134]
SIZ-7	SIV	1E3MI	CoAPO	Ionothermal	[135]
IM-16	UOS	1E3MI	$(Ge,Si)O_2$	Ge, F^-	[136]
IM-20	UWY	1Bz3MI	$(Ge,Si)O_2$	Ge, F^-	[137]
CIT-7	CSV	4b2345TMI	SiO_2 (+Al, Ti)	Highly concentrated F^- $(OH^-)^e$	[138]
CIT-13	–	oMBz23DMI, mMBz23DMI	$(Ge,Si)O_2$	Ge, F^-	[139]

[a]"True" zeolitic frameworks (i.e., three-dimensional fully connected tetrahedral (4.2) networks) but not interrupted frameworks are considered here, irrespective of their chemical composition
[b]Zeolite framework type
[c]Imidazolium used in the first report, acronyms according to Fig. 9
[d]Synthetic strategy (see text)
[e]Synthesis by the OH^- route was also possible, in a much more limited range of conditions

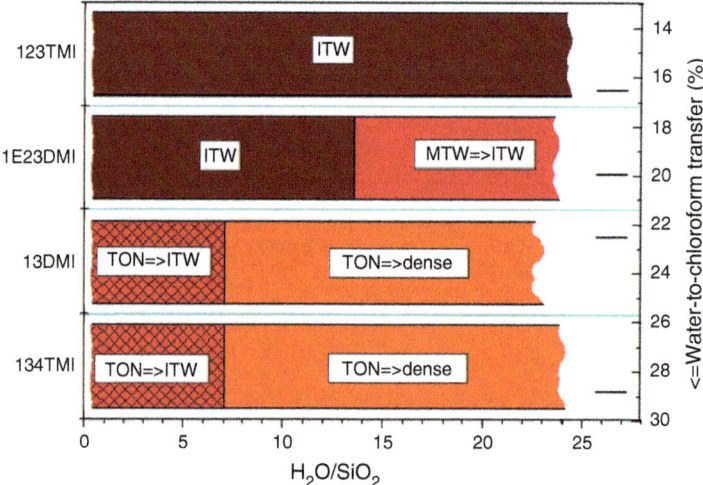

Fig. 10 Approximate crystallization fields for pure silica zeolites in fluoride media at 150°C using four imidazolium-based SDAs able to crystallize zeolite ITW at that temperature. The specificity in structure-direction towards ITW increases from bottom to top, i.e., as opposed to the % transfer from water to chloroform (shown at the *right*). Reprinted with permission from [142], Copyright 2013 American Chemical Society

bond decreases its directionality, thus relaxing the otherwise strained structure. Additionally, the combination of the fluoride and imidazolium effects was strong enough to reverse the order of stability of two SiO_2 zeolites, as proven by the in situ transformation of the initially crystallized $134TMIF-SiO_2-TON$ into $134TMIF-SiO_2-ITW$, despite the fact that TON is denser, less strained and, hence, thermodynamically more stable than ITW in the absence of guests [95].

A total of five different imidazolium cations [141–143], in combination with fluoride, were eventually found to direct to ITW with different degrees of specificity in structure-direction, following the decreasing order 123TMI >> 1E23DMI > 13DMI ~ 134TMI > 2E134TMI (see Fig. 9 for the acronyms and Fig. 10 for the specificity order). For the general conditions used, this order can be derived from the range of chemical compositions (particularly H_2O/SiO_2 ratios) and crystallization time needed to obtain the material, and from the possible prior crystallization of a different zeolite. 123TMI was effective at any of the H_2O/SiO_2 ratios tried, which included rather dilute conditions (H_2O/SiO_2 up to 24.5) with no sign of competing phases. For 1E23DMI, direct crystallization of ITW occurred only at relatively high concentrations ($H_2O/SiO_2 \leq 13$), while at more diluted conditions MTW competed. It was, however, shown that 1E23DMI-F-SiO_2-MTW could be transformed in situ to ITW at high concentrations (while the opposite, i.e., the transformation of 1E23DMI-F-ITW into MTW at high dilution didn't occur), in an additional example of "dense and stable" to "open and strained" zeolite transformation driven by host–guest interactions [142]. On the other hand, 13DMI and 134TMI needed increasing concentrations to crystallize ITW, with TON competing (but finally transforming to ITW, as stated above) [95]. The less specific SDA for ITW was 2E134TMI, a cation significantly larger in size that only worked at higher temperatures (175 or 185°C instead of 150°C for the rest of the examples) and that produced HPM-1 (the pure silica version of the chiral STW zeolite, see below) [144] at rather high concentrations and HPM-2 (a layered precursor to zeolite MTF) [145] at intermediate dilutions. Interestingly, both phases were unstable under the crystallization conditions and eventually transformed to ITW under much prolonged heating (well over 30 days). The possibility that a small degradation product acted as the true SDA, rather than the rather large 2E134TMI, had to be considered because of the prolonged heating at high temperature and the "degraded SDA" examples given above. However, ^{13}C MAS NMR of the zeolite demonstrated 2E134TMI was occluded intact into the zeolite [143]. The STW to ITW zeolite transformation is not as remarkable as the TON to ITW or MTW to ITW transformations commented above, since it probably doesn't revert the thermodynamic stability difference of the SiO_2 frameworks involved: STW is less dense (framework density 15.2 compared to 18.1 $Si/1{,}000$ Å^3) and likely more strained (higher concentration of D4R units) than ITW.

Table 2 collects the percentage transfer of iodide salts from water to chloroform solutions, proposed as a qualitative measure of the hydrophobicity of the cations [51], for several imidazolium and quaternary ammonium cations. The different behavior between both types of cations is clearly evident. The five small imidazolium cations that are able to produce ITW appear to be significantly more hydrophobic than quaternary ammonium cations within a similar range of C and N atoms per charge. Interestingly, also, the molecular shape apparently contributes largely to the observed hydrophobicity, since the three isomers with formula $C_6N_2H_{11}$ (1E3MI, 123TMI, and 134TMI) display significantly different percentage transfers. On the other hand, the most specific structure-direction effects among this imidazolium series are provided by cations with an intermediate hydrophobicity,

Table 2 Literature data for the transfer of imidazolium [142, 146] and quaternary ammonium [51] SDA iodide salts from water to chloroform solution[a]

Salt	(C + N)/q	% transfer
13DMI$^+$I$^-$	7	22.5
134TMI$^+$I$^-$	8	28.8
123TMI$^+$I$^-$	8	16.5
1E3MI$^+$I$^-$	8	8.1
1E23DMI$^+$I$^-$	9	17.7
2E134TMI$^+$I$^-$	10	21.6
TMA$^+$I$^-$	5	0
TEA$^+$I$^-$	9	0
TPA$^+$I$^-$	13	30
TBA$^+$I$^-$	17	90

[a]Acronyms for the imidazolium cations are given in Fig. 9. The quaternary ammonium cations are tetramethyl- (TMA$^+$), tetraethyl- (TEA$^+$), tetrapropyl- (TPA$^+$), and tetrabutylammonium (TBA$^+$)

which is, however, significantly lower than that proposed by Kubota and coworkers for quaternary ammonium SDAs that work well. The less hydrophobic cation (1E3MI) is a rather unselective SDA producing zeolites not highly demanding of strong structure-direction (mainly TON and, in more restricted conditions, MFI) [69]. However, the next most hydrophobic cations (123TMI and 1E23DMI) are the most specific SDAs. It has in fact been argued that a relatively high hydrophilic character may improve the ionization-based flexibility enhancement observed in these fluoride-containing silica materials [141, 142]. 1E3MI may be either too hydrophilic or too flexible [142], when compared to the rest of these imidazolium cations. Finally, it is worth mentioning that, during all these studies, the Villaescusa rule was strictly observed.

A more recent example of a new zeolite topology discovered through the use of imidazolium cations is CIT-7 (CSV) [138]. The most specific SDA used is a dication with a tetramethylene linker (4b2345TMI) although the trimethylene and pentamethylene variants and a particular monocation (2E13DMI) also worked with a lesser strength. Importantly, most of the synthesis described used the fluoride route in concentrated conditions but the hydroxide route appears also to work in this case, although in a much more limited extent (only one unseeded synthesis was reported). The structure of CIT-7 comprises a two-dimensional system of interconnected 10 and 8MR pores with a relatively large $[4^8 5^4 6^8 8^2 10^2]$ cavity at the crossing. Very interestingly is, in our opinion, the presence in CIT-7 of a small $[4^4 5^2]$ cage which resembles the $[4^6]$ or D4R cage and that can, in fact, be formally considered as derived from that cage by insertion of an additional Si atom in between two corners of the D4R unit. With regard to structure-direction in CIT-7 it may be highly relevant to find out if fluoride is occluded in the $[4^4 5^2]$ cage and the precise location and conformation of the organic SDA, so further studies are well merited.

5.1.2   SIV: Ionothermal Synthesis

The so-called ionothermal synthesis of zeolite-like analogues was developed by Morris's group at St. Andrews and consists in the use of ionic liquids or deep-eutectic mixtures as solvents and as "templates" or structure-directing agents [147–149]. In the case of eutectic mixtures they don't really act as templates but as template-delivery agents by the in situ degradation of the urea derivative of the mixture, as commented above [71]. In the case of ionic liquids, which are frequently salts of imidazolium or pyridinium derivatives, the cation in the salt acts as SDA, most frequently as itself but occasionally also after degradation and rearrangement [70].

While it may have additional advantages (like a very small vapor pressure that doesn't build up a high autogenous pressure) the idea behind the ionothermal synthesis is that it should avoid the competition between solvent and SDA for interacting with the inorganic components of the mixture. In fact, this may profoundly transform the mechanism of structure-direction, since no "hydrophobic hydration sphere" can be involved in this case. Ionic liquids are frequently highly hygroscopic and, hence, are never fully dry and water (and fluoride) may also be added as mineralizers or be introduced together with the reagents, so one could wonder if the frontiers between the ionothermal synthesis and the hydrothermal synthesis with fluoride in concentrated conditions may blurry for extremely low H_2O/SiO_2 ratios. There are, however, important differences that distinguish both routes. First, in the ionothermal synthesis the ionic liquids are, of course, in "solvent" amounts (i.e., many times the molar amount of tetrahedral atoms), while in the hydrothermal synthesis they usually do not exceed a SDA/SiO_2 ratio of 0.5 or 0.6. Second, even in wet ionic liquids water seems to be "deactivated," meaning it displays a reduced chemical reactivity. Apparently, this is so as far as water remains molecularly dispersed or in small clusters and with strong anion–water interactions that reduce the nucleophilicity of water [148]. Also at variance with the hydrothermal imidazolium fluoride synthesis that works very well for fully connected silica zeolites, the ionothermal route works better for phosphate-based materials that, most typically, have hanging P–O bonds, i.e., frequently they are not fully connected but either interrupted frameworks or layered materials, possible due to a slow rate of hydrolysis caused by the decreased water reactivity commented above.

The ionothermal route has been successful in crystallizing many new phosphate-based materials, including one fully connected material with a new zeolite topology: SIZ-7 (SIV) [135]. This is a cobalt aluminophosphate with a three-dimensional system of small pores built from double crankshaft chains and was prepared with 1E3MI$^+$ bromide as solvent. We have considered above that 1E3MI is a poor SDA in hydrothermal syntheses and, while it can act differently in an ionothermal synthesis it still doesn't seem to be a very specific SDA. For instance, with a CoAlPO composition it only produced physical mixtures of SIZ-7 (SIV), SIZ-8 (framework type AEI), and SIZ-9 (SOD), while in pure $AlPO_4$ compositions it yielded SIZ-1 (an interrupted framework), SIZ-3(AEL), SIZ-4 (CHA), or SIZ-5 (AFO) depending on the presence and amount of water and fluoride.

It is interesting that the ionothermal synthesis doesn't work well for silica-based materials. So far, only SIZ-12, a fully connected but non-porous SiO_2 phase with a framework density similar to cristobalite, has been prepared as isolated crystals among a lot of amorphous solid [148, 149]. The synthesis conditions have so far not been revealed but, since the material is dense and doesn't occlude any organics, structure-direction in this case didn't proceed through the typical organic–inorganic host–guest interactions.

5.1.3   UOS, UWY, and CIT-13: Imidazolium, Fluoride, and Germanium

By combining the structure-directing effects of imidazolium cations, fluoride anions, and germanium, three new zeolite structures were discovered: IM-16 (UOS) using 1E3MI [136], IM-20 (UWY) with 1B3MI [137], and CIT-13 using mMBz23DMI (and also oMBz23DMI in a more restricted compositional window [139]). As expected, because of the F and Ge roles, the three of them contain D4R units and this is likely the main effect affording their discovery. In fact, as we have commented before, 1E3MI shows no specificity at all in the absence of Ge. Because of its similarity with 1E3MI, the slightly longer and more flexible 1B3MI is also unlikely to exert a strong structure-direction of its own. Finally, CIT-13 is described in terms of dense cfi layers (first found in CFI [150], a one-dimensional 14MR material) connected through the D4R units, resulting in the first zeolite containing a bidimensional system of extra-large 14MR and medium 10MR pores. The relatively large benzylimidazolium derivatives that afford CIT-13 (who were later expanded to the slightly smaller mMBz3MI and oMBz3MI) [151] would help stabilize the relatively large volume by host–guest interactions, while most Ge was found in the D4R units. Thus, in all these three cases the main actor that directs the crystallization is likely Ge (and fluoride).

5.2   New Compositions Attained with Imidazolium Cations

In addition to the discovery of new zeolite structures, imidazolium cations have afforded the preparation of new zeolite compositions. Among these, the synthesis in pure form of the chiral zeolite HPM-1 (STW) is highly relevant [144] because of its implication in possible applications. This zeolite contains a helicoidal medium pore that could be useful in asymmetric catalysis and adsorption applications (see below). The STW structure type was first discovered in SU-32 [152], a germanosilicate prepared with diisopropylamine and with the help of the structure-directing effects of Ge and fluoride (STW contains a large concentration of D4R). All these combined effects were not strong enough to produce pure SU-32 but mixtures with SU-15, GeO_2, and/or amorphous solids. Unfortunately, SU-32 suffers from the stability problems typically inherent to germanosilicates with a large Ge content, which tend not only to collapse upon calcination but also by hydrolysis with ambient moisture at room temperature. By contrast pure silica HPM-1 has the outstanding thermal and hydrothermal stability typical of SiO_2 polymorphs with no connectivity defects.

The synthesis of HPM-1 in silica form made use of 2E134TMI with fluoride in highly concentrated conditions and produced the zeolite free of any impurities, despite prior theoretical predictions that such a silica polymorph could not be synthesized. The predictions were based in the lack of a "flexibility window" for STW [153] or in the too strained structure for silica [154] but both considered as relevant material the "clean," calcined SiO_2 structure. A broad flexibility window (the ability of the structure to deform without too large an energetic cost) [155] is considered necessary for a structure to be feasible, which has been rationalized as a way to improve the likelihood of a good host–guest match [156]. However, from the point of view of being able to crystallize a material, the important structure is not the clean calcined material but, obviously, the as-made one. As stated above, host–guest interactions, particularly with fluoride and imidazolium cations, may increase the ionicity of the framework [94, 95], and this has been shown to be the case also for HPM-1: The structural solution of the as-made material showed a large deformation of the $[SiO_{4/2}]$ tetrahedra [157], indeed supporting an enhanced flexibility. Fluoride was found in the D4R cages and the imidazolium cations in the larger $[4^6 5^8 8^2 10^2]$ cavity in a "locked" conformation, in a further example of a strong structure-direction by the combined effects of fluoride and imidazolium cations. This synthetic route also afforded the incorporation of Al into HPM-1, and the material was found to be an outstanding catalyst for the skeletal isomerization of normal butene [158].

Later on, Schmidt and coworkers also synthesized pure silica HPM-1 using 12345PMI by the concentrated fluoride route [111] following a study based on a computational method for determining chemically feasible SDAs for specific structures [159]. The method considers not only the host–guest interaction energy but also the possibility of synthesizing the SDA by known chemical transformations of a library of commercially available organic precursors. When the same cation was used for the synthesis of aluminosilicate zeolites, however, the aluminosilicate version of zeolite RTH was obtained instead [160]. The same aluminosilicate zeolite was obtained with a range of different imidazolium cations [161, 162]. A layered precursor of RTH, termed CIT-10, was also synthesized using an imidazolium-based dication with a pentamethylene linker, 5b2345TeMI [163].

5.3   Achieving Homochirality

The discovery that a dication, 4b2345TeMI, was able to crystallize STW with the imidazolium moieties occupying two different cages, first reported as supplementary information to the synthesis of CIT-7 [138], suggested a way to control the crystallization of this chiral zeolite towards a scalemic conglomerate. This zeolite in principle crystallizes as a racemic conglomerate, i.e., as a mixture of equal amounts of homochiral crystals of both hands. By introducing a rigid cyclopropane moiety in the middle of the spacer and using enantiopure versions of the resulting chiral dications, R- and S-cprbMbTeMI (Fig. 9, bottom right corner), Brand and coworkers were able to produce very recently enantioenriched germanosilicate and aluminogermanosilicate [132] versions of this zeolite. Presumably, the linker's chirality imposed in some extent the spatial disposition of consecutive cages, directing the crystallization preferentially towards crystals of one hand. The ring opening of 1,2-epoxyalkanes in aluminium-containing enantioenriched STW showed significant enantiomeric excess, while adsorption of 2-butanol in enantioenriched germanosilicates also proved a certain degree of enantioselectivity.

6 Conclusions

The use of organic structure-directing agents in the synthesis of zeolite materials has rendered a tremendous impact on the discovery of new zeolite topologies as well as new chemical compositions of the zeolite networks, which have greatly expanded the potential of application of the resulting zeolitic materials. To be efficient structure-directing agents, organic species have to fulfill a series of chemical conditions, in particular display a high hydrothermal stability, and particular size/shape, hydrophobic/hydrophilic, and rigidity/flexibility properties, which have been analyzed in this introductory chapter, together with some applications directly derived from these particular properties. Nonetheless, the extensive investigation carried out during the last decades on the use of new organic species as structure-directing agents has enlarged the range of conditions where they efficiently direct the crystallization of zeolitic materials, rendering new fascinating materials through a rational selection of the organic species to be used. The following chapters of this Special Issue will be devoted to analyze in detail recent trends in the science of structure-direction by organic species.

Acknowledgements Funding from the Spanish Ministry of Economy, Industry, and Competitiveness (through projects MAT2015-65767-P and MAT2015-71117-R) is acknowledged.

References

1. Cejka J, Corma A, Zones SI (2010) Zeolites and catalysis: synthesis reactions and applications. Wiley, Weinheim
2. Davis ME (2002) Ordered porous materials for emerging applications. Nature 417:813–821
3. Barrer RM (1948) Synthesis of a zeolitic mineral with chabazite-like sorptive properties. J Chem Soc 2:127–132

4. Rabo JA, Schoonover MW (2001) Early discoveries in zeolite chemistry and catalysis at union carbide, and follow-up in industrial catalysis. Appl Catal A 222:261–275
5. Breck DW, Eversole EG, Milton RM (1956) New synthetic crystalline zeolites. J Am Chem Soc 78:2338–2339
6. Sherman JD (1999) Synthetic zeolites and other microporous oxide molecular sieves. Proc Natl Acad Sci 96:3471–3478
7. Barrer RM, Denny PJ (1961) Hydrothermal chemistry of the silicates. Part IX.* Nitrogenous aluminosilicates. J Chem Soc:971–982
8. Kerr GT, Kokotailo GT (1961) Sodium zeolite ZK-4, a new synthetic crystalline aluminosilicate. J Am Chem Soc 83:4675–4675
9. Kerr GT (1966) Chemistry of crystalline aluminosilicates. II. The synthesis and properties of zeolite ZK-4. Inorg Chem 5:1537–1539
10. Wadlinger RL, Kerr GT, Rosinski EJ (1967) US Patent 3,308,069
11. Argauer RJ, Landolt GR (1972) US Patent 3,702,886
12. Kokotailo GT, Chu P, Lawton SL, et al. (1978) Synthesis and structure of synthetic zeolite ZSM-11. Nature 275:119–120
13. Kerr GT (1963) Zeolite ZK-5: A new molecular sieve. Science 140:1412
14. Wilson ST, Lok BM, Flanigen EM (1982) US Patent 4,310,440
15. Wilson ST, Lok BM, Messina CA, et al. (1982) Aluminophosphate molecular sieves: a new class of microporous crystalline inorganic solids. J Am Chem Soc 104:1146–1147
16. Lok BM, Cannan TR, Messina CA (1983) The role of organic molecules in molecular sieve synthesis. Zeolites 3:282–291
17. Kundy CS, Cox PA (2005) The hydrothermal synthesis of zeolites: precursors, intermediates and reaction mechanism. Microporous Mesoporous Mater 82:1–78
18. Gies H, Marler M (1992) The structure-controlling role of organic templates for the synthesis of porosils in the system SiO$_2$/template/H$_2$O. Zeolites 12:42–49
19. Moliner M, Rey F, Corma A (2013) Towards the rational design of efficient organic structure-directing agents for zeolite synthesis. Angew Chem Int Ed 52:13880–13889
20. Burton AW, Zones SI (2007) Organic molecules in zeolite synthesis: their preparation and structure-directing effects. Stud Surf Sci Catal 168:137–179
21. Nakagawa Y (1994) US Patent 5,281,407
22. Xie D, McCusker LB, Barlocher C, et al. (2013) SSZ-52, a zeolite with an 18-layer aluminosilicate framework structure related to that of the DeNOx catalyst Cu-SSZ-13. J Am Chem Soc 135:10519–10524
23. Elomari S (2003) US Patent 6,616,911
24. Lew CM, Davis TM, Elomari S (2016) Synthesis of new molecular sieves using novel structure-directing agents (Chapter 2). In: Mintova S (ed) Verified syntheses of zeolitic materials, 3rd revised edition. XRD Patterns: N. Barrier. Published on behalf of the Synthesis Commission of the International Zeolite Association 2016, pp 29–35. ISBN: 978-0-692-68539-6
25. Moliner M, Rey F, Corma A (2016) Synthesis design of new molecular sieves (Chapter 3). In: Mintova S (ed) Verified syntheses of zeolitic materials, 3rd revised edition. XRD Patterns: N. Barrier. Published on behalf of the the Synthesis Commission of the International Zeolite Association 2016, pp 36–41. ISBN: 978-0-692-68539-6
26. Pérez-Pariente J, Gómez-Hortigüela L (2008) The role of templates in the synthesis of zeolites. In: Čejka J, Peréz-Pariente J, Roth WJ (eds) Zeolites: from model materials to industrial catalysts. Transworld Research Network, pp 33–62. ISBN: 978-81-7895-330-4
27. Davis ME, Lobo R (1992) Zeolite and molecular sieve synthesis. Chem Mater 4:756–768
28. Ikuno T, Chaikittisilp W, Liu Z, et al. (2015) Structure-directing behaviors of tetraethylammonium cations toward zeolite beta revealed by the evolution of aluminosilicate species formed during the crystallization process. J Am Chem Soc 137:14533–14544
29. Zones SI, Burton AW, Lee GS, et al. (2007) A study of piperidinium structure-directing agents in the synthesis of silica molecular sieves under fluoride-based conditions. J Am Chem Soc 129:9066–9079

30. Jiang J, Yu J, Corma A (2010) Extra-large-pore zeolites: bridging the gap between micro and mesoporous structures. Angew Chem Int Ed 49:3120–3145
31. Li J, Corma A, Yu J (2015) Synthesis of new zeolite structures. Chem Soc Rev 44:7112–7127
32. Jiang J, Xu Y, Cheng P, et al. (2011) Investigation of extra-large pore zeolite synthesis by a high-throughput approach. Chem Mater 23:4709–4715
33. Li Y, Yu J (2014) New stories of zeolite structures: their descriptions, determinations, predictions, and evaluations. Chem Rev 114:7268–7316
34. Willhammar T, Sun J, Wan W, et al. (2012) Structure and catalytic properties of the most complex intergrown zeolite ITQ-39 determined by electron crystallography. Nat Chem 4:188–194
35. Moliner M, González J, Portilla MT, et al. (2011) A new aluminosilicate molecular sieve with a system of pores between those of ZSM-5 and beta zeolite. J Am Chem Soc 133:9497–9505
36. Moliner M, Willhammar T, Wan W, et al. (2012) Synthesis design and structure of a multipore zeolite with interconnected 12- and 10-MR channels. J Am Chem Soc 134:6473–6478
37. Moliner M, Martínez C, Corma A (2015) Multipore zeolites: synthesis and catalytic applications. Angew Chem Int Ed 54:3560–3579
38. Wang Z, Yu J, Xu R (2012) Needs and trends in rational synthesis of zeolitic materials. Chem Soc Rev 41:1729–1741
39. Burton AW, Zones SI, Elomari S (2005) The chemistry of phase selectivity in the synthesis of high-silica zeolites. Curr Opin Colloid Interface Sci 10:211–219
40. Boal BW, Zones SI, Davis ME (2015) Triptycene structure-directing agents in aluminophosphate synthesis. Microporous Mesoporous Mater 208:203–211
41. Jackowski A, Zones SI, Hwang SJ, et al. (2009) Diquaternary ammonium compounds in zeolite synthesis: cyclic and polycyclic N-heterocycles connected by methylene chains. J Am Chem Soc 131:1092–1100
42. Shvets O, Kasian N, Zukal A, et al. (2010) The role of template structure and synergism between inorganic and organic structure directing agents in the synthesis of UTL zeolite. Chem Mater 22:3482–3495
43. Davis ME (2014) Zeolites from a materials chemistry perspective. Chem Mater 26:239–245
44. Tsuji K, Beck LW, Davis ME (1999) Synthesis of 4,4'-trimethylenebis(1-benzyl-1-methylpiperidinium) diastereomers and their use as structure-directing agents in pure-silica molecular sieves syntheses. Microporous Mesoporous Mater 28:519–530
45. Lee G, Nakagawa Y, Hwang S, et al. (2002) Organocations in zeolite synthesis: fused bicyclo [l.m.0] cations and the discovery of zeolite SSZ-48. J Am Chem Soc 124:7024–7034
46. García R, Gómez-Hortigüela L, Sánchez F, et al. (2010) Diasteroselective structure directing effect of (1S,2S)-2-Hydroxymethyl-1-benzyl-1-methylpyrrolidinium in the synthesis of ZSM-12. Chem Mater 22:2276–2286
47. Gallego EM, Portilla MT, Paris C (2017) "Ab initio" synthesis of zeolites for preestablished catalytic reactions. Science 355:1051–1054
48. Burkett SL, Davis ME (1994) Mechanism of structure direction in the synthesis of Si-ZSM-5: an investigation by intermolecular ^{1}H-^{29}Si CP MAS NMR. J Phys Chem 98:4647–4653
49. Burkett SL, Davis ME (1995) Mechanisms of structure direction in the synthesis of pure-silica zeolites. 1. Synthesis of TPA/Si-ZSM-5. Chem Mater 7:920–928
50. Burkett SL, Davis ME (1995) Mechanism of structure direction in the synthesis of pure-silica zeolites. 2. Hydrophobic hydration and structural specificity. Chem Mater 7:1453–1463
51. Kubota Y, Helmkamp MM, Zones SI, et al. (1996) Properties of organic cations that lead to the structure-direction of high-silica molecular sieves. Microporous Mater 6:213–229
52. Jo C, Lee S, Cho SJ, et al. (2015) Synthesis of silicate zeolite analogues using organic sulfonium compounds as structure-directing agents. Angew Chem Int Ed 54:12805–12808
53. Delprato F, Delmotte L, Guth JL, et al. (1990) Synthesis of new silica-rich cubic and hexagonal faujasites using crown-ether based supramolecules as templates. Zeolites 10: 546–552
54. Balkus KJ, Hargis CD, Kowalak S (1992) Synthesis of NaX zeolites with metallophthalocyanines. ACS Symp Ser 499:347–354

55. Dorset DL, Kennedy GJ, Strohmaier KG, et al. (2006) P-derived organic cations as structure-directing agents: synthesis of a high-silica zeolite (ITQ-27) with a two-dimensional 12-ring channel system. J Am Chem Soc 128:8862–8867
56. Wan Y, Zhao D (2007) On the controllable soft-templating approach to mesoporous silicates. Chem Rev 107:2821–2860
57. Moteki T, Keoh SH, Okubo T (2014) Synthesis of zeolites using highly amphiphilic cations as structure-directing agents by hydrothermal treatment of a dense silicate gel. Chem Commun 50:1330–1333
58. Cho M, Na K, Kim J, et al. (2009) Stable single-unit-cell nanosheets of zeolite MFI as active and long-lived catalysts. Nature 461:246–249
59. Kim W, Kim JC, Kim J, et al. (2013) External surface catalytic sites of surfactant-tailored nano-morphic zeolites for benzene isopropylation to cumene. ACS Catal 3:192–195
60. Luo HY, Michaelis VK, Hodges S, et al. (2015) One-pot synthesis of MWW zeolite nanosheets using a rationally designed organic structure-directing agent. Chem Sci 6:6320–6324
61. Seo Y, Lee S, Jo C, et al. (2013) Microporous aluminophosphate nanosheets and their nano-morphic zeolite analogues tailored by hierarchical structure-directing amines. J Am Chem Soc 135:8806–8809
62. Corma A, Díaz-Cabañas MJ, Jorda JL, et al. (2008) A zeolitic structure (ITQ-34) with connected 9- and 10-ring channels obtained with phosphonium cations as structure directing agents. J Am Chem Soc 130:16482–16483
63. Rey F, Simancas J (2017) Beyond nitrogen OSDAs. Struct Bond. https://doi.org/10.1007/430_2017_13 (in this volume)
64. Villaescusa LA, Camblor MA (2016) Time evolution of an aluminogermanate zeolite synthesis: segregation of two closely similar phases with the same structure type. Chem Mater 28:3090–3098
65. Caullet P, Guth JL, Hazm J, et al. (1991) Synthesis, characterization and crystal-structure of the new clathrasil phase octadecasil. Eur J Solid State Inorg Chem 28:345–361
66. Villaescusa LA, Barrett PA, Camblor MA (1998) Calcination of octadecasil: fluoride removal and symmetry of the pure SiO_2 host. Chem Mater 10:3966–3973
67. Caullet P, Paillaud JL, Mathieu Y, et al. (2007) Synthesis of zeolites in the presence of diquaternary alkylammonium ions as structure-directing agents. Oil Gas Sci Technol 62:819–825
68. Lee SH, Shin CH, Yang DK, et al. (2004) Reinvestigation into the synthesis of zeolites using diquaternary alkylammonium ions $(CH_3)_3N^+(CH_2)_nN^+(CH_3)_3$ with n = 3–10 as structure-directing agents. Microporous Mesoporous Mater 68:97–104
69. Rojas A, Gómez-Hortigüela L, Camblor MA (2013) Benzylimidazolium cations as zeolite structure-directing agents. Differences in performance brought about by a small change in size. Dalton Trans 42:2562–2571
70. Parnham ER, Morris RE (2006) 1-Alkyl-3-methyl imidazolium bromide ionic liquids in the ionothermal synthesis of aluminium phosphate molecular sieves. Chem Mater 18:4882–4887
71. Parnham ER, Drylie EA, Wheatley PS, et al. (2006) Ionothermal materials synthesis using unstable deep-eutectic solvents as template-delivery agents. Angew Chem 118:2084–5088
72. Lee H, Zones SI, Davis ME (2003) A combustion-free methodology for synthesizing zeolites and zeolite-like materials. Nature 425:385–388
73. Lee H, Zones SI, Davis ME (2005) Zeolite synthesis using degradable structure-directing agents and pore-filling agents. J Phys Chem B 109:2187–2191
74. Lee H, Zones SI, Davis ME (2006) Synthesis of molecular sieves using ketal structure-directing agents and their degradation inside the pore space. Microporous Mesoporous Mater 88:266–274
75. O'Brien MG, Beale AM, Catlow CRA, et al. (2006) Unique organic-inorganic interactions leading to a structure-directed microporous aluminophosphate crystallization as observed with in situ Raman spectroscopy. J Am Chem Soc 128:11744–11745

76. Sánchez-Sánchez M, Sankar G, Gómez-Hortigüela L (2008) NMR evidence of different conformations of structure-directing cyclohexylamine in high-doped AlPO$_4$-44 materials. Microporous Mesoporous Mater 114:485–494
77. Bernardo-Maestro B, López-Arbeloa F, Pérez-Pariente J, et al. (2015) Supramolecular chemistry controlled by conformational space during structure-direction of nanoporous materials: self-assembly of ephedrine and pseudoephedrine. J Phys Chem C 119:28214–28225
78. Takekiyo T, Yoshimura Y (2006) Raman spectroscopic study on the hydration structures of tetraethylammonium cation in water. J Phys Chem A 110:10829–10833
79. Schmidt JE, Fu D, Deem MW, et al. (2016) Template–framework interactions in tetraethylammonium-directed zeolite synthesis. Angew Chem Int Ed 55:16044–16048
80. Hong SB, Min HK, Shin CH, et al. (2007) Synthesis, crystal structure, characterization, and catalytic properties of TNU-9. J Am Chem Soc 129:10870–10885
81. Gramm F, Baerlocher C, McCusker LB, et al. (2006) Complex zeolite structure solved by combining powder diffraction and electron microscopy. Nature 444:79–81
82. Hong SB, Lear EG, Wright PA, et al. (2004) Synthesis, structure solution, characterization, and catalytic properties of TNU-10: a high-silica zeolite with the STI topology. J Am Chem Soc 126:5817–5826
83. Bernardo-Maestro B, López-Arbeloa F, Pérez-Pariente J et al (2017) Comparison of the structure-directing effect of ephedrine and pseudoephedrine during crystallization of nanoporous aluminophosphates. Microporous Mesoporous Mater, published in web. https://doi.org/10.1016/j.micromeso.2017.04.008 (in press)
84. Camblor MA, Hong SB (2011) Synthetic silicate zeolites: diverse materials accessible through geoinspiration. In: Bruce DW, O'Hare D, Walton IR (eds) Porous materials. Wiley, Chichester
85. Flanigen EM, Patton RL (1978) US Patent 4,073,865
86. Guth JL, Kessler K, Higel JM, et al. (1989) Zeolite synthesis in the presence of fluoride ions. ACS Symp Ser 398:176–195
87. Guth J, Kessler H, Caullet P et al (1993) F$^-$: a multifunctional tool for microporous solids a) mineralizing, structure directing and templating effects in the synthesis. In: von Ballmoos R, Higgins J, Treacy M (eds) Proceedings of the 9th international zeolite conference, London, pp 215–222
88. Caullet P, Paillaud JL, Simon-Masseron A, et al. (2005) The fluoride route: a strategy to crystalline porous material. C R Chim 8:245–266
89. Koller H, Lobo RF, Burkett SL, et al. (1995) SiO$^-$···HOSi hydrogen bonds in as-synthesized high-silica zeolites. J Phys Chem 99:12588–12596
90. Blasco T, Camblor MA, Corma A, et al. (1998) Direct synthesis and characterization of hydrophobic aluminum-free Ti-Beta zeolite. J Phys Chem 102:75–88
91. Camblor MA, Corma A, Iborra S, et al. (1997) Beta zeolite as a catalyst for the preparation of alkyl glucoside surfactants: the role of crystal size and hydrophobicity. J Catal 172:76–84
92. Eroshenko V, Regis RC, Soulard M, et al. (2001) Energetics: a new field of applications for hydrophobic zeolites. J Am Chem Soc 123:8129–8130
93. Villaescusa LA, Camblor MA (2003) The fluoride route to new zeolites. Recent Res Dev Chem 1:93–141
94. Zicovich-Wilson CM, San-Román ML, Camblor MA, et al. (2007) Structure, vibrational analysis, and insights into host-guest interactions in as-synthesized pure silica ITQ-12 zeolite by periodic B3LYP calculations. J Am Chem Soc 129:11512–11523
95. Zicovich-Wilson CM, Gándara F, Monge A, et al. (2010) In situ transformation of TON silica zeolite into the less dense ITW: structure-direction overcoming framework instability in the synthesis of SiO$_2$ zeolites. J Am Chem Soc 132:3461–3471
96. Camblor MA, Villaescusa LA, Díaz-Cabañas MJ (1999) Synthesis of all-silica and high-silica molecular sieves in fluoride. Top Catal 9:59–76

97. Camblor MA, Barrett PA, Díaz-Cabañas MJ, et al. (2001) High silica zeolites with three-dimensional systems of large pore channels. Microporous Mesoporous Mater 48:11–22
98. Zones SI, Darton RJ, Morris R, et al. (2005) Studies on the role of fluoride ion vs reaction concentration in zeolite synthesis. J Phys Chem B 109:652–661
99. Burton AW, Lee GS, Zones SI (2006) Phase selectivity in the syntheses of cage-based zeolite structures: an investigation of thermodynamic interactions between zeolite hosts and structure directing agents by molecular modeling. Microporous Mesoporous Mater 90:129–144
100. Zones SI, Hwang SJ, Elomari S, et al. (2005) The fluoride-based route to all-silica molecular sieves; a strategy for synthesis of new materials based upon close-packing of guest–host products. C R Chim 8:267–282
101. Camblor MA, Díaz-Cabañas MJ, Cox PA, et al. (1999) A synthesis, MAS NMR, synchrotron X-ray powder diffraction, and computational study of zeolite SSZ-23. Chem Mater 11: 2878–2885
102. Lobo RF, Zones SI, Davis ME (1995) Structure-direction in zeolite synthesis. J Incl Phenom Macrocycl Chem 21:47–78
103. Catlow CRA, Coombes DS, Lewis D, et al. (1998) Computer modeling of nucleation, growth, and templating in hydrothermal synthesis. Chem Mater 10:3249–3265
104. Jorge M, Auerbach SM, Monson PA (2005) Modeling spontaneous formation of precursor nanoparticles in clear-solution zeolite synthesis. J Am Chem Soc 127:14388–14400
105. Piccione P, Yang S, Navrotsky A, et al. (2002) Thermodynamics of pure-silica molecular sieve synthesis. J Phys Chem B 106:3629–3638
106. Piccione P, Woodfield B, Boerio-Goates J, et al. (2001) Entropy of pure-silica molecular sieves. J Phys Chem B 105(25):6025–6030
107. Khan MN, Auerbach SM, Monson PA (2015) Lattice Monte Carlo simulations in search of zeolite analogues: effects of structure directing agents. J Phys Chem C 119:28046–28054
108. Yu J, Xu R (2003) Rich structure chemistry in the aluminophosphate family. Acc Chem Res 36:481–490
109. Lewis DW, Freeman CM, Catlow CRA (1995) Predicting the templating ability of organic additives for the synthesis of microporous materials. J Phys Chem 99:11194–11202
110. Lewis DW, Willock DJ, Catlow CRA, et al. (1996) De novo design of structure-directing agents for the synthesis of microporous solids. Nature 382:604–606
111. Schmidt JE, Deem MW, Davis ME (2014) Synthesis of a specified, silica molecular sieve using computationally predicted organic structure-directing agents. Angew Chem Int Ed 126: 8512–8514
112. Brunklaus G, Koller H, Zones SI (2016) Defect models of as-made high-silica zeolites: clusters of hydrogen-bonds and their interaction with the organic structure-directing agents determined from ^{1}H double and triple quantum NMR spectroscopy. Angew Chem Int Ed 55: 14459–14463
113. Dib E, Grand J, Mintova S, et al. (2015) Structure-directing agent governs the location of silanol defects in zeolites. Chem Mater 27:7577–7579
114. Gómez-Hortigüela L, Pinar AB, Corà F, et al. (2010) Dopant-siting selectivity in nanoporous catalysts: control of proton accessibility in zeolite catalysts through the rational use of templates. Chem Commun 46:2073–2075
115. Román-Leshkov Y, Moliner M, Davis ME (2011) Impact of controlling the site distribution of Al atoms on catalytic properties in ferrierite-type zeolites. J Phys Chem C 115:1096–1102
116. Yokoi T, Mochizuki H, Namba S, et al. (2015) Control of the Al distribution in the framework of ZSM-5 zeolite and its evaluation by solid-state NMR technique and catalytic properties. J Phys Chem C 119:15303–15315
117. Gómez-Hortigüela L, Álvaro-Muñoz T, Bernardo-Maestro B, et al. (2015) Towards chiral distributions of dopants in microporous frameworks: helicoidal supramolecular arrangement of (1R,2S)-ephedrine and transfer of chirality. Phys Chem Chem Phys 17:348–357
118. Lemishko T, Simancas J, Hernández-Rodríguez M, et al. (2016) An INS study of entrapped organic cations within the micropores of zeolite RTH. Phys Chem Chem Phys 18: 17244–17252

119. Gómez-Hortigüela L, Hamad S, Pinar AB, et al. (2009) Molecular insights into the self-aggregation of aromatic molecules in the synthesis of nanoporous aluminophosphates: a multilevel approach. J Am Chem Soc 131:16509–16524
120. Wang Y, Yu J, Li Y, et al. (2003) Chirality transfer from guest chiral metal complexes to inorganic framework: the role of hydrogen bonding. Chem Eur J 9:5048–5055
121. Gómez-Hortigüela L, Bernardo-Maestro B (2017) Chiral organic structure-directing agents. Struct Bond. https://doi.org/10.1007/430_2017_9 (in this volume)
122. Behrens P, van de Goor G, Freyhardt CC (1995) Structure-determining C-H⋯O-Si hydrogen bonds in cobaltocenium fluoride nonasil. Angew Chem Int Ed 34:2680–2682
123. Lee JK, Shin J, Ahn NH, et al. (2015) A family of molecular sieves containing framework-bound organic structure-directing agents. Angew Chem Int Ed 54:11097–11101
124. Lee JK, Lee JH, Ahn NH, et al. (2016) Solid solution of a zeolite and a framework-bound OSDA-containing molecular sieve. Chem Sci 7:5805–5814
125. Gómez-Hortigüela L, López-Arbeloa F, Corà F, et al. (2008) Supramolecular chemistry in the structure direction of microporous materials from aromatic structure-directing agents. J Am Chem Soc 130:13274–13284
126. Corma A, Rey F, Rius J, et al. (2004) Supramolecular self-assembled molecules as organic directing agent for synthesis of zeolites. Nature 431:287–290
127. Moliner M (2015) Design of zeolites with specific architectures using self-assembled aromatic organic structure-directing agents. Top Catal 25:502–512
128. Paris C, Moliner M (2017) Role of supramolecular chemistry during templating phenomenon in zeolite synthesis. Struct Bond. https://doi.org/10.1007/430_2017_11 (in this volume)
129. Corma A, Díaz-Cabañas MJ, Jordá JL, et al. (2006) High-throughput synthesis and catalytic properties of a molecular sieve with 18- and 10-member rings. Nature 443:842–845
130. Welton T (1999) Room-temperature ionic liquids. Solvents for synthesis and catalysis. Chem Rev 99:2071–2084
131. Arduengo AJ, Harlow RL, Kline M (1991) A stable crystalline carbene. J Am Chem Soc 113:361–363
132. Brand SK, Schmidt JE, Deem MW, et al. (2017) Enantiomerically enriched, polycrystalline molecular sieves. Proc Natl Acad Sci U S A 114:5101–5106
133. Zones SI (1989) Synthesis of pentasil zeolites from sodium silicate solutions in the presence of quaternary imidazole compounds. Zeolites 9:458–467
134. Barrett PA, Boix T, Puche M, et al. (2003) ITQ-12: a new microporous silica polymorph potentially useful for light hydrocarbon separations. Chem Commun:2114–2115
135. Parnham ER, Morris RE (2006) The ionothermal synthesis of cobalt aluminophosphate zeolite frameworks. J Am Chem Soc 128:2204–2205
136. Lorgouilloux Y, Dodin M, Paillaud JL, et al. (2009) IM-16: a new microporous germanosilicate with a novel framework topology containing D4R and MTW composite building units. J Solid State Chem 182:622–629
137. Dodin M, Paillaud JL, Lorgouilloux Y, et al. (2010) A zeolitic material with a three-dimensional pore system formed by straight 12- and 10-ring channels synthesized with an imidazolium derivative as structure-directing agent. J Am Chem Soc 132:10221–10223
138. Schmidt JE, Xie D, Rea T, et al. (2015) CIT-7, a crystalline, molecular sieve with pores bounded by 8 and 10-membered rings. Chem Sci 6:1728–1734
139. Boal BW, Deem MW, Xie D, et al. (2016) Synthesis of germanosilicate molecular sieves from mono- and di-quaternary ammonium OSDAs constructed from benzyl imidazolium derivatives: stabilization of large micropore volumes including new molecular sieve CIT-13. Chem Mater 28:2158–2164
140. Olson DH, Yang X, Camblor MA (2004) ITQ-12: a zeolite having temperature dependent adsorption selectivity and potential for propene separation. J Am Chem Soc 108:11044–11048
141. Rojas A, Martínez-Morales A, Zicovich-Wilson CM, Camblor MA (2012) Zeolite synthesis in fluoride media: structure direction toward ITW by small methylimidazolium cations. J Am Chem Soc 134:2255–2263

142. Rojas A, San-Roman ML, Zicovich-Wilson CM, et al. (2013) Host−guest stabilization of a zeolite strained framework: in situ transformation of zeolite MTW into the less dense and more strained ITW. Chem Mater 25:729–738
143. Rojas A, Camblor MA (2014) Structure-direction in the crystallization of ITW zeolites using 2-ethyl-1,3,4-trimethylimidazolium. Dalton Trans 43:10760–10766
144. Rojas A, Camblor MA (2012) A pure silica chiral polymorph with helical pores. Angew Chem Int Ed 51:3854–3856
145. Rojas A, Camblor MA (2014) HPM-2, the layered precursor to zeolite MTF. Chem Mater 26: 1161–1169
146. Rojas AE (2012) Dirección de estructuras en la síntesis de zeolitas usando cationes orgánicos imidazolios. PhD thesis, Universidad Autónoma de Madrid
147. Cooper ER, Andrews CD, Wheatley PS, et al. (2004) Ionic liquids and eutectic mixtures as solvent and template in synthesis of zeolite analogues. Nature 430:1012–1016
148. Parnham ER, Morris RE (2007) Ionothermal synthesis of zeolites, metal-organic frameworks, and inorganic-organic hybrids. Acc Chem Res 40:1005–1013
149. Parnham ER (2006) Ionothermal synthesis. A new synthesis methodology using ionic liquids and eutectic mixtures as both solvent and template in zeotype synthesis. PhD thesis, University of St. Andrews
150. Wagner P, Yoshikawa M, Lovallo M, et al. (1997) CIT-5: a high-silica zeolite with 14-ring pores. J. Chem. Soc, Chem. Commun 21:2179–2180
151. Kang JH, Xie D, Zones SI, et al. (2016) Synthesis and characterization of CIT-13, a germanosilicate molecular sieve with extra-large pore openings. Chem Mater 28:6250–6259
152. Tang L, Shi L, Bonneau C, et al. (2008) A zeolite family with chiral and achiral structures built from the same building layer. Nat Mater 7:381–385
153. Kapko V, Dawson C, Treacy MMJ, et al. (2010) Flexibility of ideal zeolite frameworks. Phys Chem Chem Phys 12:8531–8541
154. Sastre G, Corma A (2010) Predicting structural feasibility of silica and germania zeolites. J Phys Chem C 114:1667–1673
155. Sartbaeva A, Wells SA, Treacy MMJ, et al. (2006) The flexibility window in zeolites. Nat Mater 5:962–965
156. Medina ME, Platero-Prats AE, Snejko N, et al. (2011) Towards inorganic porous materials by design: looking for new architectures. Adv Mater 23:5283–5292
157. Rojas A, Arteaga O, Kahr B, et al. (2013) Synthesis, structure and optical activity of HPM-1, a pure silica chiral zeolite. J Am Chem Soc 135:11975–11984
158. Jo D, Hong SB, Camblor MA (2015) Monomolecular skeletal isomerization of 1-butene over selective zeolite catalysts. ACS Catal 5:2270–2274
159. Pophale R, Daeyaert F, Deem MW (2013) Computational prediction of chemically synthesizable organic structure directing agents for zeolites. J Mater Chem A 1:6750–6760
160. Schmidt JE, Deimund MA, Davis ME (2014) Facile preparation of aluminosilicate RTH across a wide composition range using a new organic structure-directing agent. Chem Mater 26:7099–7105
161. Schmidt JE, Deimund ME, Xie D, et al. (2015) Synthesis of RTH-type zeolites using a diverse library of imidazolium cations. Chem Mater 27:3756–3762
162. Jo D, Lim JB, Ryu T, et al. (2015) Unseeded hydroxide-mediated synthesis and CO_2 adsorption properties of an aluminosilicate zeolite with the RTH topology. J Mater Chem A 3: 19322–19329
163. Schmidt JE, Xie D, Davis ME (2015) Synthesis of the RTH-type layer: the first small-pore, two dimensional layered zeolite precursor. Chem Sci 6:5955–5963

Struct Bond (2018) 175: 43–74
DOI: 10.1007/430_2017_7
© Springer International Publishing AG 2017
Published online: 8 August 2017

Location of Organic Structure-Directing Agents in Zeolites Using Diffraction Techniques

Stef Smeets and Lynne B. McCusker

Abstract In this chapter, we delve into the X-ray diffraction techniques that can be used to address the question as to where the organic structure-directing agents (OSDAs) are located in the pores of a zeolite framework structure and give an overview of some of the practical issues involved. By examining the results of such investigations, we attempt to establish whether the OSDAs are really disordered, as is often claimed, or if it is the methods we use that give this impression. In fact, the non-framework species in the channels of a zeolite appear to be arranged quite logically in a chemically sensible manner. In most cases, the OSDA within the pores can be described well as a superposition of just a few discrete, symmetry-related positions, provided the discrepancies between the OSDA and framework symmetries can be resolved. On the basis of some selected examples, we show that their arrangements can be extracted from experimental data using a systematic strategy and sometimes supplementary information.

Keywords Simulated annealing • Structure-directing agents • X-ray (powder) diffraction • Zeolites

Contents

S. Smeets (✉)
Department of Materials and Environmental Chemistry, Stockholm University, Stockholm 10691, Sweden
e-mail: stef.smeets@mmk.su.se

L.B. McCusker
Department of Materials, ETH Zurich, Zurich 8093, Switzerland

Department of Chemical Engineering, University of California, Santa Barbara, CA 93106, USA

1 Introduction

Are the species in the pores of a zeolite framework structure really disordered as is often claimed? Or are they arranged logically in a chemically sensible manner and only appear to be disordered because of the methods we use to study them? If the latter is the case, can our methods be improved and/or supplemented to extract more correct information? In this chapter, we will attempt to answer these questions.

Even with the first syntheses of zeolites in the late 1940s, which sought to replicate the hydrothermal conditions that produced the natural zeolite minerals by heating mixtures of alkali and alkaline earth silicates and aluminates in sealed containers, structure analysis of the resulting product was considered to be of paramount importance. It was known that the negative charge of the aluminosilicate framework structure was balanced by the alkali and alkaline earth cations in the pores and that the pores were filled with water molecules, but why had one zeolite formed and not another? It was assumed that the answer would be revealed if the structure of the crystalline product could be determined.

When Barrer and Denny included organic cations in the synthesis mixture in 1961, with the idea that by using a larger cation, the amount of Al in the framework could be reduced [1], it signalled the beginning of a new era in zeolite synthesis. Initially, simple tetramethylammonium (TMA) ions were used, but later on, as researchers began to realize that the organic cations also act as excellent structure-directing agents, increasingly complicated quaternary ammonium ions were tried. Now organic cations play a key role in most zeolite syntheses, and can have a profound effect on the micropore topology that results [2, 3]. Several groups have performed extensive systematic studies on the effect of the charge, shape, size, and composition of the organic cation on the end product [4–8].

Given the importance of these organic structure-directing agents (OSDAs) in modern zeolite synthesis, it is essential that we understand the interplay between the OSDA and the zeolite framework to understand why and how these materials form. It would be beneficial, therefore, if the locations of the organic guest species could be retrieved experimentally. Single-crystal X-ray diffraction (XRD) methods are ideal for this purpose, but most industrially and commercially important zeolites are only produced and used in polycrystalline form. Fortunately, powder diffraction methodology has now advanced to the stage where it is generally accepted that novel framework structures can be solved from X-ray powder diffraction (XRPD) data [9], and that inorganic cations in an as-synthesized or an ion-exchanged zeolite or zeolite-like material can be located with no problem. However, this is not yet the case for OSDAs. Locating a low-symmetry OSDA in the pores of a high-symmetry zeolite framework using XRPD data is fraught with difficulties.

To illustrate the scope of the problem, we catalogued the structures of the type materials of the 103 framework types that have been added to the Database of Zeolite Structures over the last 20 years [10]. We classified each material by how the OSDA was located (Fig. 1). In only 29 cases was the OSDA located experimentally, using either single-crystal or powder diffraction data. In 12 cases, the location of the OSDA was not determined from the data directly, but estimated instead using molecular dynamics modelling. In 16 cases, diffraction data on the as-made zeolite were available, but the researchers chose not to determine the location of the OSDA, or simply placed C atoms in the channels arbitrarily to describe the diffuse electron density cloud there without interpreting it further. In a small number of cases (10), the zeolite was synthesized in non-conventional ways, for example, in the absence of an OSDA [11], or using top-down methods [12, 13]. And in the remaining 36 cases, the structural characterization was performed on a calcined zeolite, where the problem of locating the OSDA is avoided.

The key difficulty in locating organic guest species in inorganic host structures from diffraction data is that organic compounds consist of light scatterers and typically have low point symmetry, while their inorganic hosts consist of heavier

Fig. 1 Pie chart showing the distribution of methods used to locate the SDA in the last 103 framework types (up to May 2017) published in the Database of Zeolite Structures [10]. In only 29 of the type materials reported was the OSDA located from the data directly, via either single-crystal methods or Rietveld refinement

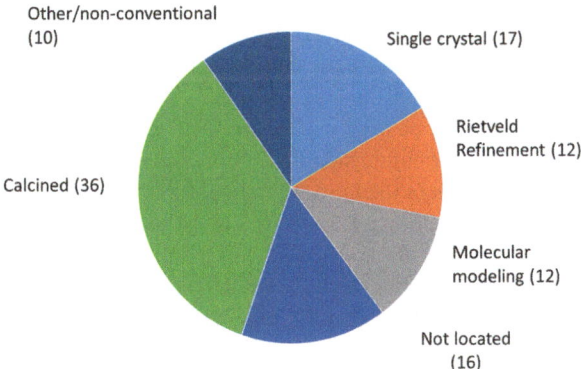

scatterers and have high symmetry. These features result in a lack of contrast that makes it difficult to *see* the organic guest, and because of the difference in symmetry, the guest species often appear to be disordered. Even when decent quality single-crystal XRD data are available, these two factors alone make retrieving the location of the organic species challenging. The problem is exacerbated if the specimen is only available in polycrystalline form, when XRPD techniques are usually the only option. The process of locating the OSDA using XRPD data is often hindered by the quality of the data that can be collected. The problem may be low data resolution (no diffraction at higher angles and/or broad peaks), or in the assignment of reflection intensities when reflections overlap, a problem that is ubiquitous in powder diffraction. This adds a second layer of complication that often results in the guest species simply being dismissed as *disordered*. Perhaps for these reasons, the location of the OSDA in many studies is avoided, and structure analysis is performed using data collected on the calcined material. This is undesirable, because calcination typically leads to a reduction in the quality of the data that can be collected, and critical information related to the synthesis is lost.

Our own experience with as-synthesized zeolite structures indicates that the guest species actually tend to be highly ordered, and that their arrangements within the pores can often be described well as a superposition of just a few discrete, symmetry-related positions, provided the discrepancies between the OSDA and framework symmetries can be resolved. In this chapter, we will introduce the X-ray diffraction techniques used for locating OSDA molecules in zeolites, give an overview of some of the practical aspects of locating OSDAs, and highlight some examples of locating the OSDA.

2   Historical Perspective

As long as there have been synthetic zeolites, there have been researchers willing to take on the challenge of locating guest species in their pores, because their positions are considered to be essential to the understanding not only of the synthesis, but also of the properties of the resulting zeolite. For example, the synthesis [14] and crystal structure [15] of zeolite A, one of the first synthetic zeolites to be reported, were published in back-to-back papers in 1956. Its framework structure had no counterpart in the natural zeolites known at the time, and was deduced using chemical reasoning and powder diffraction data recorded on photographic film. With this information, 2-dimensional electron density maps were calculated in judiciously selected slices of the unit cell (calculations had to be done by hand as there were no computers), and even the positions of some of the sodium counterions could be resolved. Not only that, the positions of the counterions in several ion-exchanged forms of the zeolite were also determined. Somewhat later, when digitized single-crystal diffraction data and modest computing facilities were available, it was also recognized that the water molecules in the large cavity of zeolite A are not randomly positioned as was first assumed, but arranged to form a hydrogen-bonded pentagon-dodecahedral cluster [16].

The conventional approach to locating the species in the pores of a known zeolite framework structure is to generate and interpret a 3-dimensional difference electron density (or Fourier) map. That is, the electron density calculated for the framework model is subtracted from the total electron density calculated from the measured reflection intensities to yield a map showing the residual density in the pores. See Sect. 4.3 for more details about how this is done in practice. These maps, while potentially extremely informative, are not always easy to interpret on the atomic level, especially if they have been generated from powder diffraction data. The interpretation is an iterative procedure that requires patience and perseverance on the part of the crystallographer and an understanding of the effects of the high symmetry of the framework on the lower symmetry OSDA. Nevertheless, the literature is full of examples of successful interpretations.

In the early studies of zeolites prepared with TMA in the synthesis mixture (late 1960s), only XRPD data could be obtained, but at least mainframe computers had become available and the cation was rigid and not too complicated. The difference Fourier approach described above was applied to TMA-sodalite, and the TMA cation was found to adopt an unexpected orientation in the sodalite cage with an unusually strong interaction between the cation and framework oxygen atoms [17]. The investigation of a second TMA zeolite was less straightforward. The presumed framework structure proved to be incorrect, so the authors considered how the four TMA cations per unit cell could be arranged, given the space group $I4_1/a$. The shape, size, and symmetry of the cation allowed only one arrangement. Then, by examining the distances between the TMA cations, it was possible to deduce where the aluminosilicate framework must be. In this way, the authors determined not only where the OSDAs were, but also that the framework structure was the same as that of the mineral gismondine [18].

The synthesis of ZSM-5 [19] was another milestone in the history of zeolite synthesis (1970s). Here, tetrapropylammonium (TPA) ions were used as the OSDA and it took years for the extremely complex high-silica framework structure to be deduced and reported [20, 21]. In a later publication, the approximate position of the TPA in the pores was found using single-crystal data from a very small crystal, but the geometry of the cation was severely distorted [22]. At this time, the Rietveld refinement technique for neutron powder diffraction data was being adapted to X-ray data, and powder diffraction methods began to evolve into a viable alternative for structure analysis. Using these new techniques and high-quality laboratory XRPD data, an improved description of the TPA could be obtained [23]. Finally, in 1987, large single crystals of ZSM-5 were produced and a definitive structure refined [24].

These early analyses of OSDAs in zeolitic materials were followed by many more, including those of zeolite-like aluminophosphates [25–27] and clathrasils [28–32] in the 1980s, and zincophosphates [33, 34] and gallophosphates [35–38] in the 1990s. Some were performed using single-crystal data and others using XRPD data. At this time, synchrotron radiation sources were beginning to become accessible to research scientists, and this provided another boost to XRPD analysis. The high intensity and parallel nature of the synchrotron X-ray beam allowed extremely highly resolved XRPD patterns to be recorded. Such data were used to investigate the structure of the aluminophosphate VPI-5, which had been shown to have the largest channels known at the time [39]. A model for its framework structure had been proposed, but its fit to the measured neutron powder diffraction pattern was poor [40] and it was not clear how the simple linear dipropylammonium (DPA) cation that was used in the synthesis could produce such a large channel. In a careful analysis using synchrotron powder diffraction data collected on the as-synthesized form of VPI-5, it became apparent that the DPA cation had not been incorporated into the final structure at all, that some of the Al atoms in the framework were octahedrally coordinated, and that there was a well-ordered triple helix of hydrogen-bonded water molecules filling the 18-ring channel [41]. All this information was derived by hand from a series of difference Fourier maps. Each improvement to the model improved the quality of the electron density map and allowed one or two more atoms to be identified and eventually the mystery of the 18-ring channel to be resolved.

There are many more examples in the literature, and just a few representative studies have been referenced here. Generating and interpreting a difference map to locate non-framework species requires care and attention, and is sometimes described in great detail, especially when researchers were limited to X-ray powder diffraction data [42, 43]. This is in stark contrast to the studies where single-crystal X-ray data were available, where the result may be reduced to a single line, if mentioned at all. Certainly having single-crystal data helps, but such data do not guarantee an easily interpretable cloud of electron density in the pores of a zeolitic material. The problem of disorder produced by the high symmetry of the framework remains, even with the best data.

With modern computers, generating a difference electron density map is quick and easy, but the interpretation step is still time consuming. More recent studies take advantage of the ever-increasing computational power that is now available, and use molecular modelling and simulated annealing algorithms to obtain an initial location of the OSDA. This has allowed a more systematic approach to the problem to be developed.

3   Brief Introduction to Powder Diffraction

As most studies of zeolitic materials require the use of powder diffraction techniques, it is important to understand the basic principles involved. A powder diffraction pattern contains a lot of information, and some of the most important features are shown in Fig. 2. While some of these readily offer qualitative information about the material, extracting quantitative information requires somewhat more expertise (see Sect. 4.1).

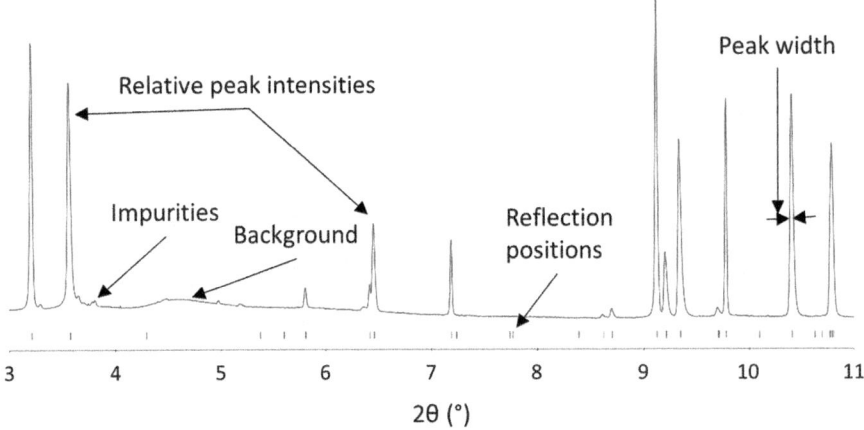

Fig. 2 A powder diffraction pattern, with the features of interest highlighted

In a powder diffraction pattern, the positions of the peaks (measured in °2θ) correspond to reflection positions. Note that a reflection does not necessarily give rise to a peak (the reflection may have an intensity of zero depending on the crystal structure), but a peak always indicates the presence of a reflection. The reflections correspond to the lattice spacings (also referred to as d-spacings) and therefore are determined solely by the size and shape of the unit cell of the crystalline phase. The unit cell is the smallest representative unit in the crystal lattice that can be translated in 3-dimensions to describe the bulk atomic arrangement of the material. Each peak corresponds to at least one reflection, and each reflection has a set of Miller indices (*hkl*) that are related to the unit cell. In a procedure called indexing, these *hkl* indices are assigned and the shape and size of the unit cell thereby determined. This is straightforward for materials with small unit cells and high symmetry, e.g. those with cubic or hexagonal crystal systems for which only one or two lattice parameters need to be determined, but is complicated by the fact that reflections that have similar d-spacings overlap. Overlapping reflections are undesirable, because they result in an ambiguity in the assignment of reflection intensities, which further complicates structure analysis. Crystalline phases with larger unit cells or lower crystallographic symmetry will typically generate more reflections and therefore more overlap. If all reflections can be indexed with a single unit cell, this is a good indication that the phase is pure. If there are reflections that are not indexed by the unit cell, this indicates that an additional crystalline phase is present or that the unit cell is incorrect. A large number of reliable auto-indexing programs are available for determining the unit cell from a list of peak positions (2θ values or d-spacings) [44].

If the positions of the reflections give information about the dimensions of the unit cell, then the relative reflection intensities describe the atomic decoration of the unit cell. In other words, the reflection intensities are related to the types and positions of the atoms (i.e. the electron density), and are the summation of all

X-ray photons diffracted by the crystals. The intensities are therefore related to the average crystal structure of the sample being studied. It is important to note that even a small change in the crystal structure (e.g. after ion exchange) will involve all reflection intensities, although some may be more affected than others.

Peak widths can be used as a first indicator of crystal quality. Narrow peaks mean small errors on the unit cell parameters, and are preferred for structural analysis, because overlapping reflections can be resolved more easily. They are dependent on the intrinsic instrumental peak width, stress or strain, and crystallite size (or, more precisely, size of the coherent domain). The smaller the crystallites, the broader the peaks become, especially for those smaller than 1 μm. For small plate-like or needle-shaped crystals, reflection broadening may occur along just one or two crystallographic directions (the short macroscopic dimensions), respectively, and this is referred to as anisotropic line broadening.

The background in a powder pattern corresponds to everything that cannot be described by the zeolite crystal structure. A high background can indicate the presence of a large amount of amorphous material (e.g. unreacted gel), but air scatter, the tail of the direct beam, or the glass capillary also contribute to the background. X-ray fluorescence from the sample may also occur and add to the background (e.g. if an Fe-containing material is measured using Cu Kα radiation). The latter can be minimized by using a different wavelength.

4 Locating the OSDA

Typically, structural characterization of zeolites using diffraction data is split into two parts, (1) determining the framework structure, and (2) locating extra-framework species in the pores. Much attention has been devoted to the framework structure determination of zeolites using a wide array of techniques. For an up-to-date overview, the reader is referred to [9].

In the following sections, we assume that the framework structure is known, and focus on the problem of locating the OSDA (more generally referred to as structure completion). Structure completion is nearly always performed as an integral part of a Rietveld refinement, so we will first summarize the theory behind Rietveld refinement, paying particular attention to the problem of locating the OSDA and using modern computational tools (simulated annealing) to do so.

4.1 Rietveld Refinement

Rietveld refinement is named after a technique devised by Hugo Rietveld in 1969 to characterize crystalline materials using neutron powder diffraction data [45]. The method uses a least-squares approach to optimize a structural model until its powder pattern matches the measured one. At the time, it signalled a significant advance in the structure analysis of powders, because it was able to deal with the problem of overlapping reflections in a reliable manner.

The theoretical intensity at step i, $Y(2\theta_i)$ can be defined as

$$Y(2\theta_i) = b(2\theta_i) + s \sum_h I_h G_h \phi(2\theta_i - 2\theta_h).$$

Here, I_h is the integrated diffracted intensity of the hth reflection, G_h is a function that corrects for preferred orientation, ϕ is the normalized profile function describing the shape of the reflections (e.g. the half-width, or peak asymmetry), b is the background function that describes the background at $2\theta_i$, and s is a scale factor. The intensities of the reflections depend on many factors, and for a more detailed description the reader is referred to a dedicated textbook such as *The Fundamentals of Crystallography* [46]. In general, it holds that the diffracted intensity is proportional to the structure factor multiplied by its complex conjugate:

$$I_h \propto F_h F_h^* = |F_h|^2.$$

Structure factors are calculated from the atomic parameters of the model using the equation

$$F_h = \sum_j n_j f_j \exp\left[2\pi i\left(hx_j + ky_j + lz_j\right)\right],$$

where F_h is the sum over all atoms in the unit cell. x_j, y_j, and z_j are the fractional coordinates, n_j is the population parameter (also referred to as the occupancy), and f_j the atomic form factor describing the shape of the observed atomic electron density of the jth atom.

In practice, Rietveld refinement can be broken down into the following steps:

1. Collection of the powder diffraction data.
2. Determination of the background function.
3. Determination of the peak shape function.
4. Evaluation of the starting values for the profile parameters.
5. Selection of the space group.
6. Refinement of the unit cell and profile parameters.
7. Addition of a (partial) structural model with geometric restraints.
8. Scaling of the calculated pattern to the observed data.
9. Structure completion.

    (a) Generation of a difference electron density map.
    (b) Interpretation of the difference electron density map.
    (c) Repetition of steps a and b until the structural model is complete.

10. Rietveld refinement of the structural and profile parameters.

Note that Rietveld refinement is initiated only in step 10, and that each of the steps preceding the Rietveld refinement is concerned with estimating the initial parameters as well as possible. This includes the crystal structure, which should be approximately correct and complete. If it is not, Rietveld refinement will simply not converge sensibly. Once the refinement is underway, there are several numerical criteria of fit that can be used to evaluate the state of the refinement, and to judge the quality of the fit. The most commonly reported one is the (weighted) profile agreement value

$$R_{\text{wp}} = \left[\frac{\sum_i w_i \left(Y_i^{\text{obs}} - Y_i^{\text{calc}} \right)^2}{\sum_i w_i \left(Y_i^{\text{obs}} \right)^2} \right]^{0.5}$$

Here, Y_i^{obs} and Y_i^{calc} are the observed and calculated profile intensities at data point i, and w_i is the weight given to each data point, conventionally taken as $w_i = 1/Y_i^{\text{obs}}$.

These steps have not changed much over the last 30 years, and are dealt with in detail in other publications. In particular for points 1–7, readers are referred to the *Rietveld refinement guidelines* [47] and *The Rietveld Method* [48]. Although only step 10 is directly related to the problem of locating the OSDA, the other steps have an important effect on the quality of the outcome. Programs that are widely used for Rietveld refinement include TOPAS [49], GSAS [50], and Fullprof [51].

The peak shape and other profile or instrument parameters are best determined beforehand, as part of a model-free whole profile fit [52, 53]. These parameters can then be kept fixed until the final stages of the Rietveld refinement. In our experience, this improves the stability of the refinement with a partial model. Background correction can also be performed as part of the model-free whole profile fit. Manual removal of the background is preferred, because of the added control it gives over the sometimes erratic and unreliable nature of fitting a polynomial, especially when the structural model is incomplete.

Rietveld refinement of zeolites is usually initiated after a geometric optimization of the framework. A program like TOPAS [49] offers the possibility of performing a geometrical refinement against the angle and distance restraints using the "penalties_only" instruction. Another option is to use a dedicated distance-least-squares algorithm as implemented in the program DLS-76 [54], or a molecular modelling program like GULP [55]. This ensures that the refinement is started with a sensible framework geometry.

4.2 Approximating the Scale Factor

One of the first tasks after the zeolite model has been introduced and profile parameters determined is to scale the calculated pattern so that it matches the observed one. This is a trivial task when the initial structural model closely matches the true one, but is complicated for as-synthesized zeolites if the position of the OSDA has not yet been accounted for. The presence or absence of non-framework species has a noticeable effect on the relative intensities in the low-angle region. A calcined material typically has much higher relative peak intensities in the low-angle region than does the as-synthesized material. The high-angle region is usually less sensitive to the presence of electron density in the zeolite channels, and more affected by small deviations in the positions in the atoms in the framework, atomic displacement (thermal vibrations of the atoms), defects, bond lengths, and the presence of heteroatoms. Figure 3 shows the powder patterns calculated for zeolite SSZ-87 with and without the OSDA included.

This gives a little bit of a chicken-and-egg problem for zeolites, because to determine the scale factor accurately, the position of the OSDA should be known, but to determine the position of the OSDA, the scale factor should be as accurate as possible. To get around this, the scale factor (for the whole pattern) is usually determined using only the high-angle data, which are less sensitive to the position of non-framework species. A reasonable estimate of the scale factor can be made by performing a few cycles of refinement with all other parameters fixed. If the profile

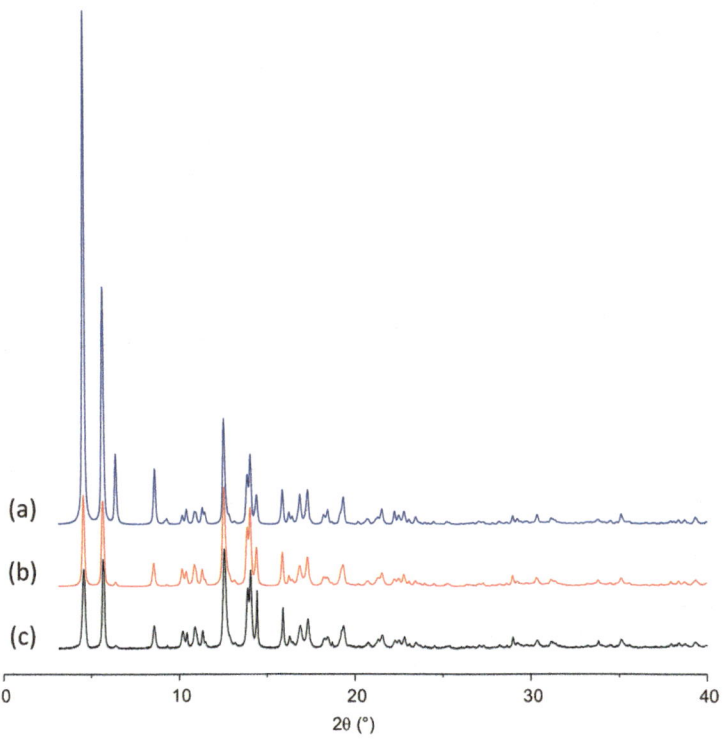

Fig. 3 Powder diffraction pattern of SSZ-87 (**a**) calculated using the framework atoms only, (**b**) calculated using the framework atoms and the OSDA, and (**c**) the observed data

fit at the higher 2θ angles is not good, the background probably needs to be adjusted. In most cases, this gives a good first approximation of the scale factor.

4.3 Generating and Interpreting a Difference Map

To get a low-resolution estimate of the location of the OSDA, a difference Fourier map, sometimes also referred to as simply a difference map or residual electron density map, is generated. These maps can be generated by subtracting the electron density map corresponding to the structural model from that calculated using the "observed" intensities. The word "observed" here is in quotation marks, because the observed intensities are actually extracted by assuming that the ratios of the reflection intensities in an overlap group calculated using the incomplete structural model are the same for the full structure. Although this usually gives a good approximation, it is important to keep in mind that the Fourier difference map is always biased towards the model used to calculate it.

A difference Fourier map thus highlights the structural difference between the observed and calculated data, and should reveal any residual electron density that has not been accounted for by the model. The algorithms needed to generate difference Fourier maps are standard in most suites of crystallographic programs, and can be visualized in programs like VESTA [56] or Chimera [57], if such functionality is not available within the Rietveld refinement program itself.

Figure 4a shows an example of a high-quality difference map obtained during a refinement of as-synthesized ZSM-5. It is trivial to recognize the shape of the TPA ion, even though there is some minor disorder in the "arms" of the cation. Figure 4b shows a more typical difference map, which corresponds to the difference between

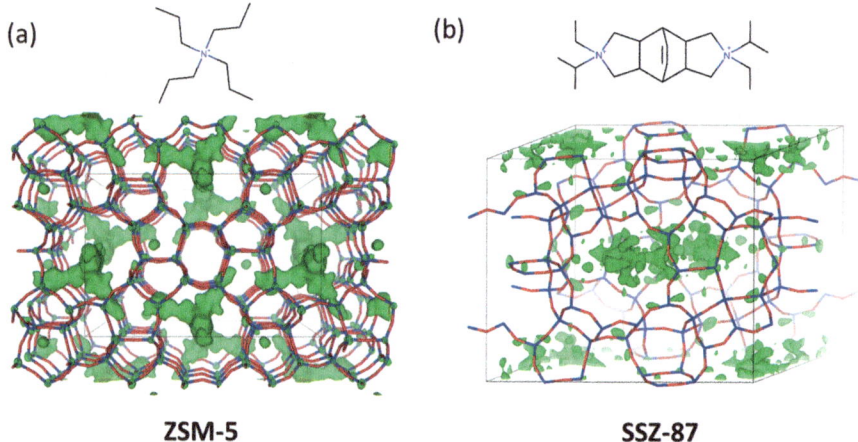

Fig. 4 Example of (**a**) a high-quality difference map for ZSM-5, and (**b**) a normal-quality difference map for SSZ-87. The corresponding OSDAs are shown above the difference maps

the observed and calculated patterns shown for as-synthesized SSZ-87 in Fig. 3. Although it is clear that the residual density describes the rough shape of the OSDA, its actual orientation cannot be discerned.

As noted above, most of the information about the residual electron density in the channel system is in the low-angle portion of the powder diffraction pattern. For the best results, it is therefore important that the scale factor be accurate, and determined using the high-angle data only. If the whole pattern is used to determine the scale factor, the differences will be too low at low angles, and too high at high angles. Figure 5 shows an example for as-synthesized ZSM-5 where the scale factor has been determined in this way. Here, the scale factor is too low by a factor of two, which results in significant electron density located on the framework atoms, and the shape of the OSDA can no longer be made out from the sliver of electron density that remains.

The crux of the problem in locating the OSDA lies in the interpretation of the electron density map. Non-framework species rarely follow the (usually higher) symmetry of their hosts. For example, an atom that is near a mirror plane of the framework (but not on it) will always be accompanied by a mirrored copy of itself on the other side (Fig. 6a). That is, it occupies two positions that are equivalent by

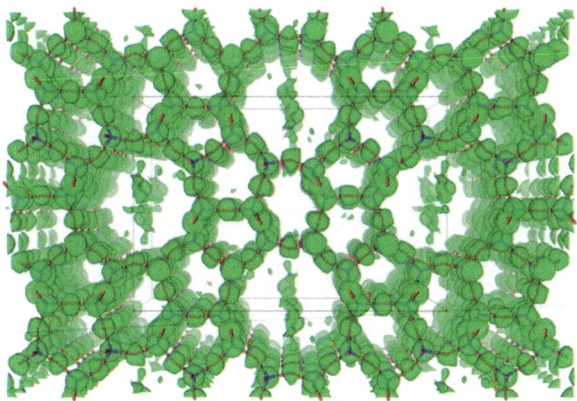

Fig. 5 Example of a difference map for ZSM-5, where the scale factor is refined against the entire diffraction pattern

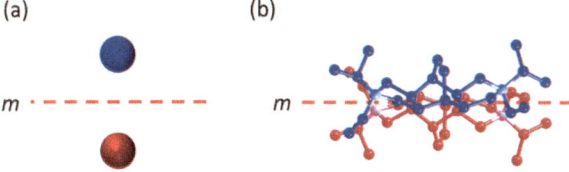

Fig. 6 Example of an (**a**) atom near a mirror plane, and (**b**) the OSDA used to synthesize SSZ-87 near a mirror plane. The equivalent species are coloured red and blue to distinguish them. In both cases, the maximum occupancy is 0.5

symmetry. Because the two positions are too close to be populated at the same time, the maximum occupancy of this position is 0.5. Interpretation of an electron density map corresponding to such a situation is manageable in simple cases, such as a Na^+ ion or a water molecule, but is much more difficult for a complicated OSDA, especially as positions of fourfold symmetry or higher are not uncommon (Fig. 6b). This problem of partial occupancy, or disorder of the OSDA, is common for zeolites, and greatly hampers the interpretation of the difference map. With powder diffraction data, the problem is exacerbated, because the observed reflection intensities are only approximated, and this causes the map to be somewhat more diffuse.

Even with high-quality single-crystal XRD data, individual atoms can rarely be discerned. Although the better intensities alleviate some of the problems related to generating and interpreting a difference map, they do not solve the disorder problem and it is rare that individual atoms can be seen when a symmetry element is nearby. With powder diffraction data, of course, the effect is even more pronounced.

4.4   Simulated Annealing

As part of our own research, we have found the simulated annealing (SA) routine to be very effective in locating organic guest species in zeolites from XRPD data [58]. SA has gained most attention in the area of crystallography as the method of choice for determining organic structures from powder diffraction data [59], but it is flexible enough to deal with various types of other materials also. Indeed, it was used initially as a method for determining zeolite framework structures from powder data [60], and has since been used to locate adsorbed species in zeolites [61–64], the OSDA in germanates [65, 66] and the organic linkers in MOFs [67–69]. Simulated annealing is a global optimization algorithm that is used in a wide variety of computational problems. In the context of crystal structure determination, it belongs to the direct-space class of methods [59]. Direct-space methods were introduced almost 30 years ago, but have been significantly facilitated by the increase in computational power that has become available over the last 20 years. They have now matured to a degree where they are widely implemented in Rietveld refinement suites like TOPAS [49], GSAS [50], and Fullprof [51], or dedicated programs like FOX [70] or DASH [71]. The idea behind direct-space methods is conceptually simple and effective when some prior information about the system, such as the chemical composition, geometry, or connectivity, is known.

Expected atoms, molecules, or fragments are defined as rigid bodies, and placed in the unit cell in a random arrangement. During an iterative procedure, their positions, orientations, and any free torsion angles are modified. After each rearrangement, the corresponding diffraction pattern is calculated and compared with the observed one. If the fit is better than the previous one, the move is accepted; if it is worse, the decision as to whether or not to accept the move is made by a simulated annealing algorithm (accepting more "false" moves initially and fewer as the procedure progresses). After convergence has been reached (based on R_{wp} or some other criterion), the process is repeated until a satisfactory model that fits the data emerges.

Direct-space methods are widely applied for structure determination of pharmaceutical and organic compounds, for which the molecular connectivities are usually known. This also makes them ideally suited for tackling the problem of locating organic guests in inorganic host materials, for which the chemical composition and connectivity are usually known, and can be confirmed using elemental analysis and solid state ^{13}C NMR, respectively. Because the algorithm uses the experimental diffraction pattern directly, the data quality is reflected in the profile parameters used to calculate the pattern for each model generated, so the problem of reflection overlap is circumvented. Furthermore, complicated partial occupancies arising from the OSDA being near or on a high-symmetry site are taken into account automatically.

Simulated annealing therefore offers a straightforward way of interpreting the inevitably low-resolution electron density clouds representing the OSDA in a difference Fourier map in an objective manner. There are several ways of introducing the OSDA molecule as a rigid-body model into the structural model. If the structure of the OSDA has been determined before, online databases such as the Cambridge structural database [72] are obvious sources for obtaining a model of the OSDA. An approximate model can also be generated from scratch, and optimized using molecular modelling programs like Avogadro [73] or Jmol [74]. These allow a three-dimensional model of the OSDA to be constructed interactively or via the SMILES syntax [75]. How the model of the OSDA should be introduced into the Rietveld refinement, depends on the program used. FOX, DASH, Fullprof, GSAS, and TOPAS all expect the rigid-body to be in the Fenske-Hall Z-matrix format, or a modified version thereof. An example of the Z-matrix format used by TOPAS is given for TPA in Fig. 7. The program Babel [76] is useful for handling some of the necessary coordinate transformations.

A rigid-body model has three rotational and three translational parameters associated with it. Additional internal rotations of rigid subgroups may also be

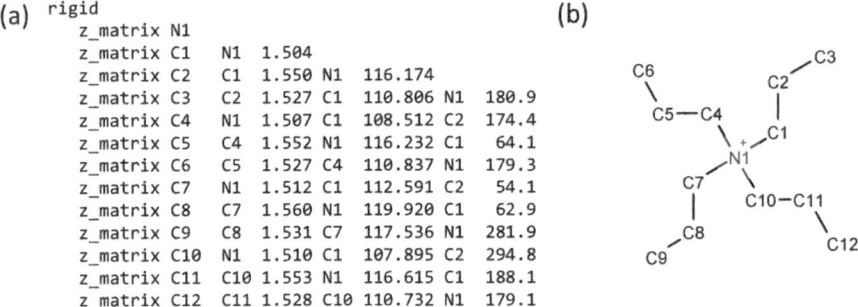

Fig. 7 (a) Example of a Z-matrix in TOPAS, corresponding to (b) tetrapropylammonium (TPA)

introduced, at the cost of more degrees of freedom. After the OSDA is introduced into the model, the global optimization can be initiated. During the optimization process, all parameters, except the six parameters describing the location and orientation of the OSDA, should remain fixed. The occupancy of the OSDA can usually be estimated from the location of the electron density cloud in the difference map, but sometimes allowing the global optimization to find a good value is helpful. Sometimes it is worthwhile to allow the scale factor to refine as well, because it can influence the fitting of the OSDA.

In our experience with TOPAS, we have not found it necessary to deviate from the default settings. Because the problem of locating the OSDA is usually well defined, the simulated annealing procedure takes no longer than a few minutes to determine a reliable starting position for further refinement. If it does not find a solution in that time, it is usually not worth continuing for a longer period. Time is better spent by changing the strategy. For example, the occupancy of the OSDA can be allowed to refine, the background and/or scale factor can be re-estimated, or "anti-bump" restraints can be introduced. If necessary, H atoms can be modelled by scaling the occupancy of the parent atom to account for the additional electron density (see Sect. 4.5). Once a suitable location for the OSDA has been found, the rigid-body parameters can also be refined further. Usually we keep the rigid-body model to refine the position of the OSDA for a couple of cycles of Rietveld refinement, and only after the refinement has converged sufficiently, do we exchange the rigid-body model for a restrained one (e.g. using the tabulated distances and angles in the International Tables for Crystallography, Vol. C, Chapter 9.5 [77] as a guide).

It is useful to know the convention the program uses to relate the rigid-body model to its placement in the unit cell, and how to take advantage of this. For example, for a rigid-body in the program TOPAS, the first atom in the Z-matrix is taken as the centre of the molecule. The second atom defines the z-axis, and the third atom the xz-plane. The x-axis is in the same direction as the a lattice vector, and y is in the ab-plane. By choosing the first three atoms carefully, the placement of the molecule can sometimes be directed by aligning it with a symmetry element. In some cases, a "dummy" atom with occupancy equal to zero is needed to achieve this. In doing so, some of the translations and/or rotations can be fixed, and the search space for the position of the OSDA reduced considerably. Note that the maximum occupancy of the OSDA should be reduced accordingly.

4.5   Hydrogen in OSDAs

XRPD data are not very sensitive to the position of hydrogen atoms as they only contribute one electron and that is delocalized in bonding. However, the contribution of a large number of hydrogen atoms adds up, and can make a small, but noticeable contribution to the observed reflection intensities. For OSDAs, hydrogen can easily contribute up to 30% of the electron density, and therefore needs to be accounted for in one way or another. Some programs allow H atoms to be added in the geometrically expected positions. Otherwise, the population of the parent atom can be inflated to account for the additional electron density. For example, the electron count of a $-CH_3$ group can be modelled by omitting the H atoms, but inflating the population of the C atom by a factor of 1.5 to account for the nine electrons in $-CH_3$.

4.6 R-values, Difference Plots, and Finalizing Matters

The progress of the Rietveld refinement is usually monitored by following the trend in the agreement values (i.e. R_{wp}). Unfortunately, there are no absolute thresholds for these agreement values that signal that a refinement is finished. It is important to keep in mind that the value of R_{wp}, while internally consistent, merely represents the quality of the fit of the model to the data. Therefore, it says absolutely nothing about the quality of the structure.

Therefore, criteria of fit only tell part of the story, and other criteria are equally, if not more, important. A visual representation of the fit of Y^{calc} to Y^{obs} along with a plot of the differences ($Y^{obs} - Y^{calc}$) can reveal problems with the profile parameters, such as peak shape, background, zero correction, sample displacement, and unit cell. The physical meaning of mismatches in intensities is best visualized by calculating a difference electron density map. A difference map can uncover problems with the structure, the position of the OSDA, or with the symmetry, but it is important to remember that the difference map is biased towards the model. Positive residual electron density peaks can indicate missing atoms, while negative peaks may indicate atoms that are only partially occupied or absent. A good indication of the quality of the refinement is how clean the residual density map is (i.e. the map is featureless).

Finally, all of this is in vain if the refined structure does not make chemical sense. It is necessary to monitor the bond angles and distances, intramolecular distances, occupancies, and atomic displacement parameters during the course of the refinement to verify that they correspond to reasonable values.

1. *Is the geometry of the framework reasonable?* For high-silica zeolites, the criteria we use are as follows:

$$1.55 \text{ Å} \leq d(\text{Si} - \text{O}) \leq 1.65 \text{ Å}$$
$$104.0° \leq \sphericalangle(\text{O} - \text{Si} - \text{O}) \leq 114.0°$$
$$\sphericalangle(\text{Si} - \text{O} - \text{Si}) \geq 135.0°$$

2. *Are the intermolecular distances between the framework and any non-framework species reasonable?* For example, for the OSDA, we try to

maintain a framework-to-OSDA distance of at least 3.0 Å unless there is a chemical bond involved.

3. *Are the interatomic distances within the OSDA sensible?* In a restrained refinement these do not usually deviate much from the expected values (i.e. ±0.02 Å).

4. *Are the atomic displacement parameters sensible?* As a rule of thumb, the B_{iso} values we use are for Si: 1.0–2.0, for O: 2.0–3.0, and for the OSDA or extra-framework water: 3.0–5.0.

5. *Are the occupancies sensible?* For example, the occupancy of an OSDA on a position of fourfold symmetry can never exceed 0.25, or the population sum of two atoms occupying the same site (like Ge and Si in a germanosilicate) has to equal one.

Only if all qualitative criteria are fulfilled, and quantitative criteria are stable and as low as possible, can the refinement be considered finished.

5 Examples from Literature

It is clear that locating the OSDA from XRPD data requires a great deal of care and attention in all but the most trivial cases. Perhaps for this reason, the process of locating the OSDA is sometimes described in great detail. This section is intended to highlight a few of those studies where the location of the OSDA from diffraction data is central.

Table 1 shows a summary of studies in which the OSDA has been located over the last 20 years, and includes the 29 cases from the Database of Zeolite Structures mentioned in the introduction, as well as some selected recent studies.

5.1 SSZ-87

Cell: $C2/m$, $a = 21.1727$ Å, $b = 17.8092$ Å, $c = 12.2869$ Å, $\beta = 124.79°$
Composition: $|(C_{22}H_{42}N_2)_2|[Si_{64}O_{128}]$

The borosilicate SSZ-87 (**IFW**) was found as a product in a new synthesis approach for silica-based zeolites using boric acid and ammonium fluoride [78]. The framework structure of SSZ-87 was determined using electron diffraction data, and revealed a framework with large cages interconnected by 8- and 10-ring channels, giving rise to a three-dimensional channel system. After the geometry of the framework had been optimized, the scale factor was estimated and a difference map generated using the method described in Sect. 4.3. Most of the residual electron density is located in the large $[10^2 8^4 6^8 5^8 4^8]$ cavity (Fig. 4b). Although the electron density cloud has the rough shape of the OSDA, its actual orientation is difficult to discern, presumably because of the high degree of reflection overlap (93%) and the

Table 1 List of some zeolites in which the location of the OSDA has been determined using diffraction techniques

Material	Code	Year	Data	Reference	Method[a]
SSZ-23	**STT**	1998	Single-crystal XRD	[88]	DiffMap
UCSB-15GaGe	**BOF**	1998	Single-crystal XRD	[89]	DiffMap
UCSB-7	**BSV**	1998	Single-crystal XRD	[89]	DiffMap
UCSB-9	**SBN**	1998	Single-crystal XRD	[90]	DiffMap
Mu-18	**UEI**	2001	Single-crystal XRD	[91]	DiffMap
UiO-28	**OWE**	2001	Single-crystal XRD	[92]	DiffMap
RUB-10	**RUT**	2001	Powder XRD	[93]	DiffMap
AlPO-SAS	**SAS**	2002	Single-crystal XRD	[94]	DiffMap
AlPO-CHA	**CHA**	2002	Single-crystal XRD	[94]	DiffMap
IST-1	**PON**	2003	Powder XRD	[95]	DiffMap
ITQ-12	**ITW**	2004	Powder XRD	[96]	DiffMap
Nu-6	**NSI**	2004	Powder XRD	[97]	DiffMap
SU-16	**SOS**	2005	Single-crystal XRD	[98]	DiffMap
SIZ-10	**CHA**	2006	Single-crystal XRD	[99]	DiffMap
SU-15	**SOF**	2008	Single-crystal XRD	[100]	DiffMap
SU-32	**STW**	2008	Single-crystal XRD	[100]	DiffMap
SSZ-74	**-SVR**	2008	Powder XRD	[80]	Modelling
PKU-9	**PUN**	2009	Single-crystal XRD	[101]	DiffMap
LSJ-10	**JOZ**	2010	Single-crystal XRD	[102]	DiffMap
RUB-50	**LEV**	2010	Powder XRD	[82]	MEM
STA-2	**SAT**	2010	Powder XRD	[86]	Modelling
Linde type J	**LTJ**	2011	Powder XRD	[103]	DiffMap
CoAPO-CJ69	**JSN**	2012	Single-crystal XRD	[104]	DiffMap
CoAPO-CJ62	**JSW**	2012	Single-crystal XRD	[105]	DiffMap
SSZ-52	**SFW**	2013	Powder XRD	[106]	Modelling
JU-92-300	**JNT**	2013	Single-crystal XRD	[107]	DiffMap
ZnAlPO-57	**AFV**	2014	Powder XRD	[108]	DiffMap
ZnAlPO-59	**AVL**	2014	Powder XRD	[108]	DiffMap
SSZ-45	**EEI**	2014	Powder XRD	[109]	SAnnealing
SSZ-61	***-SSO**	2014	Powder XRD	[79]	SAnnealing
EMM-23	***-EWT**	2014	Powder XRD	[110]	Modelling
SSZ-87	**IFW**	2015	Powder XRD	[78]	SAnnealing
Ge-BEC	**BEC**	2015	Powder XRD	[111]	SAnnealing
DAF-1	**DFO**	2015	Powder XRD	[43]	DiffMap
ITQ-24	**IWR**	2015	Powder XRD	[42]	DiffMap
EU-12	**ETL**	2016	Powder XRD	[112]	Not reported
SSZ-53	**SFH**	2016	Powder XRD	[58]	SAnnealing
SSZ-55	**ATS**	2016	Powder XRD	[58]	SAnnealing
SSZ-56	**SFS**	2016	Powder XRD	[58]	SAnnealing
SSZ-58	**SFG**	2016	Powder XRD	[58]	SAnnealing
SSZ-59	**SFN**	2016	Powder XRD	[58]	SAnnealing
CIT-13	–	2016	Powder XRD	[113]	SAnnealing

(continued)

Table 1 (continued)

Material	Code	Year	Data	Reference	Method[a]
EMM-26	–	2016	Powder XRD	[114]	DiffMap
ITQ-37	**-ITV**	2016	Single-crystal XRD	[115]	DiffMap

[a]*DiffMap* interpretation of the difference map, *SAnnealing* simulated annealing, *Modelling* molecular modelling, *MEM* maximum entropy method

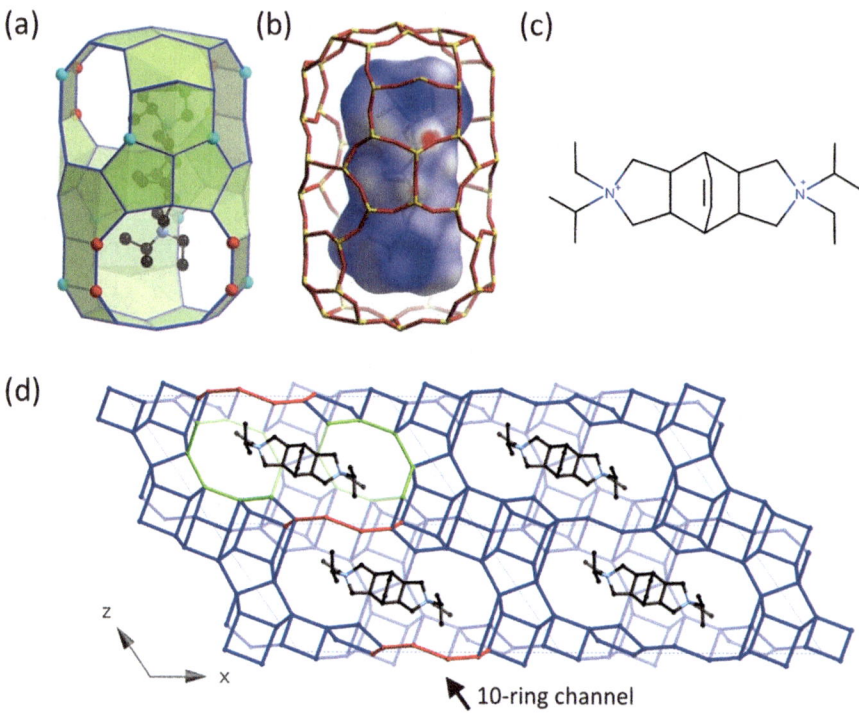

Fig. 8 Cavity of SSZ-87 showing (**a**) the refined position of the OSDA and the location of the two T-sites containing B in red and cyan, and (**b**) the Hirshfeld surface of the OSDA [87]. (**c**) Scheme of the OSDA. (**d**) Projection of the structure of SSZ-87 showing the arrangement of the large cavities and the position of the OSDA. Reprinted with permission from Ref. [78]. Copyright 2015 American Chemical Society

fourfold symmetry of the pore. Therefore, a model of the OSDA (Fig. 8c) was created and optimized using the energy minimization routine in Jmol [74]. It was not clear from the synthesis which configuration the OSDA would adopt, so XRD data on a single crystal of the OSDA were collected, revealing that the OSDA had twofold symmetry with the terminal isopropyl groups in a *cis* configuration. The OSDA was added to the structural model as a rigid-body, and its initial location and orientation were then optimized using simulated annealing. In this process, the OSDA settled on a position of fourfold disorder (point symmetry 2/*m*). The twofold rotation axis of the OSDA appeared to align itself with the twofold rotation axis of the framework, so the description of the OSDA was changed to align it with this axis, reducing the disorder by a factor of two. In the final stages of the refinement, the rigid-body description was replaced by a geometrically restrained one, and B could be located in the framework.

The refined structure reveals that the cavity wraps comfortably around the OSDA, with a minimum distance between the framework and the OSDA of 3.28 Å. All other distances are well over 3.6 Å. Interestingly, the two positively charged N atoms are located near the ends of the OSDA, close to the T-sites partially occupied by B (Fig. 8).

5.2 SSZ-61

Cell: $P2_1/c$, $a = 19.7601$ Å, $b = 10.0747$ Å, $c = 25.2192$ Å, $\beta = 106.92°$
Composition: $|H_4(C_{16}H_{26}N)_4|[Si_{80}O_{164}]$

SSZ-61 (*-SSO) is a high-silica zeolite with large one-dimensional, dumbbell-shaped 18-ring channels, and an interrupted framework structure that is closely related to that of **MTW** and **SFN** [79]. The location of the OSDA in this particular structure was not only determined from the powder diffraction data, it was used as an argument in the structure determination process. Structure analysis of SSZ-61 was initially attempted using a C-centred cell ($a = 19.76$ Å, $b = 5.04$ Å, $c = 25.22$ Å, $\beta = 106.9°$), and resulted in only a partial structure. However, to accommodate the bulky OSDA (7.1 Å × 2.3 Å × 4.2 Å; Fig. 9b), the unit cell had to be doubled along b. The reasoning was that, in this way, two OSDA molecules could be arranged side by side, with their main axes parallel to the channel direction. With the expanded unit cell, the partial framework structure for SSZ-61 could be completed and confirmed against the XRPD data. However, the initial residual electron density map did not reveal any sign of the OSDA in the 18-ring pores, despite the fact that ^{13}C NMR clearly showed that the OSDA was intact (Fig. 9a). This could simply be an effect of the data quality, or problems with the estimation of the scale factor or background function. Therefore, an idealized model of the OSDA was constructed and added as a rigid-body. The initial location and orientation could then be found by applying the simulated annealing routine. This model was converted to a flexible model with geometric restraints for further refinement, which confirmed the framework structure and the location of the OSDA. In the refined structure, each half of the 18-ring channel in SSZ-61 accommodates one OSDA cation to give a total of four per unit cell (Fig. 9c, d). The dumbbell-shaped pore provides room for the bulky part of the OSDA, and allows the positively charged N atom to lie near two terminal O atoms.

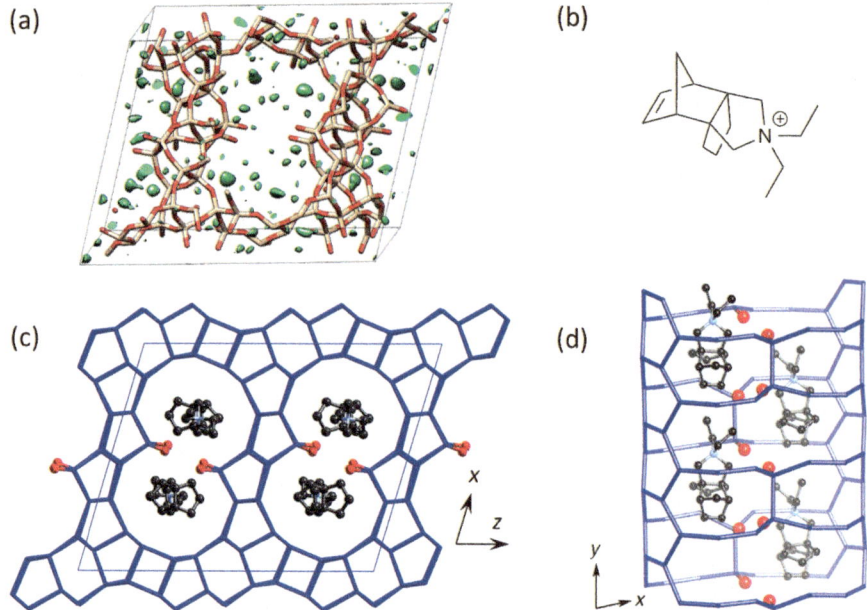

Fig. 9 (**a**) Initial difference electron density map calculated with only the framework structure of SSZ-61 along the channel direction. (**b**) Scheme of the OSDA. Framework structure of SSZ-61 showing the 18-ring channels and the location of the OSDAs viewed (**c**) down the channel and (**d**) from the side. The terminal O atoms are shown in *red* and other O and H atoms have been omitted for clarity

5.3 DAF-1

Cell: $P6/mmc$, $a = 22.2244$ Å, $b = 42.3293$ Å
Composition: $|(C_9H_{17}N_2)_{17.2}F_4(H_2O,OH,F)_{29.8}|[Zn_{6.1}Al_{125.9}P_{132}O_{528}]$

In a study exploring the use of ionic liquid reactions based on imidazolium halides, Pinar et al. found a new zincoaluminophosphate zeolite with the **DFO** framework type that they termed Zn-DAF-1 [43]. **DFO** has an open framework structure with a three-dimensional channel system consisting of two separate parallel 12-ring channels along the c-axis, linked via perpendicular 10- and 8-ring channels. With the goal of locating the Zn in the framework and the N,N'-di-isopropyl-imidazolium (DIPI) ions (Fig. 10c) in the pores, a full structure analysis was performed. Rietveld refinement was initiated using the published coordinates for Mg-DAF-1, and after a scale factor had been estimated using the steps described in Sect. 4, a difference map was generated. Pinar et al. write that although the difference map revealed a few small clouds of electron density in the void volume of the structure (Fig. 10a, b), the high symmetry of the framework made it difficult to interpret these clouds. Therefore, they made an educated guess for the approximate positions of the OSDA. Although these four positions were not necessarily accurate, it did improve the structural model so that

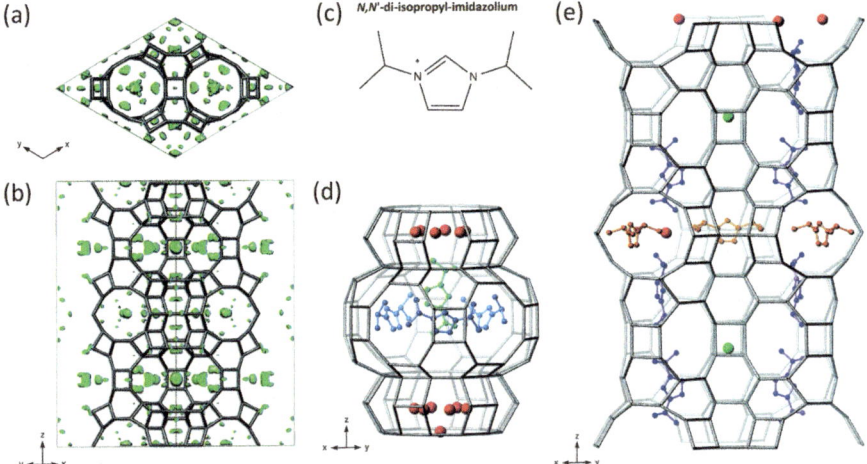

Fig. 10 Initial difference electron density map calculated with only the framework structure of DAF-1 along (**a**) the [001] and (**b**) [110] directions. (**c**) Scheme of the OSDA. Possible arrangements of (**d**) DIPI-1 (*blue*) and DIPI-2 (*green*) in the large cavity and the water molecules in the smaller cavity (*red* balls), and (**e**) DIPI-3 (*purple*), DIPI-4 (*orange*), the fluoride ions (*green* balls), and the water molecules (*red* balls). Framework oxygen atoms have been omitted for clarity. Reproduced from Ref. [43] with permission from the Centre National de la Recherche Scientifique (CNRS) and The Royal Society of Chemistry

the scale factor could be refined and the Zn atoms located. The four crystallograpically independent sites (two per 12-ring channel) that could host the OSDA were found, but attempts to refine the atomic positions failed, even with strong geometric restraints. In the end, a model with hard constraints to keep the imidazolium rings planar, and the isopropyl moieties chemically sensible, led to a better positioning of the OSDA species and a cleaner final difference map (Fig. 10d, e). H atoms on the OSDAs were taken into account by increasing the population factors of the C atoms, and the occupancy of each OSDA was refined, giving a total of 17.2 OSDA atoms in the 22 positions available per unit cell. The authors comment that this model yielded a reasonable geometry and a good profile fit.

This study is interesting, because the position of four crystallographically independent OSDAs could be determined from XRPD data, and highlights the necessity for a careful, systematic approach to structure refinement. The main difficulty was that the DIPI does not follow the high symmetry of the framework and is therefore disordered. For example, one of the OSDA atoms lies close to the special position where the sixfold rotation axis and the mirror plane perpendicular to it intersect. It is therefore disordered over 12 equivalent positions, but only one of these 12 positions is occupied at a time. This not only makes it difficult to interpret the shape of the density cloud, but also dilutes its intensity by a factor 12. Another OSDA is located on a similar symmetry element, but there three out of 12 positions can be occupied simultaneously. It is worth noting that locating four OSDAs using simulated annealing would therefore be problematic.

5.4 SSZ-74

Cell: Cc, $a = 20.4756$ Å, $b = 13.3839$ Å, $c = 20.0859$ Å, $\beta = 102.1°$
Composition: $|C_{16}H_{34}N_2)_4|[Si_{92}\square_4O_{184}(OH)_8]$

SSZ-74 (**-SVR**) is a high-silica zeolite catalyst with a 3-dimensional 10-ring channel system, and ordered Si vacancies [80]. Structure analysis was performed using high-resolution powder diffraction data, collected on the as-synthesized material, because significant line broadening had been observed upon the removal of the OSDA. Although difference Fourier maps showed a cloud of electron density in the pores, the individual atoms of the 1,6-(*N*-methylpyrrolidinium)-hexane (Fig. 11d) used to synthesize the material could not be resolved. Therefore molecular modelling was used to estimate the positions using the energy-optimization docking procedure described by Burton et al. [81]. Rietveld refinement was started using this as the initial position of the OSDA, with geometric restraints imposed on the bond distances and angles of both the framework and the OSDA. The final refinement showed the OSDA taking its place in the centre of the large cavity (Fig. 11a), with the two closest contacts between the terminal oxygen atoms at the Si vacancy to the two nitrogen atoms of the doubly charged OSDA at 3.62 Å (O33···N5) and 3.58 Å (O3···N13).

The position found at the end of the refinement deviates significantly from the one determined initially with molecular modelling (Fig. 11b, c), perhaps because it was not known when the molecular modelling was performed that there was a Si vacancy in the framework. This example nicely highlights the potential of combining molecular modelling with Rietveld refinement and the fact that refinement can indeed correct deficiencies in the initial model.

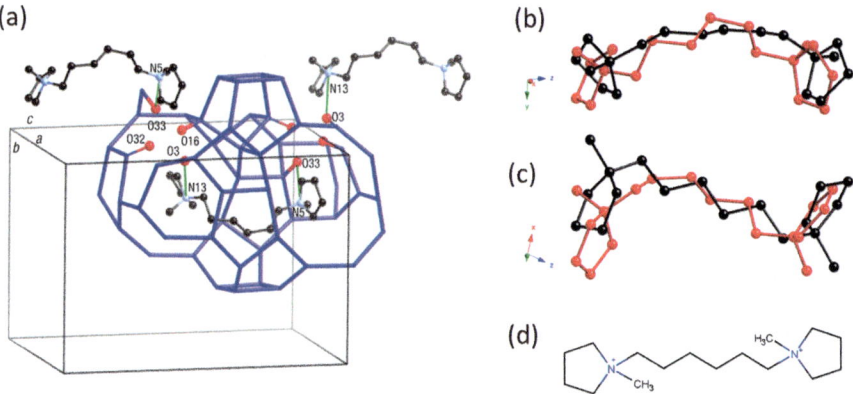

Fig. 11 (**a**) A portion of the structure of SSZ-74 showing the position of the OSDA. (**b**) Conformations of the OSDA found using molecular modelling assuming a fully connected framework structure in *black* and from the final refinement in *red*, and (**c**) the same projection rotated by 90°. (**d**) Scheme of the OSDA

5.5   RUB-50

Cell: $R\bar{3}m$, $a = 13.1090$ Å, $c = 22.4740$ Å

Composition: $|Na_{0.24}H_{4.9}\cdot(C_6H_{17}NO)_{5.8}|[Si_{48.86}Al_{5.14}O_{108}]$

RUB-50 is an aluminosilicate zeolite with a two-dimensional 8-ring channel system that can be synthesized using diethyldimethylammonium (DEDMA; Fig. 12d) as the OSDA. It has a **LEV**-type framework structure, which can be described as an AABCCABBC sequence of 6-ring layers (it is a member of the ABC-6 family of framework structures). As part of a study into the physicochemical properties of RUB-50, Yamamoto et al. performed a structure refinement of as-synthesized RUB-50 using lab XRPD data [82]. The refinement was initiated using the **LEV** framework model. The OSDA was approximated using a dummy atom with the scattering amplitude of $C_6H_{16}N$, and then Na^+ could be located in the difference map (Fig. 12a). Although the refinement converged, and resulted in a good fit to the data, the authors did not leave it at that. The disordered arrangement of the OSDA in the *lev* cavity was found from the electron density distribution determined using the maximum entropy method (MEM) [83] with the program PRIMA (now Dysnomia) [84]. MEM is itself model free and only structure factors from isolated (non-overlapping) reflections were used for the analysis. After the initial MEM analysis, the electron density distribution was redetermined using MEM-based pattern fitting (MFP), which combines MEM analysis and whole-pattern fitting, making it efficient in representing highly disordered atomic arrangements. In this way, the position of the disordered OSDA, OH^-, and water molecules could be found (Fig. 12b). Their results suggest that the electron density distribution of the OSDA is anisotropically elongated to form a "garlic" shape (Fig. 12b). Yamamoto et al. note that this is probably because the OSDA is disordered around the threefold axis and because hydroxyl ions or water molecules are close to the OSDA. These findings were then fed back into the Rietveld refinement to confirm that the location of the OSDA is sensible.

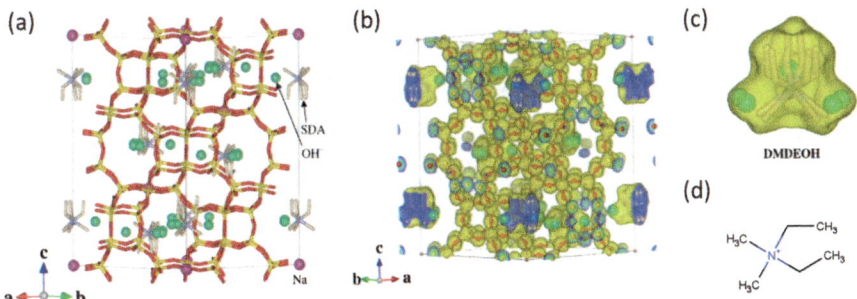

Fig. 12 (**a**) Structure of the as-made RUB-50 obtained from the Rietveld refinement, and electron density maps of (**b**) the structure and (**c**) the disordered OSDA, both obtained using MPF refinement. (**d**) Scheme of the OSDA. Reprinted from Ref. [82] with permission from Elsevier

5.6 STA-2

Cell: $R\bar{3}$, $a = 12.726$ Å, $c = 30.939$ Å
Composition: $|(BDAB)_3|[Al_{12}P_{12}O_{24}]$, BDAB $= C_{18}H_{32}N_2$

The aluminophosphate zeolite STA-2 (**SAT**) was initially prepared and character-ized by Noble et al. in 1997 using 1,4-bis-*N*-quinuclidiniumbutane (BQNB; Fig. 13c) as the OSDA [85]. Its framework structure could be determined using synchrotron microcrystal XRD, and was found to be a member of the ABC-6 family with an ABBBCBCCACAAB stacking sequence of 6-ring layers, giving it a three-dimensional 8-ring channel system. All atoms of the OSDA could be located from the difference map, confirming its location along the threefold axis (Fig. 13a). The authors note that although there is a disorder in the positions of the atoms within the tetramethylene chain, as seen in the larger temperature factors and chemically inaccurate bond lengths for C–N and C–C, the positions of the quinuclidinium fragments are particularly well described. This is probably because the quinu-clidinium fragments have a threefold axis that can line up with that of the frame-work, so only the atoms in the connecting methylene chain are disordered.

A follow-up study appeared in 2010, with the aim of finding a cheaper OSDA to produce STA-2. To do this, Castro et al. investigated a series DABCO-derived OSDAs (DABCO = diazabicyclooctane) using molecular modelling, and found bis-diazabicyclooctane-butane (BDAB; Fig. 13d) to give the lowest framework stabilization energy (Fig. 13b) [86]. This is not surprising, because BDAB is essentially identical to BQNB, with the exception that two tertiary C atoms at either end are replaced by two tertiary amine N atoms. They were able to produce

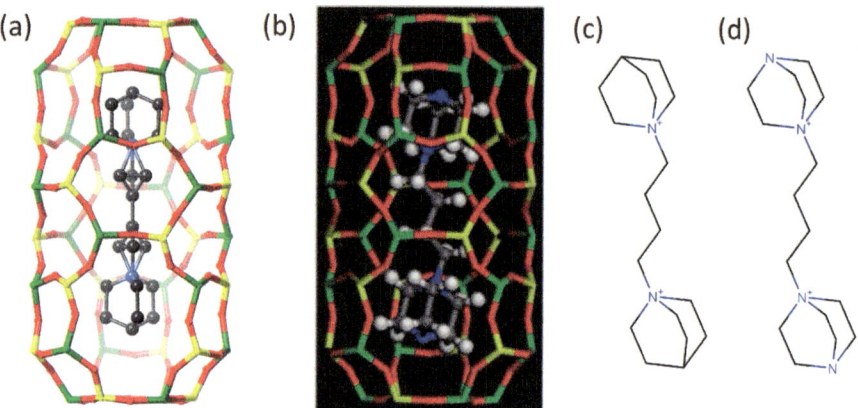

Fig. 13 *SAT*-cage including the position of the OSDA as determined using (**a**) microcrystal diffraction data and (**b**) molecular modelling. Scheme of (**c**) 1,4-bis-*N*-quinuclidiniumbutane (BQNB) and (**d**) bis-diazabicyclooctane-butane (BDAB), representing the OSDAs in (**a**) and (**b**), respectively. (**a**) Reprinted (adapted) with permission from [86]. Copyright 2010 American Chemical Society

STA-2 using BDAB as the OSDA, and performed a Rietveld refinement with the XRPD data to confirm the structure of STA-2.

6 Discussion/Conclusion

From our work on as-synthesized zeolites, we find that the OSDA is often highly ordered, despite the fact that their locations usually have to be described as a superposition of multiple discrete positions. Indeed, they probably do occupy each of these symmetry-related positions in different unit cells in the crystal. The OSDA just does not have the fully symmetry of the zeolite framework structure, and therefore appears to be more disordered than it is. The examples presented in this chapter demonstrate that given enough attention, the OSDA can be located systematically using diffraction methods. Difficulties in locating the OSDA can often be attributed to the lack of high-quality data, in particular when working with XRPD data, for which reflection overlap obfuscates the true reflection intensities. Part of the information that is lost can be overcome by using modern approaches like simulated annealing and/or by using supplementary information gained from other measurements. MAS NMR is particularly useful in determining whether the OSDA is intact and/or protonated.

The reliability of the structure analysis will also depend upon the specific problem at hand. For example, a rigid OSDA will generally be easier to locate than a flexible one, an OSDA with at least a subset of the symmetry of the framework will be easier to refine than one with no symmetry relationship, and an OSDA at a position of low symmetry will be easier to locate and recognize than one near a symmetry element.

Despite the constant improvement in the quality of diffraction data, and structure determination and refinement techniques, locating an OSDA from diffraction data cannot yet be taken for granted. We hope that some of the information provided here will provide the reader with some insight into how the problem can be approached in a systematic manner and where some of the pitfalls lie.

Acknowledgements S.S. thanks the Swiss National Science Foundation for financial support (project number: 165282) and L.M. Chevron ETC.

References

1. Barrer RM, Denny PJ (1961) J Chem Soc 971–982. doi:10.1039/JR9610000971
2. Lok BM, Cannan TR, Messina CA (1983) Zeolites 3(4):282–291
3. Lobo RF, Zones SI, Davis ME (1995) J Incl Phenom Macrocycl Chem 21(1–4):47–78
4. Gies H, Marler B (1992) Zeolites 12(1):42–49
5. Gies H (1994) In: Jansen JC, Stöcker M, Karge HG, Weitkamp J (eds) Advanced zeolite science and applications, vol 85, Elsevier, pp 295–327

6. Kubota Y, Helmkamp MM, Zones SI, Davis ME (1996) Microporous Mater 6(4):213–229
7. Millini R, Carluccio L, Frigerio F, O'Neil Parker W, Bellussi G (1998) Micropor Mesopor Mat 24(4–6):199–211
8. Wagner P, Nakagawa Y, Lee GS, Davis ME, Elomari S, Medrud RC, Zones SI (2000) J Am Chem Soc 122(2):263–273
9. McCusker LB, Baerlocher C (in press) In: Vol H, Gilmore CJ, Kaduk J, Schenk H (eds) International tables for crystallography, Wiley
10. Baerlocher C, McCusker LB. Database for zeolite structures. http://www.iza-structure.org/databases/
11. Jordá JL, Rey F, Sastre G, Valencia S, Palomino M, Corma A, Segura A, Errandonea D, Lacomba R, Manjón FJ, Gomis Ó, Kleppe AK, Jephcoat AP, Amboage M, Rodríguez-Velamazán JA (2013) Angew Chem 125(40):10652–10656
12. Roth WJ, Nachtigall P, Morris RE, Wheatley PS, Seymour VR, Ashbrook SE, Chlubná P, Grajciar L, Položij M, Zukal A, Shvets O, Čejka J (2013) Nat Chem 5(7):628–633
13. Chlubná-Eliásová P, Tian Y, Pinar AB, Kubr UM, Cejka J, Morris RE (2014) Angew Chem 126(27):7168–7172
14. Breck DW, Eversole WG, Milton RM, Reed TB, Thomas TL (1956) J Am Chem Soc 78 (23):5963–5972
15. Reed TB, Breck DW (1956) J Am Chem Soc 78(23):5972–5977
16. Gramlich V, Meier WM (1971) Z Kristallografiya 133(1–6):134–149
17. Baerlocher C, Meier WM (1969) Helv Chim Acta 52(7):1853–1860
18. Baerlocher C, Meier WM (1970) Helv Chim Acta 53(6):1285–1293
19. Argauer RJ, Landolt GR (1972) Crystalline zeolite ZSM-5 and method of preparing the same. US3702886 A
20. Flanigen EM, Bennett JM, Grose RW, Cohen JP, Patton RL, Kirchner RM, Smith JV (1978) Nature 271(5645):512–516
21. Kokotailo GT, Lawton SL, Olson DH, Meier WM (1978) Nature 272(5652):437–438
22. Price GD, Pluth JJ, Smith JV, Araki T, Bennett JM (1981) Nature 292(5826):818–819
23. Baerlocher C (1984) In: Olson DH, Bisio A (eds) 6th Int Zeolite Conf.; Guildford Butterworths: Reno, pp 823–833
24. van Koningsveld H, van Bekkum H, Jansen JC (1987) Acta Cryst B 43(2):127–132
25. Bennett JM, Cohen JP, Flanigen EM, Pluth JJ, Smith JV (1983) In: Intrazeolite chemistry, ACS Symposium Series, vol 218, American Chemical Society, pp 109–118
26. Parise JB (1984) J Chem Soc Chem Commun 21:1449–1450
27. Parise JB (1984) Acta Crystallogr C 40(10):1641–1643
28. Gies H (1983) Z Kristallogr 164(3–4):247–257
29. Gerke H, Gies H (1984) Z Kristallogr 166(1–4):11–22
30. Gies H (1984) Z Kristallogr 167(1–4):73–82
31. Gies H (1986) Z Kristallogr 175(1–4):93–104
32. McCusker L (1988) J Appl Cryst 21(4):305–310
33. Harrison WTA, Martin TE, Gier TE, Stucky GD (1992) J Mater Chem 2(2):175–181
34. Harrison WTA, Nenoff TM, Eddy MM, Martin TE, Stucky GD (1992) J Mater Chem 2 (11):1127–1134
35. Parise JB (1986) Acta Cryst C 42(6):670–673
36. Loiseau T, Férey G (1992) J Chem Soc Chem Commun 17:1197–1198
37. Weigel SJ, Morris RE, Stucky GD, Cheetham AK (1998) J Mater Chem 8(7):1607–1611
38. Wragg DS, Bull I, Hix GB, Morris RE (1999) Chem Commun 20:2037–2038
39. Davis ME, Saldarriaga C, Montes C, Garces J, Crowder C (1988) Zeolites 8(5):362–366
40. Richardson JW, Smith JV, Pluth JJ (1989) J Phys Chem 93(25):8212–8219
41. McCusker LB, Baerlocher C, Jahn E, Bülow M (1991) Zeolites 11(4):308–313
42. Pinar AB, McCusker LB, Baerlocher C, Schmidt J, Hwang S-J, Davis ME, Zones SI (2015) Dalton Trans 44(13):6288–6295

43. Pinar AB, McCusker LB, Baerlocher C, Hwang S-J, Xie D, Benin AI, Zones SI (2016) New J Chem 40(5):4160–4166
44. Bergmann J, Le Bail A, Shirley R, Zlokazov V (2004) Z Kristallogr Cryst Mater 219 (12):783–790
45. Rietveld HM (1969) J Appl Cryst 2(2):65–71
46. Giacovazzo C, Monaco HL, Artioli G, Viterbo D, Ferraris G, Gilli G, Zanotti G, Catti M (2002) Fundamentals of crystallography. In: Giaccovazzo C Series Ed., Oxford Science Publications
47. McCusker LB, Von Dreele RB, Cox DE, Louër D, Scardi P (1999) J Appl Crystallogr 32 (1):36–50
48. Young RA (1993) The rietveld method. In: Young RA, Series Ed, Oxford University Press
49. Coelho AA (2012) TOPAS-ACADEMIC v5.0
50. Toby BH, Von Dreele RB (2013) J Appl Crystallogr 46(2):544–549
51. Rodríguez-Carvajal J (1990) In: Galy J, Louër D (eds) Abstracts of the meeting on powder diffraction (Toulouse, France), pp 127–128
52. Le Bail A, Duroy H, Fourquet JL (1988) Mater Res Bull 23(3):447–452
53. Pawley GS (1981) J Appl Crystallogr 14(6):357–361
54. Baerlocher C, Hepp A, Meier WM (1976) DLS-76
55. Gale JD, Rohl AL (2003) Mol Simul 29(5):291–341
56. Momma K, Izumi F (2011) J Appl Crystallogr 44(6):1272–1276
57. Pettersen EF, Goddard TD, Huang CC, Couch GS, Greenblatt DM, Meng EC, Ferrin TE (2004) J Comput Chem 25(13):1605–1612
58. Smeets S, McCusker LB, Baerlocher C, Elomari S, Xie D, Zones SI (2016) J Am Chem Soc 138(22):7099–7106
59. David WIF, Shankland K (2008) Acta Cryst A 64(1):52–64
60. Deem MW, Newsam JM (1989) Nature 342(6247):260–262
61. Porcher F, Borissenko E, Souhassou M, Takata M, Kato K, Rodriguez-Carvajal J, Lecomte C (2008) Acta Crystallogr B 64(6):713–724
62. Fyfe CA, Lee JSJ, Cranswick LMD, Swainson I (2008) Micropor Mesopor Mat 112 (1–3):299–307
63. Meilikhov M, Yusenko K, Fischer RA (2010) Dalton Trans 39(45):10990–10999
64. Dejoie C, Martinetto P, Tamura N, Kunz M, Porcher F, Bordat P, Brown R, Dooryhée E, Anne M, McCusker LB (2014) J Phys Chem C 118(48):28032–28042
65. Inge AK, Huang S, Chen H, Moraga F, Sun J, Zou X (2012) Cryst Growth Des 12 (10):4853–4860
66. Xu Y, Liu L, Chevrier DM, Sun J, Zhang P, Yu J (2013) Inorg Chem 52(18):10238–10244
67. Chen R, Yao J, Gu Q, Smeets S, Baerlocher C, Gu H, Zhu D, Morris W, Yaghi OM, Wang H (2013) Chem Commun 49(82):9500–9502
68. Reimer N, Reinsch H, Inge AK, Stock N (2015) Inorg Chem 54(2):492–501
69. Halis S, Inge AK, Dehning N, Weyrich T, Reinsch H, Stock N (2016) Inorg Chem 55 (15):7425–7431
70. Favre-Nicolin V, Černý R (2002) J Appl Crystallogr 35(6):734–743
71. David WIF, Shankland K, van de Streek J, Pidcock E, Motherwell WDS, Cole JC (2006) J Appl Crystallogr 39(6):910–915
72. Allen FH (2002) Acta Crystallogr B 58(3):380–388
73. Hanwell MD, Curtis DE, Lonie DC, Vandermeersch T, Zurek E, Hutchison GR (2012) J ChemInform 4(1):17–17
74. Hanson RM (2010) J Appl Cryst 43(5):1250–1260
75. Weininger D (1988) J Chem Inf Comput Sci 28(1):31–36
76. O'Boyle NM, Banck M, James CA, Morley C, Vandermeersch T, Hutchison GR (2011) J Chem 3(1):33
77. Prince E (ed) (2006) International tables for crystallography: mathematical, physical and chemical tables, vol C, 1st edn. Fuess H, Hahn T, Wondratschek H, Müller U, Shmueli U,

Prince E, Authier A, Kopský V, Litvin DB, Rossmann MG, Arnold E, Hall S, McMahon B, Series Eds, International tables for crystallography; International Union of Crystallography, Chester, England

78. Smeets S, McCusker LB, Baerlocher C, Xie D, Chen C-Y, Zones SI (2015) J Am Chem Soc 137(5):2015–2020
79. Smeets S, Xie D, Baerlocher C, McCusker LB, Wan W, Zou X, Zones SI (2014) Angew Chem 126(39):10566–10570
80. Baerlocher C, Xie D, McCusker LB, Hwang S-J, Chan IY, Ong K, Burton AW, Zones SI (2008) Nat Mater 7(8):631–635
81. Burton A, Lee GS, Zones SI (2006) Micropor Mesopor Mat 90(1–3):129–144
82. Yamamoto K, Ikeda T, Onodera M, Muramatsu A, Mizukami F, Wang Y, Gies H (2010) Micropor Mesopor Mat 128(1–3):150–157
83. Collins DM (1982) Nature 298(5869):49–51
84. Momma K, Ikeda T, Belik AA, Izumi F (2013) Powder Diffract 28(3):184–193
85. Noble GW, Wright PA, Kvick Å (1997) Dalton Trans (23):4485–4490
86. Castro M, Seymour VR, Carnevale D, Griffin JM, Ashbrook SE, Wright PA, Apperley DC, Parker JE, Thompson SP, Fecant A, Bats N (2010) J Phys Chem C 114(29):12698–12710
87. Spackman MA, Jayatilaka D (2009) CrstEngComm 11(1):19–32
88. Camblor MA, Díaz-Cabañas M-J, Perez-Pariente J, Teat SJ, Clegg W, Shannon IJ, Lightfoot P, Wright PA, Morris RE (1998) Angew Chem Int Ed 37(15):2122–2126
89. Bu X, Feng P, Gier TE, Zhao D, Stucky GD (1998) J Am Chem Soc 120(51):13389–13397
90. Bu X, Feng P, Stucky GD (1998) J Am Chem Soc 120(43):11204–11205
91. Josien L, Simon A, Gramlich V, Patarin J (2001) Chem Mater 13(4):1305–1311
92. Ove Kongshaug K, Fjellvåg H, Petter Lillerud K (2001) J Mater Chem 11(4):1242–1247
93. Marler B, Werthmann U, Gies H (2001) Micropor Mesopor Mat 43(3):329–340
94. Wheatley PS, Morris RE (2002) J Solid State Chem 167(2):267–273
95. Jordá JL, McCusker LB, Baerlocher C, Morais CM, Rocha J, Fernandez C, Borges C, Lourenco JP, Ribeiro MF, Gabelica Z (2003) Micropor Mesopor Mat 65(1):43–57
96. Yang X, Camblor MA, Lee Y, Liu H, Olson DH (2004) J Am Chem Soc 126 (33):10403–10409
97. Zanardi S, Alberti A, Cruciani G, Corma A, Fornés V, Brunelli M (2004) Angew Chem Int Ed 43(37):4933–4937
98. Li Y, Zou X (2005) Angew Chem Int Ed 44(13):2012–2015
99. Parnham ER, Morris RE (2006) Chem Mater 18(20):4882–4887
100. Tang L, Shi L, Bonneau C, Sun J, Yue H, Ojuva A, Lee B-L, Kritikos M, Bell RG, Bacsik Z, Mink J, Zou X (2008) Nat Mater 7(5):381–385
101. Su J, Wang Y, Wang Z, Lin J (2009) J Am Chem Soc 131(17):6080–6081
102. Armstrong JA, Weller MT (2010) J Am Chem Soc 132(44):15679–15686
103. Broach RW, Kirchner RM (2011) Micropor Mesopor Mat 143(2–3):398–400
104. Liu Z, Song X, Li J, Li Y, Yu J, Xu R (2012) Inorg Chem 51(3):1969–1974
105. Shao L, Li Y, Yu J, Xu R (2012) Inorg Chem 51(1):225–229
106. Xie D, McCusker LB, Baerlocher C, Zones SI, Wan W, Zou X (2013) J Am Chem Soc 135 (28):10519–10524
107. Wang Y, Li Y, Yan Y, Xu J, Guan B, Wang Q, Li J, Yu J (2013) Chem Commun 49 (79):9006–9008
108. Broach RW, Greenlay N, Jakubczak P, Knight LM, Miller SR, Mowat JPS, Stanczyk J, Lewis GJ (2014) Micropor Mesopor Mat 189:49–63
109. Smeets S, Xie D, McCusker LB, Baerlocher C, Zones SI, Thompson JA, Lacheen HS, Huang H-M (2014) Chem Mater 26(13):3909–3913
110. Willhammar T, Burton AW, Yun Y, Sun J, Afeworki M, Strohmaier KG, Vroman H, Zou X (2014) J Am Chem Soc 136(39):13570–13573
111. Smeets S, Koch L, Mascello N, Sesseg J, Hernández-Rodríguez M, Mitchell S, Pérez-Ramírez J (2015) CrstEngComm 17(26):4865–4870

112. Bae J, Cho J, Lee JH, Seo SM, Hong SB (2016) Angew Chem 128(26):7495–7499
113. Kang JH, Xie D, Zones SI, Smeets S, McCusker LB, Davis ME (2016) Chem Mater 28 (17):6250–6259
114. Guo P, Strohmaier K, Vroman H, Afeworki M, Ravikovitch PI, Paur CS, Sun J, Burton A, Zou X (2016) Inorg Chem Front 3(11):1444–1448
115. Chen F-J, Gao Z-H, Liang L-L, Zhang J, Du H-B (2016) CrstEngComm 18(15):2735–2741

Struct Bond (2018) 175: 75–102
DOI: 10.1007/430_2017_16
© Springer International Publishing AG 2017
Published online: 15 December 2017

Molecular Modelling of Structure Direction Phenomena

Check for updates

Alessandro Turrina and Paul A. Cox

Abstract Organic structure-directing agents (OSDAs) are widely used in the synthesis of zeolitic materials. Molecular modelling methods are playing a key part in helping to establish the role of the OSDA in the synthesis process. Moreover, modelling is increasingly being used to design and screen new OSDAs for specific targets. This review aims to provide an overview of the methods used to investigate the relationship between OSDAs and their zeolitic products and to provide a series of examples to highlight the important contribution that modelling is making in this field.

Keywords De novo method • High-throughput screening • OSDA • Template modelling • Zeolite

Contents

A. Turrina
Johnson Matthey Technology Centre, Billingham, UK
e-mail: alessandro.turrina@matthey.com

P.A. Cox (✉)
School of Pharmacy and Biomedical Sciences, University of Portsmouth, Portsmouth, UK
e-mail: paul.cox@port.ac.uk

1   Introduction

A large number of synthetic parameters may affect the hydrothermal synthesis of zeolite and zeotype materials. They can be divided into two categories: gel composition (sources of inorganic reagents, framework cations, organic additive, mineraliser, solvent and seeds) and synthesis conditions (sequence of mixing, temperature, time, pH, concentration and agitation). All are important, and small variations can lead to different products. One important objective during zeolite synthesis is to achieve control of the pore dimensions and their connectivity. At the present time, this is best accomplished through the use of additives referred to as organic structure-directing agents (OSDAs). Hence, an understanding of the relationship between OSDAs and their zeolitic products is of fundamental importance in understanding the role of the OSDA in the crystallisation process.

Molecular modelling methods have been used extensively to investigate the structure-directing role of these OSDAs. The key goals of the calculations have been to:

1. Identify the location and energetics of the OSDA inside the zeolite product. This type of work has played an important role in helping us to understand the fundamental relationship between OSDAs and the product they form.
2. Understand the role of the OSDA and structure-building units (SBUs) in the detailed energetics and pathways of the synthesis process.
3. Determine the role of the OSDA in influencing the location and loading of aluminium and other heteroatoms within the framework. The distribution of these ions can have an important influence on the catalytic behaviour of the framework.
4. Rationally design new templates in order to synthesise new zeolites or to identify cheaper or more selective OSDAs for the preparation of existing structures.

The aim of this chapter is to summarise the key methodologies that have been used to investigate the relationship between the OSDAs and their product frameworks, followed by selected illustrations to demonstrate the type of information that these calculations can provide.

2 The Basic Methodologies

2.1 Molecular Mechanics

The majority of calculations reported in the literature involving zeolites and their OSDAs have used the molecular mechanics (MM) method [1]. The MM method is based on a very simple approach that allows calculations involving hundreds, or even thousands, of atoms to be completed quickly using fairly modest computational resources. The basis of this method is that each atom is represented by a hard sphere and each bond is represented by a spring. The classical laws of physics are then applied to calculate the energy of the system under investigation. The total energy is determined by a force field which comprises a number of component energy terms that are summed together to give the total energy. A simple force field has the following terms:

$$E_{total} = E_{stretch} + E_{bend} + E_{torsion} + E_{VDW} + E_{electrostatic} \tag{1}$$

The first term on the right-hand side of the equation, $E_{stretch}$, looks at every pair of atoms and calculates an energy according to the bond length of the model, r. A very simple way to represent this is via Hooke's law (Eq. 2) which tells us that the energy required to stretch or compress a bond is proportional to the square of the extension (or compression) of the bond:

$$E_{stretch}(r) = \frac{1}{2}k(r - r_0)^2 \tag{2}$$

where r is the current bond distance, r_0 is the equilibrium bond distance, and k is the force constant.

In practice, more complex expressions than this are used, either involving higher-order terms or a Morse function [2]. However, looking at Hooke's equation reveals an important feature of the MM approach: the need for parameters, such as k and r_0. A force constant, k, which represents how easy it is for the bond to be stretched or compressed, is required for each pair of atoms. An equilibrium bond length, r_0, is also required. The assumption when using a MM force field is that these parameters are applicable to every system studied. This assumption is less likely to be true when, for example, systems with unusual chemistry are investigated.

The second term, E_{bend}, looks at every set of three connected atoms. Here, an energy penalty is applied when the angle between the three atoms deviates away from its equilibrium value. An equation similar to one used for the $E_{stretch}$ term can be applied:

$$E_{bend}(\theta) = \frac{1}{2}k(\theta - \theta_0)^2 \tag{3}$$

Again, more complex functions are usually used, but again a value for θ_0 is required before the energy of a model can be evaluated.

The $E_{torsion}$ term looks at every set of four connected atoms. In this case, an energy penalty is applied when the torsion energy between the four atoms deviates away from its equilibrium value:

$$E_{torsion}(\omega) = \tfrac{1}{2}V_n \cos(n\omega) \tag{4}$$

where V_n is the torsional rotation force constant and ω is the current torsional angle.

The final two terms in the force field are referred to as 'nonbonded' terms because the atoms do not need to be directly bonded together in order for the interaction to contribute to the total energy. The electrostatic term, $E_{electrostatic}$, is the total electrostatic energy for the system. This comes about from the charges and partial charges associated with the ions and atoms that are present in the system.

The electrostatic interaction between charged atoms is modelled by a Coulomb potential:

$$E_{electrostatic} = \frac{1}{4\pi\varepsilon_0} \frac{q_i q_j}{r_{ij}} \tag{5}$$

where q_i and q_j are the partial charges on atoms i and j, r_{ij} is the distance the charges are apart, and ε_0 is the relative permittivity of the medium.

The electrostatic energy is a long-range term which means that it is not normally appropriate to ignore the interaction between charged particles that are a long way apart. As a result, Ewald summation [3] is usually used to calculate the total electrostatic energy. This is a mathematical method used to compute electrostatic interactions in periodic systems over infinite distances.

The partial charges for atoms in MM calculations are usually derived by 'bond increments' associated with the force field. For every pair of atom types within the molecule, a pair of opposite values is added to the charges of the atoms in the bond. For example, a 'bond increment' of 0.1268 for atom types C and H would lead to a charge of +0.1268 being added to the charge of the carbon atom and -0.1268 added to the charge of the hydrogen atom. Several other methods of calculating charges are available including Gasteiger and Marsili [4] and QEq [5]. The assignment of charges for framework-OSDA calculations is often a potential source of difficulty, particularly when the location of Al atoms in the framework is not known. Moreover, charge balancing cations are often omitted from calculations.

The final term, the van der Waals energy, E_{VDW}, encompasses all the interactions between atoms (or molecules) that are not covered by the electrostatic interaction. These include dispersion, repulsion and induction. One of the most common forms used for E_{VDW} is the Lennard-Jones (LJ) potential. At small distances (r), atomic repulsion leads to large positive values of E_{VDW}. At larger distances, the value of E_{VDW} becomes small and negative (i.e. favourable) before going to zero as the distance goes to infinity. The form of the LJ potential for two atoms i and j is:

$$E_{VDW} = \frac{A_{ij}}{r_{ij}^{12}} - \frac{C_{ij}}{r_{ij}^{6}} \tag{6}$$

where r_{ij} is the distance apart of the two charged particles and A_{ij} and C_{ij} are suitable constants.

There are several force fields that have been used for zeolite-OSDA calculations. These include the CVFF [6], Universal [7] and COMPASS [8] force fields, which are available within the program Materials Studio [9]. These force fields have parameters specifically derived for silicon, aluminium and oxygen in zeolite frameworks. One particular feature of the Universal force field is that it contains parameters for tetrahedral phosphorous. GULP [10] also allows for force field-type calculations and utilises interatomic Buckingham potentials [11] to simulate the zeolitic framework. This approach has been very successful in predicting the structures and energies of zeolitic frameworks [12, 13]. The OSDA and its interaction with the framework are modelled with the inclusion of more general force fields such as the ones by Kiselev et al. [14] and Oie et al. [15].

2.2 Energy Minimisation and OSDA Placement

The goal for most calculations is to find the lowest (minimum) energy configuration for the OSDA inside its zeolite host. One problem with energy minimisation is that it can be hard for the algorithms used to find the minimum energy (such as steepest descents and Newton-Raphson methods [1]) and avoid locating a local minimum rather than the global minimum energy configuration, particularly if the OSDA starts at a point that is a long way from the global minimum. The problem is exacerbated further when the calculation involves the use of multiple OSDAs.

Two techniques are used to help with this problem. The first one is the use of the molecular dynamics (MD) technique [16]. MD simulates the OSDA (and/or zeolite framework) at a particular temperature by providing additional thermal energy, allowing the movement of the OSDA to be simulated as a function of time. By performing this type of calculation at an elevated temperature, typically 500 K, 'snapshots' of the OSDA's position at different time points during the simulation are saved to generate a series of different positions to use as the starting point for subsequent energy minimisation runs. The hope is that at least one of these configurations will have a starting point that enables the global minimum energy to be found. Another, related approach, is that of simulated annealing (SA). The SA method uses the MD method to simulate the OSDA at elevated temperature, followed by successive cycles of simulation at gradually reduced temperatures. A final energy minimisation cycle completes the process. With the SA approach, the aim is that the thermal energy of the OSDA will drive it over energy barriers and the gradual reduction of temperature will allow the OSDA to follow a low-energy pathway to the global minimum, in much the same way as slow annealing can lead to more highly ordered structures in lab experiments. The usefulness of a simulated annealing protocol is illustrated in Fig. 1. Six OSDA molecules optimised in the LEV framework from the same starting point achieve a lower-energy, more ordered

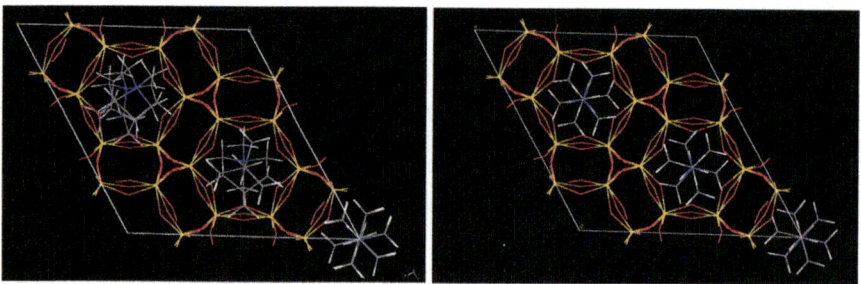

Fig. 1 Six *N*-methyl quinuclidinium molecules in the LEV framework optimised via energy minimisation (left) and via a simulated annealing protocol (right)

arrangement of the OSDAs using a SA algorithm compared with standard energy minimisation approaches.

The generation of a random starting position for the OSDAs in the host framework prior to any energy minimisation run is useful to prevent any preconceptions about their position or the way multiple OSDAs pack inside a framework. This is generally achieved by the application of a Monte Carlo (MC)-type approach which places an OSDA at a random location and orientation inside the zeolite framework. The energy is then calculated, and if it falls below a specified threshold energy, then the position is accepted. If not, then the process is repeated until the condition is met. This method can be used to predict how many OSDAs will pack within a certain volume of the framework by continuing to dock OSDAs inside the framework until a certain number of unsuccessful attempts have been made. The MC and SA approaches are often combined to first dock and then optimise the energy of the templates in a framework [17].

2.3 Quantum Mechanical Methods

Molecular mechanics calculations have the advantage that they are quick to perform, and it has been established that they can yield results in excellent agreement with experiment. However, their dependence on the transferability of parameters from one system to another, coupled with other deficiencies such as the fact that charges on the atoms remain the same throughout the calculation, means that they will be subject to error. Quantum mechanical (QM) approaches do not rely on derived parameters and are likely to be more chemically accurate, but they are very time-consuming and therefore limited in the number of atoms that can routinely be included in this type of calculation. This means that relatively few QM studies on zeolite-OSDA systems have appeared in the literature so far, although more are likely to appear as computers continue to become more powerful.

QM calculations rely on the solution of the Schrodinger equation:

$$H\psi = E\psi \tag{7}$$

where H is the Hamiltonian operator, E is the total energy of the system, and ψ is the wave function.

In theory, solution of the Schrodinger equation to yield the wave function for the system should lead to the determination of accurate properties for any chemical system. In practice, an exact solution is only possible for the simplest of systems and certainly not for something as large as a zeolite-OSDA calculation.

QM approaches can be split into two types: semiempirical and ab initio methods. Ab initio methods do make approximations in the solution of the wave function but do not rely on the use of empirical parameters. The main type of ab initio method is the Hartree-Fock (HF) approach [18]. The key approximation made here is that correlated electron-electron repulsion is not taken into account; instead this is replaced by its average effect. However, calculations of this type are still computationally very time consuming, which often means that calculations are restricted to looking at just fragments of the framework structure. Another ab initio approach that is proving more popular in the study of zeolite-OSDA interaction is density functional theory (DFT) [19]. In DFT, the wave function is bypassed in favour of the electron density of the system, which can be used to determine the properties of the system. Unlike the HF approach, DFT includes an approximate treatment of electron correlation, which makes it potentially more accurate. Moreover, it is much less time demanding computationally, allowing periodic systems to be investigated successfully using high-end computers.

The semiempirical methods (often referred to as AM1, PM3, INDO, MINDO etc.) are based on the Hartree-Fock approach but use different approximations [20]. Only the valence electrons are considered (core electrons are treated together with the nucleus). Some quantities, such as electron repulsion integrals, are taken from experiment. Other quantities are estimated by fitting to experimental data, whereas some (small) quantities are neglected altogether. While this type of calculation has been widely used in connection with zeolite catalysis, there seems to be little benefit to using these methods for OSDA-zeolite investigations rather than the faster MM approach.

3 Modelling Case Studies

3.1 Prediction of OSDA Location

An early example of the use of molecular modelling to successfully determine the location of an OSDA molecule inside a zeolite host is provided by the work on hexamethonium in EU-1 [21]. Force field-based energy minimisation was used to predict that the hexamethonium ions reside in the pockets that alternate along the

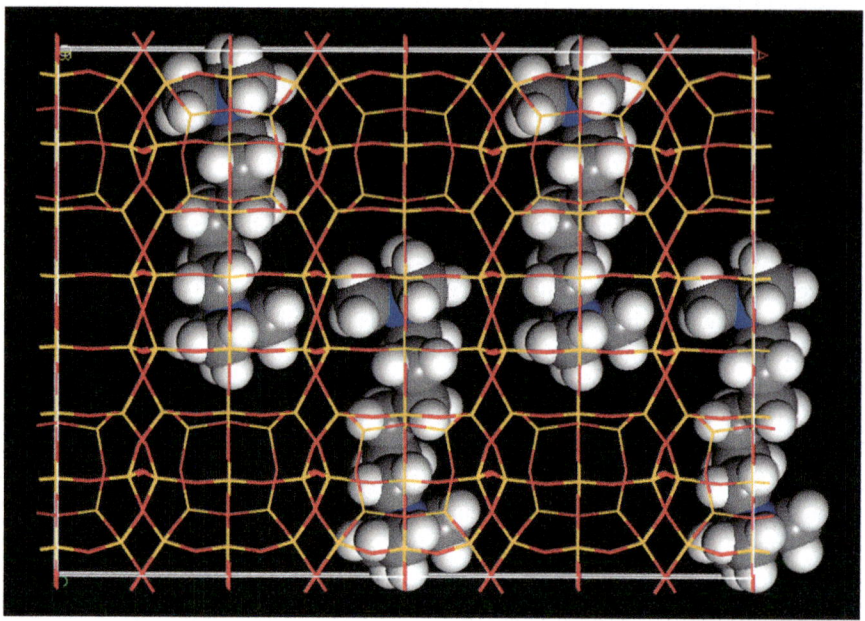

Fig. 2 The optimised location of hexamethonium ions in the EU-1 structure

framework's ten-ring channels (Fig. 2). This example demonstrates the complementary nature of theory and experiment. X-ray diffraction (XRD) does have the
ability to locate OSDA molecules inside a framework structure, but the Rietveld
refinement procedure often benefits enormously from a good starting model, which
modelling can often provide. In this particular case, Fig. 3a shows the experimental
X-ray diffraction pattern for the framework and the OSDA compared with the
X-ray diffraction pattern for just the empty framework. Clear differences can be
seen between the two patterns (the line underneath shows the difference between
them), and these differences are due to the presence of the OSDA. Figure 3b shows
a comparison between the X-ray diffraction pattern obtained from experiment
compared with a model of the framework and the OSDAs located at their predicted
location from the theoretical calculation. There is now excellent agreement between
the two patterns, providing clear evidence in support of the predicted location for
the OSDA, which was confirmed by subsequent Rietveld refinement.

Modelling also made an important contribution to confirming the as-made
structure of the zeolite ZSM-25 [22] by identifying the most likely positions of
the ca 40 tetraethylammonium (TEA^+) OSDA cations in the unit cell, which are
present as extra-framework species, along with Na^+ cations and water molecules.
The structure of ZSM-25 is tremendously complex, with over 4,500 atoms in the
unit cell and five different types of cages present (Fig. 4).

With a large number of guest molecules/cations in the structure, it is necessary to
have a good starting model for the NaTEA-ZSM-25 structure prior to undertaking

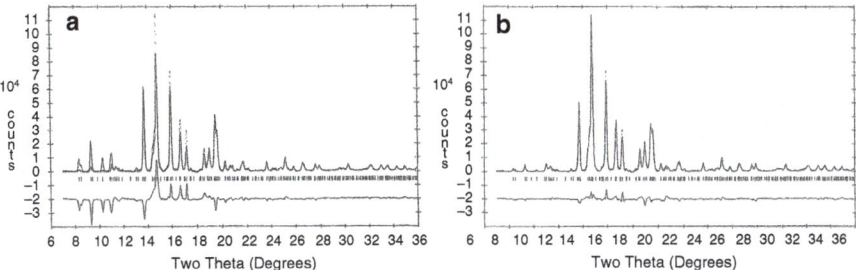

Fig. 3 (a) Comparison of the XRD patterns for EU-1 with and without hexamethonium and (b) comparison of experimental XRD pattern for EU-1 with hexamethonium and the pattern generated from the modelled location of hexamethonium in EU-1 [21]

Fig. 4 Polyhedral representation of ZSM-25

Rietveld refinement of the X-ray data. The most favourable sites were identified using a force field approach which showed that the most likely sites for the TEA^+ cations in the structure are the *pau* [$4^6 6^2 8^6$] and [$4^{12} 6^8 8^6$] (*lta*) cages, each of which is large enough that the TEA^+ cation can be included without unfavourable close contacts. The modelled positions of the TEA^+ inside the different cages are shown in Fig. 5, and the interaction energies are shown in Table 1. For the cages showing favourable energies, the interaction is significantly more favourable with the slightly smaller *pau* and [$4^6 6^2 8^6$] cages than with the larger [$4^{12} 6^8 8^6$] *lta* cage. The interaction energies with the smaller [$4^7 8^5$] and [$4^6 8^4$] cages are much less favourable,

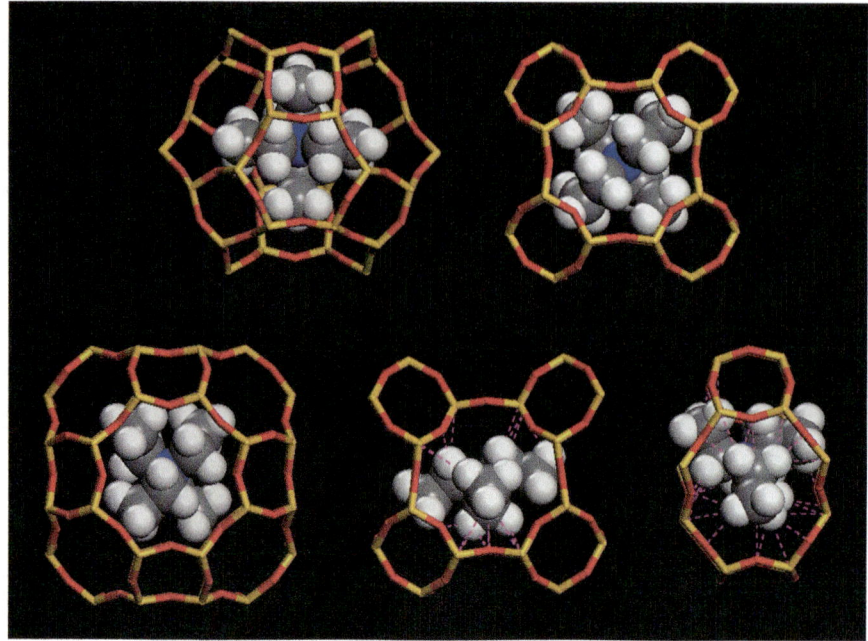

Fig. 5 Energy-minimised location of TEA$^+$ cations in (clockwise, from top left) [$4^6 6^2 8^6$] (*t-plg*), [$4^{12} 8^6$] (pau), [$4^6 8^4$] (*t-gsm*), [$4^7 8^5$] (*t-phi*) and [$4^{12} 6^8 8^6$] (*lta*) cages in the ZSM-25 framework. Dashed lines indicate energetically unfavourable close contacts

Table 1 Calculated interaction energies of TEA$^+$ cations in different cages of the ZSM-25 framework

Cage type	Interaction energy of TEA$^+$/kcal mol^{-1}
[$4^6 6^2 8^6$]	−52.7
[$4^{12} 8^6$] (*pau*)	−51.8
[$4^{12} 6^8 8^6$] (*lta*)	−39.6
[$4^7 8^5$]	−3.7
[$4^7 8^5$]	+8.6

i.e. more positive (Table 1) because they are smaller and so close contacts arise with TEA$^+$. So, it could be concluded from the modelling that the TEA$^+$ cations would adopt sites in the [$4^6 6^2 8^6$], [$4^{12} 8^6$] *pau* and [$4^{12} 6^8 8^6$] *lta* cages. Since there are two *lta*, 18 *pau* and 24 [$4^6 6^2 8^6$] cages and ca. 40 TEA$^+$ cations per unit cell, it was concluded that most of the *pau* and [$4^6 6^2 8^6$] cages are occupied by TEA$^+$ cations. Having established that it was likely that TEA$^+$ cations occupy the [$4^{12} 8^6$] *pau* and [$4^6 6^2 8^6$] cages, these were included in these cages in the starting structural model for ZSM-25, with one TEA$^+$ cation in each of the *pau* and [$4^6 6^2 8^6$] cages. This model was used as the starting point for the X-ray refinement of the ZSM-25 structure and yielded a model that was in excellent agreement with experiment, again confirming the usefulness of modelling in successfully locating OSDAs within the host structure.

3.2 Zeolite-OSDA Energetics

Calculated interaction energies between the OSDA and the framework can be used to interpret and predict the products obtained from synthesis reactions. A good example is provided by Rollmann et al. [23] who investigated the relationship between small amines and the zeolite product they synthesise. Table 2 shows the interaction energy for different amines modelled in four different zeolites.

The product(s) synthesised by each of the OSDAs is shown by the interaction energy in bold. In this case, it can be seen that the changes in the interaction energy nicely track the changes in zeolite product observed. Not only does this help to rationalise the observed trends, but more importantly it demonstrates the potential of this type of calculation to predict new OSDAs for known and hypothetical frameworks.

Wagner et al. [24] used modelling to rationalise the synthesis of three zeolite structures, SSZ-35, SSZ-36 and SSZ-39. One particularly interesting feature to emerge from this investigation is the importance of thoroughly investigating competing phases. Their experimental work found that the [4.1.1] octane molecule had a strong specificity for the synthesis of the SSZ-35 framework. Modelling results confirmed the stabilisation of the SSZ-35 cage structure by −11.83 kcal per cage. However, adding a single methyl group to the ring has a dramatic effect on the binding energy, making the interaction with the SSZ-35 unfavourable (+0.69 kcal per cage); experimentally the methylated OSDA results in the formation of the competing SSZ-36 phase. In general, one of the difficulties associated with the design of novel OSDAs using computational techniques is screening them in potential competing phases; a molecule may have a good interaction energy in a targeted framework, but it may have a better binding energy in a competing phase.

Modelling OSDAs used in the formation of MFI and MEL illustrate how modelling has evolved to investigate more subtle effects associated with the synthesis process. Initial work by Lewis et al. [25] used force field calculations to show the importance of considering more than one template to rationalise why tetrapropylammonium (TPA) synthesises the MFI structure, while tetrabutylammonium (TBA) synthesises MEL. A

Table 2 Interaction energies calculated for cyclic amine-zeolite pairs

Amine	ZSM-35	MCM-22	ZSM-5	ZSM-12
Pyrrolidine	**−15.3**	−13.5	−14.0	−13.5
Pyrrolidine-H$^+$	**−14.4**	−12.6	−12.6	−12.4
Piperidine	**−17.6**	−16.6	−14.9	−16.2
Piperidine-H$^+$	**−15.9**	−15.4	−13.9	−15.1
Piperazine	**−18.1**	−16.4	−15.6	−16.3
Hexamethyleneimine	−17.5	**−18.7**[a]	−17.4	**−18.5**[a]
Hexamethyleneimine-H$^+$	−16.0	**−17.7**[a]	−16.4	**−17.3**[a]
Homopiperazine	−18.4	**−18.7**[a]	−16.8	**−18.7**[a]

All values are given in kcal mol^{-1}
Effective amine-zeolite combinations, as observed experimentally, are shown in boldface type
[a]Product shifts from MCM-22 to ZSM-12 at different SiO$_2$/Al$_2$O$_3$ ratios [23]

single TBA cation has a good interaction energy with the MFI framework (Table 3). However, this result does not take into account whether the OSDAs can pack efficiently into each of the two different structure types. When additional OSDAs are included in the calculation, it becomes clear that a second TBA cation destabilises the first one due to steric considerations and they cannot pack efficiently in the MFI structure without incurring a significant energetic penalty (Table 4), thus highlighting the importance of investigating packing interactions.

Shen and Bell [26] observed similar trends using a similar force field-based approach but by loading different numbers of template molecules into the unit cell.

Szyja et al. [27] carried out a combined DFT and molecular mechanics investigation to investigate the early-stage synthesis of the MFI and MEL structures. This is a good example of the use of modelling to investigate the interaction between the OSDA and silicate oligomer precursors taking part in the synthesis process. Such calculations, performed on relatively small numbers of atoms, are ideal for higher levels of theory, such as DFT. The calculations yield the geometries of the OSDA clusters as well as their interaction energies. One of the key findings of this investigation is that TBA is found to stabilise one of the key (Si_{22}) precursors required for formation of the MEL structure, emphasising the importance of the interaction between OSDA and silicate precursors in determining the outcome of the synthesis.

Recent work by Sánchez et al. [28] focused on investigating different conformations for TPA and TBA (Fig. 6). Dynamics studies show that TPA has higher numbers of molecules in an angular conformation at experimental temperatures than TBA. This angular conformation for TPA fits inside the MFI structure very favourably. In contrast, for TBA, the semi-planar conformation is more stable. Hence, the conformation adopted by the OSDA at the synthesis temperature has an important effect on the product.

Table 3 Nonbonded energies for single molecules of TPA and TBA in the MFI and MEL frameworks [25]

Template/framework	Stabilisation energy/kJ mol^{-1}
TPA/MFI	−133.9
TBA/MFI	−165.5
TBA/MEL	−159.5
TPA/MEL	−119.9

Table 4 Stabilisation energy of template/framework combinations when two OSDA molecules are included at adjacent intersection sites [25]

Template/framework	Stabilisation energy/kJ mol^{-1}
TPA/MFI	−29.7
TBA/MFI	+14.9
TBA/MEL	−18.3
TPA/MEL	−8.5

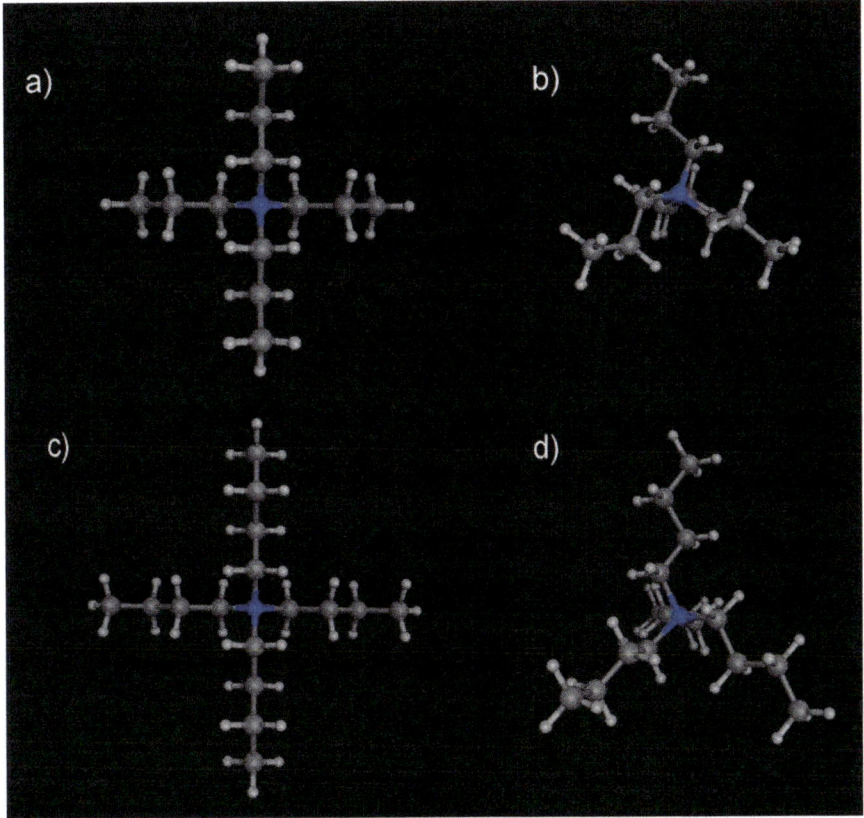

Fig. 6 TPA (**a**, **b**) and TBA (**c**, **d**) in their semi-planar and mean angular conformations, respectively

3.3 The Influence of the OSDA on the Location of Al and Protons

One important structural feature of any zeolite framework is the distribution of the aluminium in the framework. The distribution of Al and other heteroatoms can have an important influence on the behaviour of a catalyst. Modelling calculations have helped to yield an important insight into the role of the OSDA in controlling the location of Al and proton siting.

The value of modelling in this area is nicely illustrated by the work of Sastre et al. [29] In their work they used force field simulations using the program GULP to investigate the Al distribution in the zeolite ITQ-7. These calculations showed that the OSDAs located inside the zeolite's channels, along with the fluorine ions that occupy the D4R units, combine to make some of the potential sites for Al atoms become more energetically favoured. Hence, rather than just the energy of the

ITQ-7 framework influencing the Al distribution in the structure, it is the combined effect of the interaction that occurs between the zeolite and OSDA during the crystallisation process that determines where the Al is located. Once the location of the Al has been determined, then the preferred location of the proton, once the OSDA has been removed, can also be predicted.

Gómez-Hortigüela et al. [30] performed a computational study using a similar approach to investigate the siting of Al in the FER framework. This work also established the role of the OSDA in influencing the Al distribution in FER. Moreover, the ability of a template to direct Al to its vicinity during synthesis was shown to be related to the ability of the OSDA to form strong H-bond interactions with the O atoms neighbouring Al in the framework. In particular, it was concluded that OSDAs with acid protons, including primary or secondary amines in the outmost regions of the molecule, provide a strong driving force for the incorporation of Al in the vicinity. This work was followed up by a further investigation that elegantly showed how the knowledge gained from this type of computational study could be exploited in the design of template systems that yield different Al distributions in the FER framework [31].

3.4   Synthesis of Different Products from a Single OSDA

Many OSDAs can direct the formation of several different framework types when synthesis conditions are altered. One example of this is the hexamethonium ion $(CH_3)_3N^+(CH_2)_6N^+(CH_3)_3$ which can direct the formation of EU-1, ITQ-13, ITQ-22 and ITQ-24 when germanium is included in the synthesis gel. Sastre et al. studied this phenomenon using the GULP code [32]. They concluded that the relatively low 'match' between the OSDA and each of the four different frameworks meant that the product formed was very sensitive to the reaction conditions. The experimental trends observed could be rationalised by taking into account different energy terms associated with the zeolite and the OSDA. These are the structural stability of the zeolite, the interaction between the OSDA and the zeolite and, finally, the strain of the OSDA when it is occluded inside the zeolites pore system.

3.5   Template Flexibility

A recent computational study of the use of rigid/flexible templates has adopted a combined force field/DFT approach to probe the subtle effects brought about by small changes to the OSDA [33]. The two OSDAs shown in Fig. 7 are both specifically designed for the synthesis of multipore zeolites with interconnected large and medium pores. Both OSDAs have a bulky piperidinium-derived body included to template large pores, in combination with alkyl chains that direct towards ten-ring channels in zeolites, such as ZSM-5. The second OSDA, SDAEt, has two shorter ethyl chains. In a

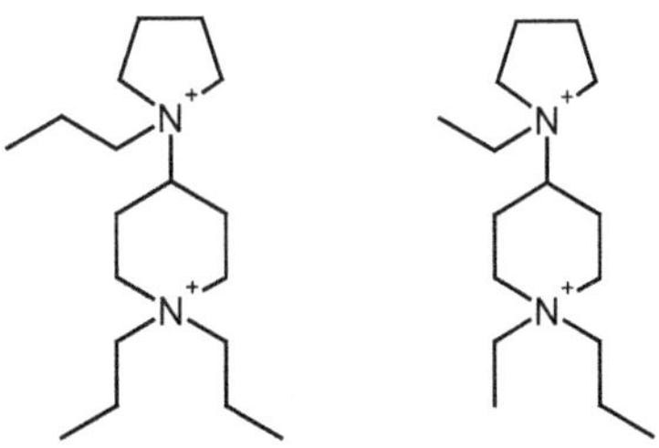

Fig. 7 SDAPr (left) and
SDAEt (right)

pure silica system, SDAEt synthesises the zeolite ITQ-39, an interesting structure that comprises three polymorphs, although its channel system can be considered to be composed of straight 12-ring channels interconnected by 10-ring zigzag channels. In contrast, in a pure silica system, SDAPr produces the MFI structure, a common synthesis product, which comprises straight ten-ring channels connected by ten-ring zigzag channels. However, increasing the amount of aluminium in the synthesis mixture means that both OSDAs can be used to produce the ITQ-39 framework. The approach used in this study was to use force field calculations to 'prescreen' potential docking sites for both OSDA molecules in the ITQ-39 and MFI structures. Periodic DFT was then used as a more accurate and reliable methodology to probe the subtle differences between the OSDAs and the two framework structures. The calculations show that the highest stabilisation energy for SDAPr and SDAEt is in MFI and ITQ-39, respectively, in agreement with experimental observations. One interesting outcome of the calculations is that the optimised location for the two OSDAs within the ITQ-39 framework is in a distinctly different part of the framework. This is potentially significant because it suggests that samples of ITQ-39 made with these different OSDAs may have subtle differences in their Al/H distributions. This is a nice illustration of how modelling can lead to new avenues of investigation for experimental work.

Sastre et al. have also investigated why changing Si/Al ratios in the synthesis gel leads to the formation of different zeolite products [34]. This study involved the determination of interaction energies for cyclohexyl alkyl pyrrolidinium salts inside EU-1, ZSM-11, ZSM-12 and β-zeolites. One of the nice features of this work is that the authors were able to rationalise why different structures are formed at different Si/Al ratios. For example, in the case of the N-Cyclohexyl-N-methyl-pyrrolidinium cation, they found that the larger porosity of the EU-1 structure allows more OSDA cations to enter the pores than in a less open structure such as ZSM-12. Hence, at higher Al content, the calculated final energies show that EU-1 becomes more stable than the product formed at lower Si/Al ratios, ZSM-12.

3.6   Inclusion of Water

As well as the importance of the OSDA, water is known to have an influence on the crystallisation mechanism in both pure-silica and aluminophosphate synthesis.

Gómez-Hortigüela et al. have developed a protocol to give an insight in the competition between the OSDA and water during synthesis [35]. The approach involves:

1. Loading the framework with the maximum amount of OSDA via a Monte Carlo approach.
2. Starting from the maximum loading obtained from step 1, the number of OSDA molecules is decreased until sections of empty space appear in the structure. OSDA locations are optimised by simulated annealing and energy minimisation.
3. For each of the different OSDA loadings investigated, water is added via a second Monte Carlo step. Again, water and OSDA locations are optimised via a simulated annealing and energy minimisation step.

Initial studies focused on the incorporation of three different OSDAs used in the synthesis of AFI, triethylamine (TEA), benzylpyrrolidine (BP) and (S)-(-)-N-benzylpyrrolidine-2-methanol (BPM). The results obtained are in good agreement with experiment for the relative amounts of water and OSDA content for these different systems. In the case of TEA, higher amounts of water are incorporated inside the framework. This is attributed to the relatively low interaction between TEA and the framework, leading to higher incorporation of water molecules to provide the necessary additional thermodynamic stabilisation needed for the structure to crystallise. In effect, water is acting cooperatively with the OSDA to direct formation of the product, thus highlighting the importance of the inclusion of water for certain systems.

3.7   Co-templating

Co-templating involves the use of more than one OSDA, each one used to generate a different cage or part of the channel system within the structure. Modelling, based on a force field approach, was used to rationalise the synthesis of SAPO materials STA-7 (SAV) and STA-14 (KFI) [36]. In the case of STA-14 (KFI), modelling was used to screen candidate OSDAs for the formation of the *mer* cage in the KFI structure; the other, larger, co-templating OSDA was identified from its known ability to synthesise the other cage in the framework. Calculations correctly predicted that the tetraethylammonium (TEA^+) ion was the most likely to succeed from a list of ten potential candidates. Indeed, all other OSDAs failed to make the desired KFI product. The predicted location of TEA^+ inside the *mer* cage was subsequently found to be in excellent agreement with the one obtained from experiment (Fig. 8).

Almeida et al. [37] also used modelling to rationalise co-templating in the formation of ferrierite by 1,6-bis-(N-methylpyrrolidinium)hexane (MPH) in combination with tetramethylammonium (TMA). The MPH molecules are found to pack along the ten-ring channels in the structure, while the smaller TMA molecules occupy the cavity bridging the ten-ring channels.

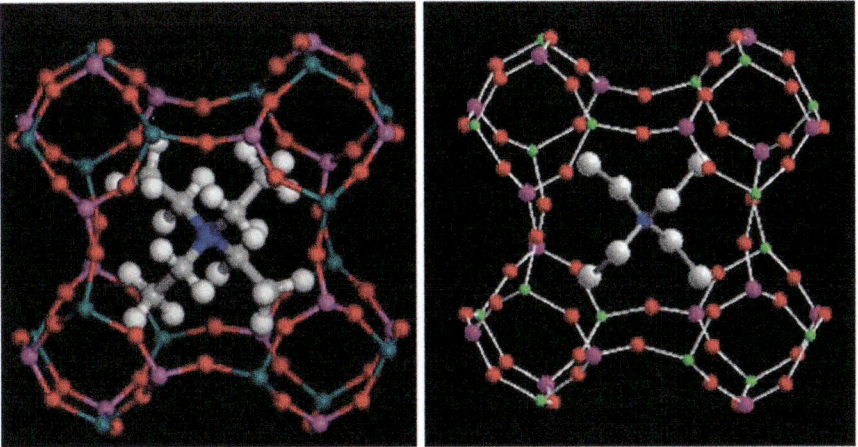

Fig. 8 The modelled location of the TEA⁺ OSDA within the *mer* cages of the KFI structure (left) and the experimentally observed location (right) [36]

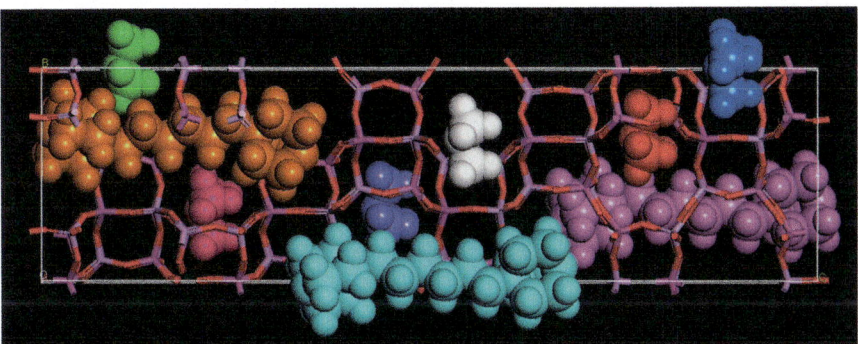

Fig. 9 TrMA and DABCO-C6 OSDAs optimised inside the SFW framework

Modelling was used in the design of co-templates for three related SAPO versions of AFX (SAPO-56), SFW (STA-18) and GME (STA-19) [38]. The protonated trimethylammonium (TrMA) cation was selected to be the best OSDA for templating the small *gme* cages present in all three of the structures. Bisdiazabicyclooctane (DABCO) alkane cations and quaternary ammonium oligomers of DABCO with connecting polymethylene chain lengths of 4–8 methylene units acted as templates for the additional channels or cages, respectively. The approach used was validated by the successful synthesis of each of these frameworks, and predicted locations for the co-templates inside each of the three frameworks were found to be in excellent agreement with those obtained from experiment. The modelled positions of TrMA and diDABCO-C6 within the six *gme* and three *sfw* cages per SFW unit cell are shown in Fig. 9.

The cooperative way in which two OSDAs act during nucleation and growth is intriguing and is something that requires further investigation for this approach to realise its full potential.

3.8 Self-Assembled Molecules as OSDAs

This approach involves the use of a supramolecular OSDA formed by molecules that self-assemble in aqueous solution prior to the crystallisation of the product. It allows more complex, larger zeolites to be created from relatively simple, potentially cheaper molecules. This concept was impressively demonstrated experimentally by Corma et al. who used an organic cation that self-assembles through π–π interactions to form a dimer that fits nicely in the LTA cage of the ITQ-29 structure [39]. This self-assembly approach has since been investigated by modelling. Modelling can help in two ways. The aggregation of the initial molecules can be investigated by molecular dynamics studies. The aggregates can then be geometry optimised inside the host structure to investigate their fit and energetics. This modelling approach is illustrated nicely by the work of Álvaro-Muñoz et al. [40] who have investigated the use of (1*R*,2*S*)-Ephedrine as a self-assembly OSDA for the synthesis of the AFI framework. The simulations identified strong H-bond interaction between the O atom of the –OH group and an H atom of an amino atom on an adjacent molecule. These interactions are maintained inside the AFI framework (Fig. 10). The theoretical results are in excellent agreement with experimental observations.

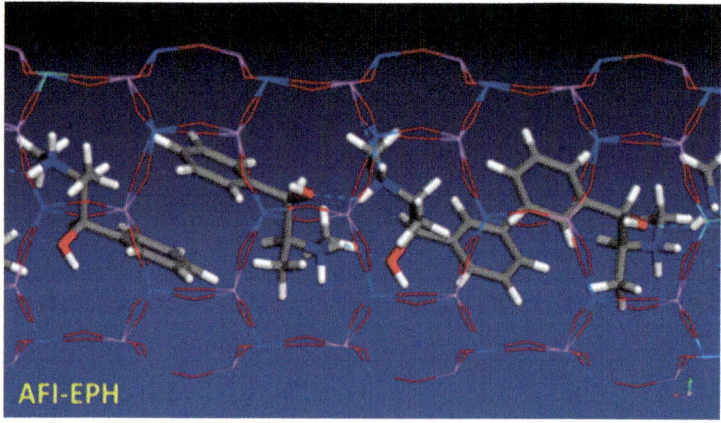

Fig. 10 Geometry-optimised structure of EPH aggregates within the AFI structure [40]

3.9 Zeolite Polymorphs and Chiral Structures

One area where modelling has been used successfully is in the synthesis of specific zeolite polymorphs. Early work on the synthesis of zeolite beta (BEA) resulted in the formation of an intergrowth of different polymorphs. Pure polymorphs of a zeolite structure are an attractive synthesis target because they will possess subtly different void space that may improve their suitability for certain applications. Moliner et al. [41] used the GULP code to screen a series of nine different potential OSDAs that targeted the synthesis of one of the three polymorphs of the beta structure, polymorph C. The OSDA that the calculations predicted was the best of these candidate OSDAs successfully synthesised the target structure.

One important goal for zeolite synthesis is the production of framework structures with chiral channel systems. Such materials would have the potential to possess enantiospecific catalytic pathways and separation processes. A recent study by Brand et al. [42] used modelling to select large, rigid chiral OSDAs to synthesise enantiomers of the STW framework. Calculations were carried out using a force field approach using GULP, and the interaction energies were used to select the OSDA used to perform experimental studies. The final OSDAs shown below were selected for experimental investigation because of the large differences obtained for the interaction energy between the two different enantiomers of the STW structure (Table 5).

Subsequent experimental investigation confirmed the success of this approach, with both the R- and S- forms of this OSDA producing products that showed significant enhancement of the two different enantiomers of the STW framework. The initial design of the OSDAs in this work was based on the procedure outlined in Sect. 4.2.

Table 5 Interaction energies for candidate OSDAs in the two enantiomers of the STW framework [42]

OSDA	$E_{\text{enantiomer 1}}$/kJ (mol Si)$^{-1}$	$E_{\text{enantiomer 2}}$/kJ (mol Si)$^{-1}$
	−16.32	−14.60
	−14.65	−1.37
	−15.27	−1.58

4 Automated Template Design

4.1 De Novo Methods

Automated approaches for the design of new OSDAs are particularly attractive. De novo design algorithms were initially developed to generate new molecules designed to fit inside the active site of a protein. The key concept behind the approach is to 'grow' a molecule inside the active site by the repeated addition of smaller fragments from a database. At each addition, the organic molecule is optimised within the pocket in order to maximise the scoring function (or binding energy) [43]. An atom or functional group is added to the organic molecule to achieve a more favourable interaction energy [44]. Each individual molecule can either grow or recombine with a part of another molecule in the population. A genetic algorithm is used to coordinate the evolution of the individuals in the population [45].

Lewis et al. adapted the de novo approach for the design of new OSDAs for zeolite synthesis in the program ZEBEDEE (zeolites by evolutionary de novo design) [46]. In ZEBEDEE, a seed molecule (guest) is randomly, or specifically, placed within a target space (i.e. cage or channel). A number of random actions, using the fragment library as the source of new atoms, are used to grow the seed molecule. The available actions are:

1. *Build*: a new randomly selected fragment is joined onto the existing template by forming a bond between the neighbouring atoms of selected hydrogens, with these hydrogens being deleted from the molecule.
2. *Rotate*: the last fragment added is rotated about the new bond formed.
3. *Shake*: the template is displaced along a random vector with respect to the host.
4. *Rock*: the template is randomly rotated with respect to the host.
5. *Random bond twist*: a randomly selected bond joining fragments in the template is rotated.
6. *Ring formation*: where conditions are appropriate, specific atoms are joined to form a ring.
7. *Energy minimisation*: the template geometry is optimised either in the gas phase or with respect to the fixed host (using force field-based or semiempirical methods).

The development of the new template is governed by a cost function based on overlap of van der Waals spheres:

$$f_c = \sum_t C(tz)/n \qquad (8)$$

where $C(tz)$ represents the closest contact between the template atom t and any host atom z and n is the number of atoms in the template. f_c provides a measure of the 'tightness of the fit'.

Growth is terminated when the cost function reaches a pre-ordained value, or when all the replaceable hydrogens are exhausted, or when a pre-set number of actions has been performed. ZEBEDEE allows the simultaneous growth of a number of symmetry-related templates within the unit cell. Once candidate OSDA molecules are generated then they can be screened further using any of the methods outlined earlier.

The ZEBEDEE code has enjoyed some notable successes. Examples of the successful prediction of new OSDAs, include 4-piperidinopiperidine for DAF-5 (CHA) [47] and 2-Methylcyclohexylamine for DAF-4 (LEV) [48] (Fig. 11).

One general problem with de novo design algorithms is the synthetic accessibility of the compounds that are generated. Although the new OSDAs are generated in a chemically correct way, the resulting molecules may turn out to be very hard, or impossible, to synthesise in the laboratory.

4.2   De Novo Design Incorporating Synthetic Accessibility

Deem et al. [49] have elegantly adapted the de novo approach in order to take into account the synthetic accessibility of the generated compounds. The method used in their work consists of a compound generator, a scoring function, and an optimisation algorithm.

1. *Compound generator*: a predicted compound is generated by selecting a reaction from the reaction set (84 reactions have been implemented so far with the possibility of adding further reactions) and one or more available reagents (12,838 compounds, again with the possibility of expanding this list) from a reagent database that possess the correct functionality to participate in this reaction. Once a compound is created, it can undergo further reactions, depending on its functionalities. Additional reagents are randomly selected from the reagent database where necessary.

2. *Scoring function*: is a stepwise computational protocol that first calculates a number of computationally inexpensive molecular properties that are used as filters.

   (a) *Stability under synthesis condition*: esters, amides, lactams, ketones, aldehydes, alcohols, acids and halogens are removed because they are undesirable in zeolite synthesis.

   (b) *Number of torsional degrees of freedom*: defined as the number of rotatable bonds in non-cyclic moieties of the molecule. Small end-standing symmetrical groups such as methyl, amino and nitro are not included as rotatable bonds.

   (c) *Molecular volume*: approximated as an integral over atomic cantered Gaussian functions. Volume of the OSDA is used as a criterion to eliminate quickly those OSDAs that are either too small to be effective or too large to fit within the pore of the zeolite. The number of OSDAs that can be

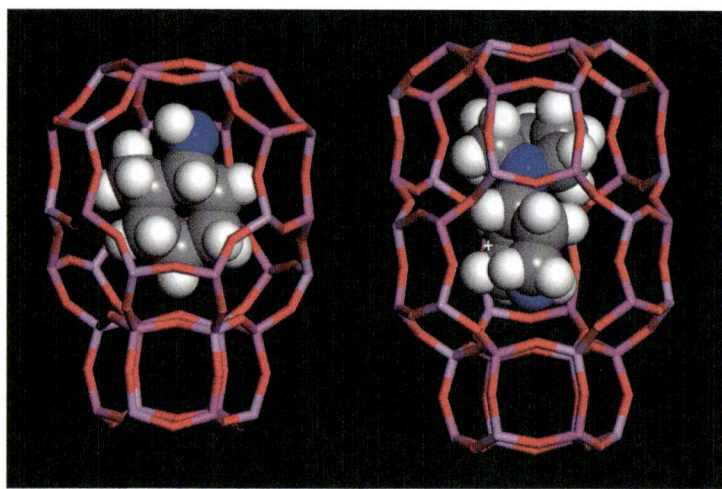

Fig. 11 New OSDA molecules designed by the ZEBEDEE code: 2-methycyclohexylamine for DAF-4 (LEV) (left) and 4-piperidinopiperidine for DAF-5 (CHA) (right)

accommodated in a unit cell of the zeolite is determined by an optimised grid search procedure.

3. *Optimisation algorithm*: allows the determination of the nonbonded, Lennard-Jones interaction energy of the OSDA with the zeolite and with other OSDAs. The energy is reported in units of kJ mol^{-1} per Si, so that the stabilisation energy per silicon atom of the zeolite is given, thus allowing comparison of different OSDAs in the same zeolite.

This method has also successfully predicted new OSDAs, namely, pentamethylimidazolium for aluminosilicate RTH in both fluoride- and hydroxide-containing syntheses and pure-silica STW [50, 51], N-ethyl-N-methyl-2,2,6,6-tetramethylpiperidine for aluminosilicate AEI [52] and N-ethyl-N-(2,4,4-trimethylcyclopenthyl) pyrrolidinium and N-ethyl-N-(3,3,5-trimethylcyclohexyl) pyrrolidinium for aluminosilicate SFW [53] shown in Fig. 12.

4.3 High-Throughput Screening

Another way to help identify potential new OSDA molecules is to use one of the large numbers of databases that are available with molecules in an electronic format. One example is the ZINC database [54] which contains over 35 million known compounds. In theory, all of these compounds could be screened in a target framework, although filters can be applied to eliminate molecules that are obviously too big or too small for a candidate structure. Alternatively, searches can be limited to chemicals available from selected suppliers.

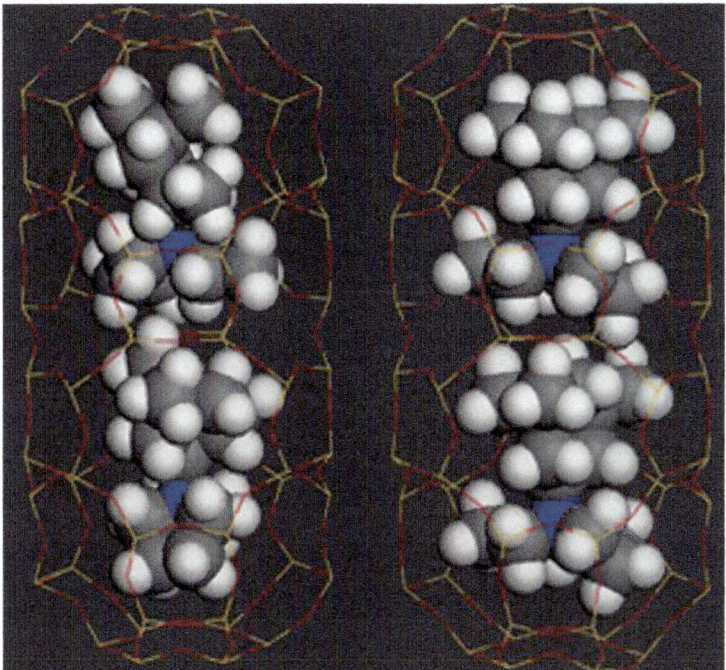

Fig. 12 Two new OSDAs successfully designed to promote the crystallisation of SSZ-52 (SFW) using the modified de novo approach [53]

In order to screen large numbers of compounds, they need to be docked into the framework at a suitable starting point, prior to energy minimisation. This can be done by using 'alpha spheres' to help define an empty region of space inside the zeolite framework. An alpha sphere is a sphere that makes contact with four framework atoms on its boundary and contains no internal atoms [55]. Each alpha sphere is unique since there is no other sphere that will contact the same four framework atoms. A collection of cavity-sized spheres describes well the shape of the zeolites' internal channels and cages. In addition, each sphere can be classified as either hydrophobic or hydrophilic depending on its local environment in the structure. The centres of the alpha spheres calculated for the LEV cage are shown below in Fig. 13.

Prior to docking, several energetically accessible conformations of the molecule are used. These conformations are then initially docked inside the zeolite by matching triangles of atoms in the molecule with triangles of alpha spheres, taking hydrophobic/hydrophilic matches into account. This is a very fast method, and it should provide a good starting position for the molecule inside the framework. This is then followed by a final energy minimisation using an appropriate force field.

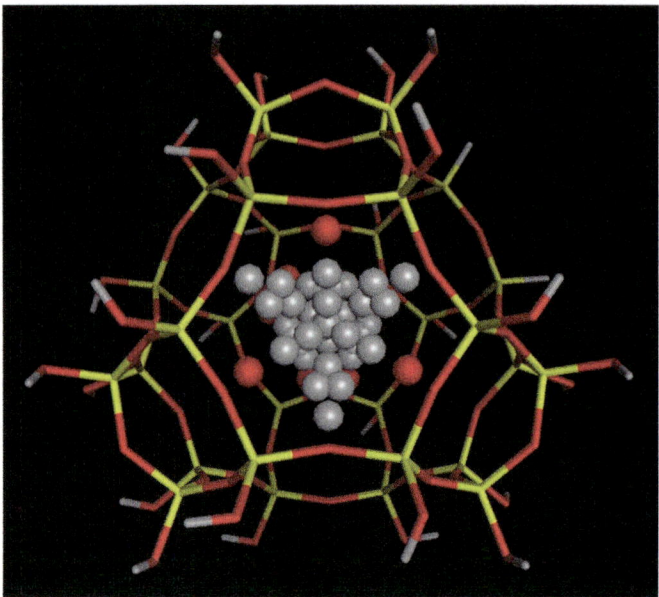

Fig. 13 Alpha sphere centres calculated for the LEV cage. White spheres are hydrophilic; red spheres are hydrophobic

As an illustration of the type of results obtained from using this approach, the molecules shown in Fig. 14 were all ranked highly from a search of over 100,000 compounds in the ZINC database for potential OSDAs for the LEV cage.

As can be seen, several of the candidate molecules are likely to require some modification before they can be suitable to use as OSDAs. However, these modifications can be made, and the molecules can be examined in depth using the procedures outlined earlier in this review. In addition, further rigorous investigation may be necessary to optimise chain lengths and functional groups. However, it is clear that some of the molecules closely resemble known templates for LEV-type frameworks, such as N-methylquinuclidinium and 1adamantanamine hydrochloride, alongside some very different molecules, demonstrating the value of the approach to generate potential new 'lead' compounds.

5 Conclusion

In a relatively short period of time, molecular modelling has made an enormous contribution to our understanding of the relationship between the OSDA and its framework product. It is a powerful tool for rationalising experimental results. Moreover, it is also showing its value as a predictive tool for guiding experiments and selecting new OSDAs for selected targets.

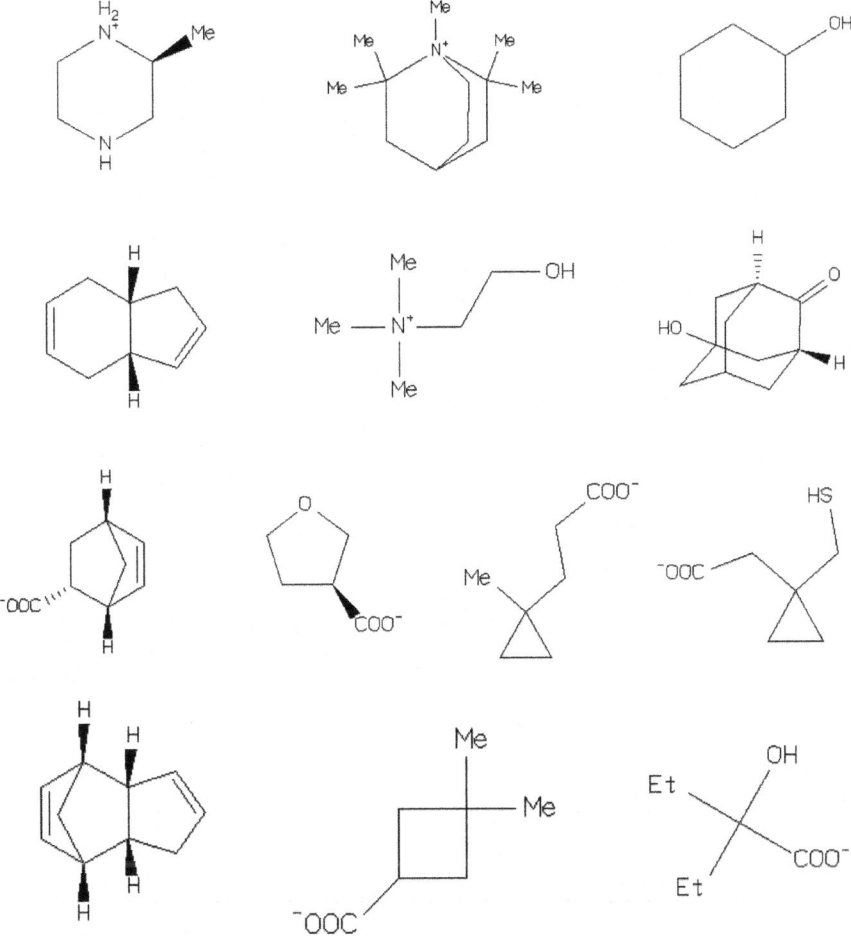

Fig. 14 Selected hits from screening over 100,000 compounds from the ZINC database in the LEV framework cage

However, there is much work still to be done before modelling can be routinely used to reliably predict the outcome for any given synthesis reaction. New methods for identifying and screening potential competing phases will be important, as well as understanding the role of other experimental variables in the synthesis process. There are many other areas for modelling to investigate, such as the role of the OSDA in the formation of intergrowths, an important and challenging target.

As the power of computers continues to increase, higher levels of theory will be used to probe the interaction between zeolite and OSDA with even greater accuracy. In the future, modelling will continue to complement experimental techniques and exert an even greater influence on the design of synthetic research programmes.

References

1. Grant GH, Richards WG (1998) Computational chemistry. Oxford University Press, Oxford
2. Leach AR (2001) Molecular modelling: principles and applications. Prentice Hall, Dorchester
3. Wells BA, Chaffee AL (2015) Ewald summation for molecular simulations. J Chem Theory Comput 11:3684–3695
4. Gasteiger J, Marsili M (1980) Iterative partial equalization of orbital electronegativity – a rapid access to atomic charges. Tetrahedron 36:3219–3228
5. Rappé AK, Goddard III WA (1991) Charge equilibration for molecular dynamics simulations. J Phys Chem 95:3358–3363
6. Dauber-Osguthorpe P, Roberts VA, Osguthorpe DJ, Wolff J, Genest M, Hagler AT (1988) Structure and energetics of ligand binding to proteins. Proteins Struct Funct Genet 4:31–47
7. Rappé AK, Casewit CJ, Colwell KS, Goddard III WA, Skiff WM (1992) UFF, a full periodic table force field for molecular mechanics and molecular dynamics simulations. J Am Chem Soc 114:10024–10035
8. Sun H (1998) COMPASS: an ab initio force-field optimized for condensed-phase applications – overview with details on alkane and benzene compounds. J Phys Chem B 102:7338–7364
9. Dassault Systèmes BIOVIA (2017) Materials studio version 2017 R2. San Diego
10. Gale JD (1997) GULP: a computer program for the symmetry-adapted simulation of solids. J Chem Soc Faraday Trans 93:629–637
11. Jensen F (2017) Introduction to computational chemistry. Wiley, Chichester
12. Jackson RA, Catlow CRA (1988) Computer simulation studies of zeolite structure. Mol Simul 1:207–224
13. Shannon MD, Casci JL, Cox PA, Andrews SJ (1991) Structure of the two-dimensional medium-pore high-silica zeolite NU-87. Nature 353:417–420
14. Kiselev AV, Lopatkin AA, Shulga AA (1985) Molecular statistical calculation of gas adsorption by silicalite. Zeolites 5:261–267
15. Oie T, Maggiora TM, Christoffersen RE, Duchamp DJ (1981) Development of a flexible intra- and intermolecular empirical potential function for large molecular systems. Int J Quantum Chem 20:1–47
16. Tildesley DJ (1993) The molecular dynamics method. In: Allen MP, Tildesley DJ (eds) Computer simulation in chemical physics, NATO ASI series 397, pp 23–47
17. Stevens AP, Gorman AM, Freeman CM, Cox PA (1996) Prediction of template location via a combined monte-carlo simulated annealing approach. J Chem Soc Faraday Trans 92:2065–2073
18. Ramachandran KI, Deepa G, Namboori K (2008) Computational chemistry and molecular modelling. Springer-Verlag, Berlin
19. Hasnip PJ, Refson K, Probert MIJ, Yates JR, Clark SJ, Pickard CJ (2014) Density functional theory in the solid state. Phil Trans R Soc A 372:20130270
20. Thiel W (2014) Semiempirical quantum-chemical methods. WIREs Comput Mol Sci 4:145–157
21. Andrews SJ, Casci JL, Cox PA, Shannon MD (1999) Determination of the location of the template molecules in zeolite EU-1 via a combined molecular modelling and X-ray diffraction approach. In: Treacy MMJ, Marcus BK, Bisher ME and Higgins JB (eds) Proceedings of the 12th international zeolite conference, Materials Research Society, Warrendale, pp 2355–2360
22. Guo P, Shin J, Greenaway AG, Min JG, Su J, Choi HJ, Liu LF, Cox PA, Hong SB, Wright PA, Zou XD (2015) A zeolite family with expanding structural complexity and embedded isoreticular structures. Nature 524:74–78
23. Rollmann LD, Schlenker JL, Lawton SL, Kennedy CL, Kennedy GJ, Doren DJ (1999) On the role of small amines in zeolite synthesis. J Phys Chem B 103:7175–7183

24. Wagner P, Nakagawa Y, Lee GS, Davis ME, Elomari S, Medrud RC, Zones SI (2000) Guest/host relationships in the synthesis of the novel cage-based zeolites SSZ-35, SSZ-36 and SSZ-39. J Am Chem Soc 122:263–273
25. Lewis DW, Freeman CM, Catlow CRA (1995) Predicting the templating ability of organic additives for the synthesis of microporous materials. J Phys Chem 99:11194–11202
26. Shen V, Bell AT (1996) Computer simulation of the interactions of tetraalkylammonium cations with ZSM-5 and ZSM-11. Microporous Mater 7:187–199
27. Szyja BM, Vassilev P, Trinh TT, van Santen RA, Hensen EJM (2011) The relative stability of zeolite precursor tetraalkylammonium-silicate oligomer complexes. Microporous Mesoporous Mater 146:82–87
28. Sánchez M, Diaz RD, Cordova T, Gonzalez G, Ruette F (2015) Study of template interactions in MFI and MEL zeolites using quantum methods. Microporous Mesoporous Mater 203:91–99
29. Sastre G, Fornes V, Corma A (2002) On the preferential location of Al and proton siting in zeolites: a computational and infrared study. J Phys Chem B 106:701–708
30. Gómez-Hortigüela L, Pinar AB, Cora F, Pérez-Pariente J (2010) Dopant-siting selectivity in nanoporous catalysts: control of proton accessibility in zeolite catalysts through the rational use of templates. Chem Commun 46:2073–2075
31. Pinar AB, Gómez-Hortigüela L, McCusker LB, Pérez-Pariente J (2013) Controlling the aluminium distribution in the zeolite ferrierite via the organic structure directing agent. Chem Mater 25:3654–3661
32. Sastre G, Pulido A, Castañeda R, Corma A (2004) Effect of the germanium incorporation in the synthesis of EU-1, ITQ-13, ITQ-22 and ITQ-24 zeolites. J Phys Chem B 108:8830–8835
33. Pulido A, Moliner M, Corma A (2015) Rigid/flexible organic structure directing agents for directing the synthesis of multipore zeolites: a computational approach. J Phys Chem C 119:7711–7720
34. Sastre G, Leiva S, Sabater MJ, Gimenez I, Rey F, Valencia S, Corma A (2003) Computational and experimental approach to the role of structure-directing agents in the synthesis of zeolites: the case of cyclohexyl alkyl pyrrolidinium salts in the synthesis of β, EU-1, ZSM-11 and ZSM-12 zeolites. J Phys Chem B 107:5432–5440
35. Gómez-Hortigüela L, Pérez-Pariente J, Cora F (2009) Insights into structure direction of microporous aluminophosphates: competition between organic molecules and water. Chem Eur J 15:1478–1490
36. Castro M, Garcia R, Warrender SJ, Slawin AMZ, Wright PA, Cox PA, Fecant A, Mellot-Fraznieks C, Bats N (2007) Co-templating and modelling in the rational synthesis of zeolitic solids. Chem Commun 33:3470–3472
37. Almeida RKS, Gómez-Hortigüela L, Pinar AB, Pérez-Pariente J (2016) Synthesis of ferrierite by a new combination of co-structure-directing agents: 1,6-bis(N-methylpyrrolidinium)hexane and tetramethylammonium. Microporous Mesoporous Mater 232:218–226
38. Turrina A, Garcia R, Cox PA, Casci JL, Wright PA (2016) A retrosynthetic co-templating method for the preparation of silicoaluminophosphate molecular sieves. Chem Mater 28:4998–5012
39. Corma A, Rey F, Rius J, Sabater MJ, Valencia S (2004) Supramolecular self-assembled molecules as organic directing agent for the synthesis of zeolites. Nature 431:287–290
40. Álvaro-Muñoz T, López-Arbeloa FL, Pérez-Pariente J, Gómez-Hortigüela L (2014) (1R,2S)-Ephedrine: a new self-assembling chiral template for the synthesis of aluminophosphate frameworks. J Phys Chem C 118:3069–3077
41. Moliner M, Serna P, Cantin A, Sastre G, Díaz-Cabañas MJ, Corma A (2008) Synthesis of the Ti-Silicate form of BEC polymorph of β-Zeolite assisted by molecular modeling. J Phys Chem C 112:19547–19554
42. Brand SK, Schmidt JE, Deem MW, Daeyaert F, Ma Y, Terasaki O, Orazov M, Davis ME (2017) Enantiomerically enriched, polycrystalline molecular sieves. Proc Natl Acad Sci 114:5101–5106
43. Lewis RA (1990) Automated site-directed drug design. J Comput Aided Mol Des 4:205

44. Douquet D, Munier-Lehmann H, Labesse G, Pochet S (2005) LEA3D: a computer-aided ligand design for structure-based drug design. J Med Chem 48:2457

45. Pegg SCH, Haresco JJ, Kuntz ID (2001) A genetic algorithm for structure-based de novo design. J Comput Aided Mol Des 15:911–933

46. Lewis DW, Willock DJ, Catlow CRA, Thomas JM, Hutchings GJ (1996) De novo design of structure-directing agents for the synthesis of microporous solids. Nature 382:604–606

47. Lewis DW, Sankar G, Wyles JK, Thomas JM, Catlow CRA, Willock DJ (1997) Synthesis of a small-pore microporous material using a computationally designed template. Angew Chem Int Ed Engl 36:2675–2677

48. Barrett PA, Jones RH, Thomas JM, Sankar G, Shannon IJ, Catlow CRA (1996) Rational design of a solid acid catalyst for the conversion of methanol to light alkenes: synthesis, structure and performance of DAF-4. Chem Commun 17:2001–2002

49. Pophale R, Daeyaert F, Deem MW (2013) Computational prediction of chemically synthesizable organic structure directing agents for zeolites. J Mater Chem A 1:6750–6760

50. Schmidt JE, Deimund MA, Davis ME (2014) Facile preparation of aluminosilicate RTH across a wide composition range using a new organic structure-directing agent. Chem Mater 26:7099–7105

51. Schmidt JE, Deem MW, Davis ME (2014) Synthesis of a specified, silica molecular sieve by using computationally predicted organic structure-directing agents. Angew Chem Int Ed 53:8372–8374

52. Schmidt JE, Deem MW, Lew C, Davis TM (2015) Computationally-guided synthesis of the 8-ring zeolite AEI. Top Catal 58:410–415

53. Davis TM, Liu AT, Lew CM, Xie D, Benin AI, Elomari S, Zones SI, Deem MW (2016) Computationally guided synthesis of SSZ-52: a zeolite for engine exhaust clean-up. Chem Mater 28:708–711

54. Irwin JJ, Teague S, Mysinger MM, Bolstad ES, Coleman RG (2012) ZINC: a free tool to discover chemistry for biology. J Chem Inf Model 52:1757–1768

55. Liang J, Edelsbrunner H, Woodward C (1998) Anatomy of protein pockets and cavities: measurement of binding site geometry and implications for ligand design. Protein Sci 7:1884–1189

Struct Bond (2018) 175: 103–138
DOI: 10.1007/430_2017_13
© Springer International Publishing AG 2017
Published online: 25 November 2017

Beyond Nitrogen OSDAs

Fernando Rey and Jorge Simancas

Abstract The use of organic structure-directing agents (OSDAs) is perhaps the most important factor to be considered for the synthesis of zeolitic materials. Several OSDAs had been used along the last 70 years, especially ammonium organic cations, letting the synthesis of a large number of materials. But besides ammonium cations, organic cations with different chemical natures had also been used, which resulted in the synthesis of very interesting zeolitic materials. This review includes most of the non-ammonium cations used up to date, namely, phosphorous cations, sulfonium cations, crown macrocycles and metal complexes, but also when N-containing OSDAs play a different role than in conventional zeolite syntheses, such as proton sponges, self-assembled compounds or ionic liquids.

Keywords Heterosubstituted organic cation • Structure directing agent • Synthesis • Zeolite

Contents

F. Rey (✉) and J. Simancas
Instituto de Tecnología Química, Consejo Superior de Investigaciones Científicas – Universitat
Politècnica de València (C.S.I.C. – U.P.V.), Valencia, Spain
e-mail: frey@itq.upv.es

1   Introduction

Within the materials science field, the research into the synthesis of microporous zeolitic materials has been and keeps on being a field in steady expansion since the first synthesis of zeolite was reported [1]. However, since then, the synthesis of zeolites mostly remains as a trial-and-error science, and the prediction and rational design of a given microporous material still heavily rely on previous experience in the synthesis of the chosen material. Nonetheless, progress in the area is still considerable, and thousands of publications dealing about zeolitic materials are published yearly, and among them, the number of papers discussing the synthesis of these materials is increasing every year (Fig. 1).

The successful synthesis of new zeolitic structures and compositions still remains heavily dependent on the use of OSDAs. In this way, the use of ammonium-containing OSDAs has been extensively employed because of the large knowledge about the preparation of this family of compounds [2]. The use of these OSDAs, either to substitute or to complement the use of the classical alkaline inorganic structure-directing agents, has led to the synthesis of at least 142 new zeolitic structures of the 235 known structures up to date and the widening of the chemical composition of most zeolites to improve their industrial applicability (August 2017) [3]. While the use of these OSDAs has led to the synthesis of most of the known zeolitic structures, it is important to note the large number of amine and ammonium compounds tested along all these years (Fig. 2) indicating that most of the syntheses yield already known zeolites.

Beyond the employ of nitrogen-based OSDAs (quaternary, di-quaternary and tri-quaternary alkylammonium cations, surfactants, imidazolium derivatives or amines) [4–11], there had been attempts to introduce new families of compounds suitable to be used as OSDAs. The main reason to explore new families lies

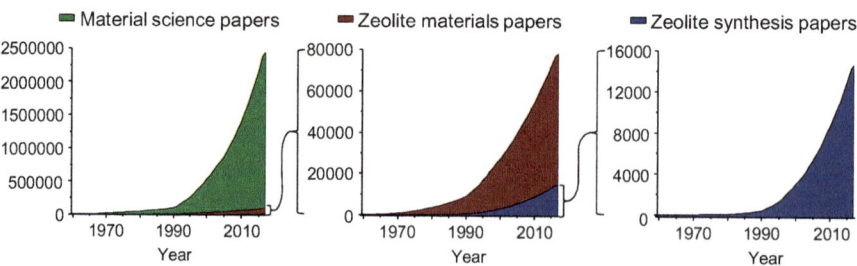

Fig. 1 Number of papers published per year in materials science (green), zeolitic materials (red), and zeolite synthesis (blue). Font: Web of Science

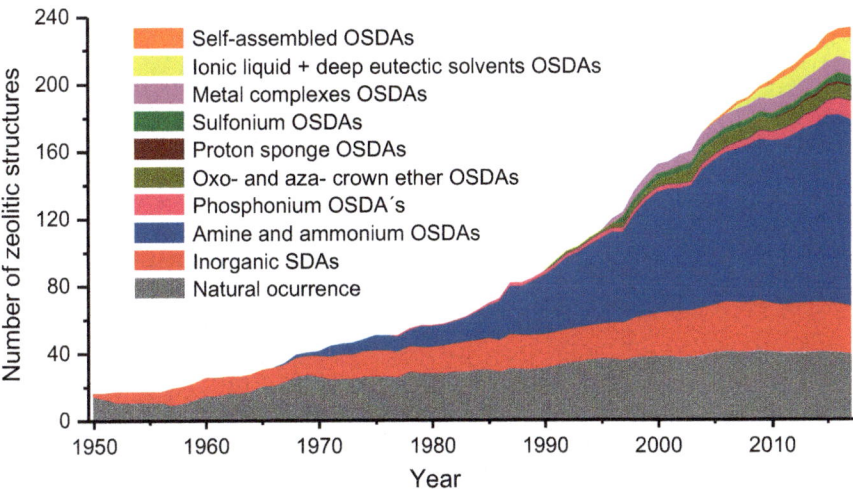

Fig. 2 Number of zeolite structures synthesized and share of each type of OSDA used per year

in the aim to obtain new zeolite structures, especially with large pore and/or multidimensional channels, or alternatively known structures with novel features as wider compositional ranges, improved textural and adsorption properties or different crystallite morphologies. These new families of compounds comprise phosphorous compounds (quaternary and di-quaternary alkylphosphonium cations, quaternary and di-quaternary alkyl-aminophosphonium cations and phosphazene bases), sulphonium cations, aza- and oxo-crown macrocycles, metal complexes, proton sponges, self-assembled compounds and ionic liquids (Fig. 3).

2 Onium OSDAs

Typically, the term onium refers to peralkylated and positively charged organic molecules. These include tetraalkylammonium, tetraalkylphosphonium and tri-alkylsulphonium cations, among others [12]. In these review, only tetraalkyl-phosphonium and trialkylsulphonium cations will be discussed.

2.1 Phosphonium Cations as OSDAs

Alkylphosphonium cations are known for a long time and are used in numerous fields of applications as a source of ylides in organic chemistry; the production of organic polymers, insecticides and fungicides; flame retardants for textiles and paper; anti-static and softening agents in textile and resins: and corrosion inhibitors

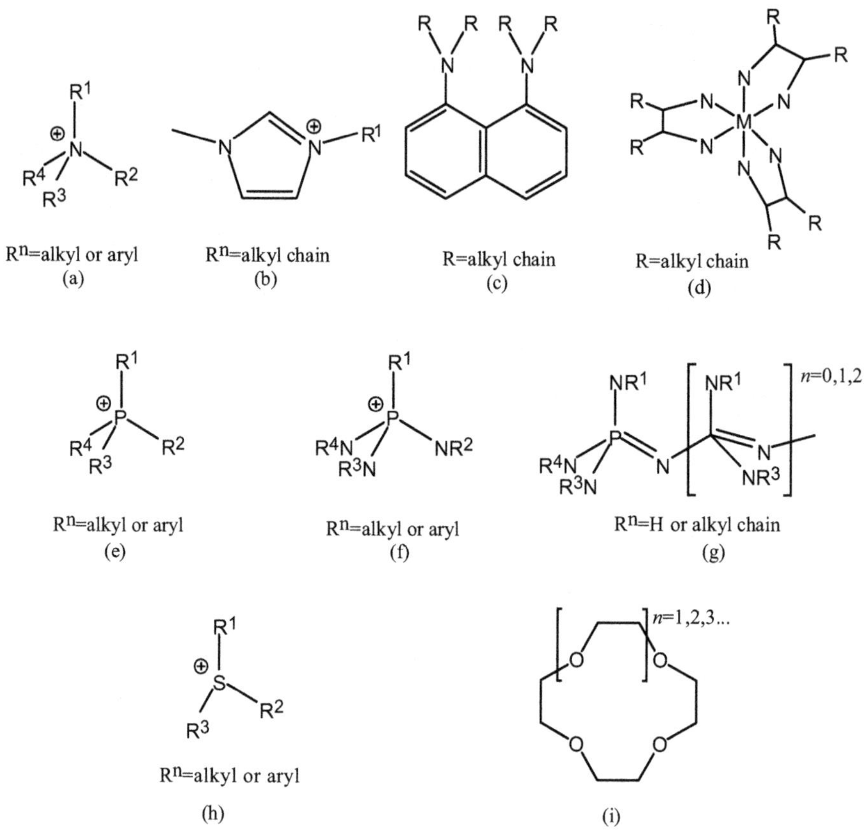

Fig. 3 Schematic chemical structures of some OSDAs applied in the synthesis of zeolites. Tetraalkylammonium cations (**a**), imidazolium derivatives (**b**), proton sponges (**c**), metal complexes (**d**), tetraalkylphosphonium cations (**e**), alkyl-triaminophosphonium cations (**f**), phosphazene bases (**g**), trialkylsulfonium cations (**h**) and crown ethers (**i**)

of photographic chemicals [13]. Recently, these compounds have received additional interest because of their possibility of forming ionic liquids with superior properties compared to their nitrogen-based analogues [14–17].

The typical synthesis procedure to obtain alkylphosphonium cations relies on the use of phosphine derivatives (Fig. 4) [13, 18–21]. First, it should be stated that special care when working with phosphines must be taken because of their very high toxicity. Phosphines are usually less stable than their amine counterparts because of their high reactivity and, especially, they are unstable in contact with water or air, yielding the corresponding oxidized compounds, although these properties are heavily linked with the compound volatility (e.g. the solid triphenylphosphine is almost inert in contact with water and air but keeps the general reactivity of phosphines) [19, 22]. Because of that, the working up to synthesize complex phosphonium derivatives turns into a demanding task to comply, especially in laboratories focused on materials science. This could be an

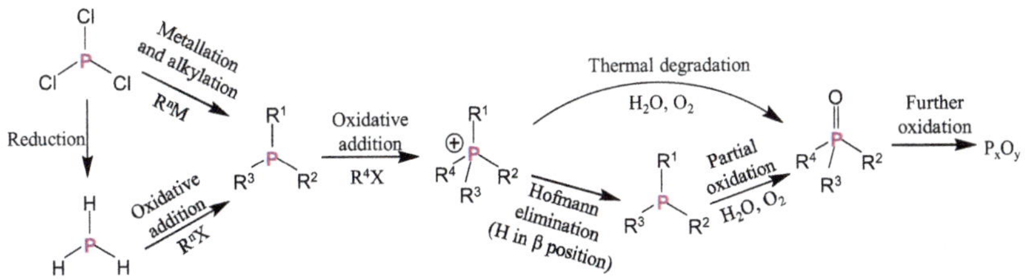

Fig. 4 General pathway for the synthesis (left side) and main degradation processes (right side) of tetraalkylphosphonium cations

important reason for the scarcity of labs and publications related to the use of the P-containing OSDAs for synthesis of zeolites, although there is an important industrial interest to obtain them because of their applicability, as will be seen in this review. However, when properly handled, the synthesis of phosphonium compounds is achieved with nearly quantitative yields with very few undesired products, as opposed to the mid-yields of the ammonium OSDAs [2, 13, 15, 18]. And more importantly, phosphonium cations exhibit a higher stability than ammonium cations against any physical condition, notably under hydrothermal conditions, lessening the occurrence of the β-Hofmann elimination (Fig. 4) [20, 23–26]. This is of paramount importance because most of the syntheses of zeolites are carried under hydrothermal conditions and the use of phosphonium cations allows to widen the range of physical conditions by increasing the synthesis temperature or the alkalinity of the media or, at least, to maintain them for very long crystallization times without any significant decrease in the concentration of the cationic species active as OSDAs.

Additionally, phosphorous compounds exhibit a more versatile chemistry than nitrogen compounds, and thus we can also find the phosphorus directly bonded to an uncharged nitrogen atom. These molecules are interesting because they allow tuning the charge distribution around the quaternized phosphorus by exchanging a carbon atom with a nitrogen atom. This property gives rise to two different compound families, aminophosphines [21, 27], which can be turned into alkyl-aminophosphonium cations. and phosphazenium cations [28, 29]. Both are readily available to be used as OSDAs, as phosphazenium cations are commercially available as phosphazene proton sponges [29] and aminophosphines are complexing agents that are less hazardous and volatile than their alkylphosphine counterparts [30, 31].

The first trials to use phosphonium cations for the synthesis of zeolites in the 1970s, 1980s and 1990s mainly came from the petrochemical industry. These syntheses involved the use of quite simple tetraalkylphosphonium cations analogous to the well-known tetraalkylammonium cations used then, namely, tetraethyl, tetrapropyl and tetrabutylphosphonium cations, which were made readily available shortly before their use as OSDAs. Several patents were issued claiming the use of phosphonium OSDAs to yield different zeolites: zeolite theta-3 (MTW) related to theta-1 and MFI zeolites and possessing a ZSM-12 structure [32], zeolite ZSM-5

(MFI) [33] and zeolite EU-13 (MTT) [34]. Next, patents and academic articles addressed the use of these phosphorus-OSDAs, yielding to a number of zeolites: ZSM-5 (MFI) [35], TS-2 (MEL) [36], TS-1 (MFI) [37], ZSM-5 (MFI) and ZSM-11 (MEL) [38] and RUB-35 (a disordered EUO-NES-NON family material) [39].

However, this first approach provided the same phase selectivity than the chemically equivalent ammonium cations, and none unknown zeolite was obtained. It was not until 2006 that a new zeolite structure was described by using a phosphorus-OSDA, the large pore zeolite ITQ-27 (IWV) [40], by using a more complex and asymmetrical tetraalkylphosphonium cation. From then on, several novel zeolite structures have been obtained with phosphonium cations, all of them included within the ITQ series of zeolites: the large pore germanosilicate ITQ-26 (IWS) [41], the medium pore germanosilicate ITQ-34 (ITR) [42], the extra-large pore germanosilicate ITQ-40 (-IRY) [43], the medium pore zeolite ITQ-45 [44], the small pore germanosilicate ITQ-49 (ITN) [45], the small and medium multi-pore borosilicate ITQ-52 (IFW) [46], the extra-large pore germanosilicate ITQ-53 (-IFT) [47] and the small pore borosilicate ITQ-58 [48] (Table 1).

Also, already known zeolites with different properties have been obtained with phosphorus-OSDAs along these years. For example, zeolite beta (BEA) with extra-framework P species can be obtained without any post-synthesis treatment [57]; the use of a phosphorus-proton sponge allowed obtaining the zeolite ITQ-47, the synthetic natural analogue of the large and medium multi-pore boggsite (BOG) zeolite [58, 59], and the extra-large pore germanosilicate ITQ-33 (ITT), allowing to introduce P without any post-synthesis treatment in this sensitive structure [60]; and the use of P-OSDA together with N-OSDA allowed obtaining the one-pot self-pillaring of the medium pore ZSM-11 (MEL) [61] and the zeolite ZSM-5 (MFI), to introduce P without any post-synthesis treatment [62], the synthesis of high-silica AEI zeolite by topotactic transformation of FAU zeolite by using a dual-template synthesis with ammonium and phosphonium cations [63], the synthesis of self-pillared zeolites ZSM-5 (MFI) and beta (BEA) and titanosilicates ETS-4 and ETS-10 [64], the synthesis of high-silica CHA zeolite by topotactic transformation of FAU zeolite by using a dual-template synthesis with ammonium and phosphonium cations [65], the synthesis of small pore all-silica zeolite RUB-13 (RTH), to introduce P as probe atom for NMR and INS techniques [66], and the synthesis of nanocrystalline zeolite SSZ-39 (AEI) by topotactic transformation of FAU zeolite using a single phosphonium OSDA [67] (Table 2).

The use of phosphorus-containing cations as OSDA is interesting because, commonly, their ammonium equivalents gave different phase selectivities (see Tables 1 and 2). Also, phosphorus is a valuable element to incorporate into zeolites, as it has been used for many years to improve the hydrothermal stability of the aluminium catalytic centres to further extent the catalytic lifetime of these materials (Fig. 5) [111–115]. Phosphorus can be introduced in the zeolite by impregnation with several compounds, as H_3PO_4, $NH_4H_2PO_4$, PCl_3, $P(OCH_3)_3$, $P(C_6H_5)_3$, etc. [112–120]. However, most of these treatments present a great disadvantage, as they are only available for medium or larger pore system zeolites. The use of

Table 1 New zeolites obtained by using P-OSDA. The middle column corresponds to the P-OSDA used to obtain the zeolite structures listed in the column to its left. Further to the left, it is listed the N-OSDA also used to obtain these zeolitic structures. The analogous N-OSDA column corresponds to the chemically equivalent of the listed P-OSDAs column (if any), and the analogous zeolites correspond to the zeolitic structures obtained with those analogous N-OSDAs

N-OSDA	Zeolitic structure	P-OSDA	Analogous N-OSDA	Analogous zeolitic structure
No N-OSDA yields IWS	ITQ-26 (IWS) [41]	[41]	[49]	ZSM-12 (MTW), beta (BEA), ITQ-17 (BEC) [49]
n = 4,5,6,8,10 [50]	ITQ-27 (IWV) [40]	[40]	[51, 52]	EU-1 (EUO) [51], beta (BEA) [52]
No N-OSDA yields ITR	ITQ-34 (ITR) [42]	[42]	[53, 54]	ZSM-23 (MTT) [53]
				Nonasil (NON) [54]
No N-OSDA yields -IRY	ITQ-40 (-IRY) [43]	[43]		No zeolite obtained with analogous N-OSDA
No N-OSDA yields ITQ-45	ITQ-45 [44]		No possible analogous N-OSDA	–
No N-OSDA yields IRN	ITQ-49 (IRN) [55]	[44, 45, 55]	No possible analogous N-OSDA	–
[56]	ITQ-52 (IFW) [46]	[46]	No possible analogous N-OSDA	–
No N-OSDA yields ITQ-53	ITQ-53 (-IFT) [47]	[47]	No possible analogous N-OSDA	–
No N-OSDA yields ITQ-58	ITQ-58 [48]	[48]	No possible analogous N-OSDA	–

Table 2 Known zeolites obtained by using P-OSDA. The middle column corresponds to the P-OSDA used to obtain the zeolite structures listed in the column to its left. Further to the left, it is listed the N-OSDA also used to obtain these zeolitic structures. The analogous N-OSDA column corresponds to the chemical equivalent of the listed P-OSDA column (if any), and the analogous zeolites correspond to the zeolitic structures obtained with those analogous N-OSDAs

N-OSDA	Zeolitic structure	P-OSDA	Analogous N-OSDA	Analogous zeolitic structure
R: Me, Et, Pr, Bu... (quaternary ammonium). Several more N-OSDAs [68–72]	ZSM-5 (MFI) [35]	[35]	No possible analogous N-OSDA	–
	TS-1 (MFI) [37] ZSM-5 (MFI) [62]	[62, 73]		BEA [74]; MTW [75]; EMT/FAU [76]; MFI [68]; MOR [77]; MTN [78]; LTL [79]; PAU [80]; LTF [81]; CHA/OFF [82]; VSV [83]; CHA [84]; AFI [85]; OSI [86]; ETR [87]; VET [88]; FAU [89]; AEI [90]; etc.
[91]	TS-2 (MEL) [36]	[36, 60, 64]		MEL [91]; MFI [92]; VFI [93]; LTL [79]; FAU [94]; LOS [95]; ITT, NUD-1 [96]; etc.
[97]	ITQ-33 (ITT) [60]			
R: Me, Et, Pr, Bu... Several more N-OSDAs for MFI and BEA [68–72]; Na⁺ and K⁺ for ETS-4 [98] and ETS-10 [99]	Self-pillared ZSM-5 (MFI), beta (BEA), ETS-4 and ETS-10 [64]			
[100]	Self-pillared ZSM-11 (MEL) [61]	[61]		

(continued)

Table 2 (continued)

N-OSDA	Zeolitic structure	P-OSDA	Analogous N-OSDA	Analogous zeolitic structure
Carbon Nanoparticles [101]	RUB-35 (EUO-NES-NON) [39]	[39]		LTA [102]; EAB, ERI, MAZ, OFF [103]; LTA, MAZ, ANA [104]; RUT [105]; FAU-EMT [106]; ETS-10 [107]; etc.
No N-OSDA yields ITQ-47	ITQ-47 (BOG) [58, 59]	[58, 59]	No possible analogous N-OSDA	–
Mixture of the two [108]	High-silica AEI [63, 67]	[63, 67]		BEA [74]; MTW [75]; EMT/FAU [76]; MFI [68]; MOR [77]; MTN [78]; LTL [79]; PAU [80]; LTF [81]; CHA/OFF [82]; VSV [83]; CHA [84]; AFI [85]; OSI [86]; ETR [87]; VET [88]; FAU [89]; AEI [90]; etc.
[109]	High-silica CHA [65]	[65]		
[110]	RUB-13 (RTH) [66]	[66]	No possible analogous N-OSDA	–

phosphonium cations proves to be a convenient way to introduce phosphorus even in small pore system zeolites [65, 67].

After the synthesis of zeolites aided by OSDAs (except in some modular and self-assembled OSDAs) [124, 125], it is required to remove the OSDA for emptying the pores of the material. This is usually made by calcination, usually under air, yielding small gaseous molecules (mostly, H_2O, CO_2, NO_x and small amines). In contrast, the calcination of P-OSDA-containing zeolites also gives rise to gaseous molecules (H_2O and CO_2), but most of the phosphorus remains inside the zeolite as extra-

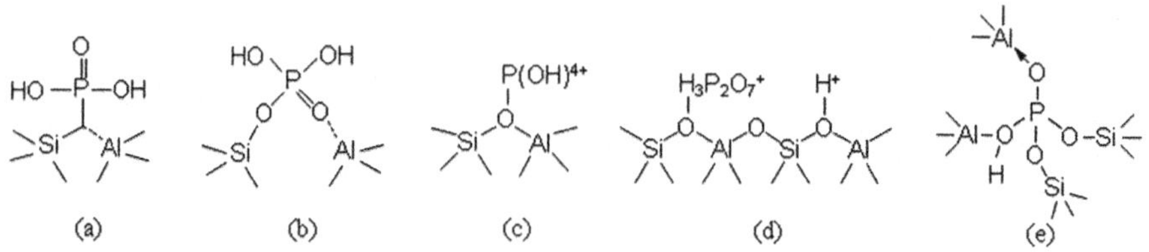

Fig. 5 Proposed models for phosphorus-zeolite interaction. (**a**) Vedrine et al. [121], (**b**) Lercher et al. [122], (**c, d**) Corma et al. [112], (**e**) Xue et al. [123]

framework oxidized phosphorous species (Fig. 6) [112, 117, 121–123]. These species are responsible for the increasing hydrothermal stability of the catalytic centres. On the other hand, P-OSDA-containing materials could be subjected to hydrogenation processes at high temperature, removing most of the phosphorus as lightweight phosphines [44, 46, 48, 58, 59].[1] However, in both processes, it is difficult to control the removal of the P species, as it depends on the temperature, heating rate and time of the thermal treatment, the composition of the zeolite and, especially, the structure, pore openings and channel dimensionality of each zeolite. A smarter way for controlling the remaining amount of P inside the zeolite would be the employ of the dual-template synthesis methodology [73, 126–128], employing a combination of N-OSDAs and P-OSDAs and optimizing the final aluminium/phosphorus ratio just by changing the ratio between the N-OSDA and P-OSDA in the synthesis stage, as most of the P will remain inside the zeolite upon calcination [65, 129].

The prospects for the use of phosphonium OSDAs are immense due to the increase in the availability and the decrease of the cost of the starting compounds (phosphines and aminophosphines) in the last decades. Despite this, they are still more expensive than the N-OSDA, so their implementation for the synthesis of materials on a large scale seems at first distant. However, their use in dual-template synthesis would allow designing materials with the properties that are of interest and taking advantage of the features that these compounds offer without incurring in the drawbacks of them. Also, these materials have shown that they are powerful OSDAs allowing the crystallization of a large number of new zeolitic structures, and therefore, the future in the development of new P-OSDA with the objective of obtaining new structures continues being shining.

2.2 Sulphonium Cations as OSDAs

Trialkylsulphonium cations have been studied in a wide range of applications, as ionic liquids [130, 131], electrolytes in solar cells [132], alkylation reagents [133] or

---

[1]Warning: these compounds are extremely hazardous, even in small quantities, and appropriated facilities for their use and disposal must be used.

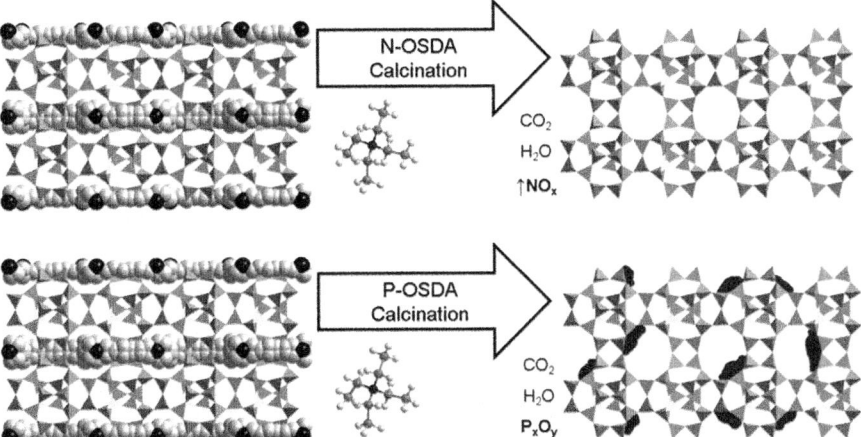

Fig. 6 Schematic view of the calcination process of N-OSDA and P-OSDA inside a zeolite structure

herbicides [134]. Trialkylsulphonium cations are commercially accessible as salts, but their availability is very low. On the other hand, the synthesis of new trialkylsulphonium cations is feasible by commonly used organic reactions (mainly by using organometallic compounds and/or oxidative addition reactions) [130, 131, 135–137].

The use of trialkylsulphonium cations as OSDAs is relatively new, and despite the low number of studies that have addressed these compounds, many aluminophosphate zeotypes such as AlPO-5, SAPO-5 and MAPO-5 (AFI); AlPO-31 (ATO), AlPO-41 and SAPO-41 (AFO); SAPO-11 and MAPO-11 (AEL); MAPO-14 and CoAPO-14 (AFN); and MAPO-34 (CHA) have been obtained [138]. In addition, the crystallization of several zeolites using trialkylsulphonium as OSDA has been reported, such as ITQ-33 (ITT), ITQ-7 (ISV) and ZSM-11 (MEL), and interestingly, a new silicogermanate, GeZA, has also been obtained [139] (Table 3).

Because of their high polarities, these compounds are water soluble even with C/S ratios as high as 18. Thus, this is still an unexplored field for the discovering of new zeolitic materials by employing trialkylsulphonium cations with different alkyl or aryl substituents providing larger and bulkier OSDAs. More interestingly, trialkylsulphonium salts (as several tetraalkylphosphonium salts) could be employed as low-temperature ionic liquids and, then, are good candidates as solvents/OSDAs for the ionothermal syntheses of zeolites. In this way, sulphonium-based ionic liquids usually have a better ionic conductivity and lower viscosity than their ammonium counterparts, which is helpful for the ionothermal synthesis of inorganic solids. However, their thermal stability is lower than ammonium cations, and their stability against β-Hoffman elimination is a matter of study yet, since it could limit their application as OSDAs [137].

Table 3 Zeolitic structures obtained by using S-OSDAs. The left column corresponds to the S-OSDA used to obtain the zeolite structures listed in the column in the middle, and the right column corresponds to the N-OSDA typically used to obtain the zeolitic structure

Sulphonium OSDA	Zeolitic structures	N-OSDA	Sulphonium OSDA	Zeolitic structures	N-OSDA
	AlPO-5, SAPO-5 and MAPO-5 (AFI) [138]	[140]		ITQ-33 (ITT) [139]	[97]
	AlPO-41 and SAPO-41 (AFO) [138]	[141]		ITQ-7 (ISV) [139]	[142]
	SAPO-11 and MAPO-11 (AEL) [138]	[143]		GeZA [139]	No N-OSDA yields GeZA
	MaPO-14 and CoAPO-14 (AFN) [138]	[144]		ZSM-11 (MEL) [139]	[145]
	MAPO-34 (CHA) [138]	[146]	[139]		
[138]					

3 Crown Macrocycles as OSDAs

The family of crown macrocycle molecules is very large and has been known for long [147], but in this review, we will discuss just about two types of crown macrocycles, the oxo-crown macrocycles (also named crown ethers) and the aza-macrocycles (as well as the mixed oxo-aza-crown macrocycles) that have been used for syntheses of zeolites.

3.1 Aza-crown Macrocycles as OSDAs

Aza-macrocycles are cyclic macromolecules containing one or more nitrogen atoms either as divalent group NH, substituting a CH_2 group, or a single trivalent nitrogen atom for the group CH. They have been extensively used in coordination chemistry [148–150]. There are several compounds commercially available, and the presence of an amine group allows their chemical modification by alkylation reactions, allowing their ammonium forms. Although several syntheses have been reported by using the ammonium forms of these compounds or azoniomacrocycles [151–153], this review will be just focused on the neutral aza-crown macrocycles as alternative OSDAs.

The use of aza-macrocycles has yielded a number of zeolitic structures. The first use of aza-macrocycles as OSDAs was described for the synthesis of aluminophosphates: the aluminophosphate AlPO-42 (LTA) was prepared using the cryptand Kryptofix 222 [154]; then, the use of aza-macrocycles has yielded the discovery of the novel gallophosphate Mu-5, an organic-inorganic hybrid zeolite-like material by using cyclam [155, 156]; zeolites X and Y (FAU) and the aluminophosphate VPI-5 (VFI) were obtained by using porphyrin molecules, with a strong affinity for these zeolites as deduced from the very high incorporation of the OSDA [157]. Also, the synthesis of novel magnesioaluminophosphate STA-6 (SAS) was reported using tmtacn aza-macrocycle as OSDA [158] as well as several aluminophosphates in the presence of Mg^{2+}, Cr^{3+}, Mn^{2+}, Fe^{2+}, Co^{2+}, Ni^{2+}, Cu^{2+} and Zn^{2+}, like the novel MAPO-cyclam-1, MAPO-5 (AFI), MAPO-18 (AEI), ALPO-21 (AWO), MAPO-42 (LTA), STA-6 (SAS) and STA-7 (SAV) by using cyclam and aza-oxacryptands tmtacn, tmtact and hmhaco [159], the aluminophosphate Mu-13 (MSO) with an aza-oxacryptand [160], the aluminophosphate analogues of CHA and SAS with cyclam [161], the all-silica ITQ-29 (LTA) with Kryptofix 222 [162] and the pure-silica FAU and EMT zeolites with di-aza-crown macrocycle OSDAs aided by molecular modelling [163] (Table 4).

3.2 Oxo-crown Macrocycles (Crown Ethers) as OSDAs

Since their discovery by Pedersen [167], crown ethers have found several applications because of their high affinity for cations in solution [168–171]. However, the first report of these compounds being used as OSDA in the synthesis of zeolites came many years later, when high-silica faujasite (FAU) with a Si/Al~5 (SAR~5) and pure hexagonal faujasite (EMT) were synthesized with 15-crown-5 and 18-crown-6, respectively [172–174]. Since then, high-silica zeolite Rho (RHO) (SAR~4.5) [175], all-silica zeolite KFI [176] and the novel aluminosilicate clathrate MCM-61 (MSO) [177] have also been synthesized with the 18-crown-6. Additionally, the use of smaller crown ethers, such as 15-crown-5, has led to the synthesis of all-silica zeolite sodalite (SOD) [178, 179]. An interesting approach have been recently taken, building a supramolecular double 18-crown-6 through interaction with Cs^+ in between, which yielded the highest-silica RHO zeolite (SAR~15) ever reported [180]. In this section, the use of mixed oxa-aza-crown macrocyclic molecules that were previously discussed could also be addressed [154, 160] (Table 5).

The main issue about macrocycle compounds is the usual need for the presence of cations in the synthesis of zeolites, either inorganic or organic, to provide the required alkalinity and, in most cases, to form charged supramolecular species that are the true active species in the synthesis process [174, 182].

Table 4 Zeolitic structures obtained by using aza-macrocycle OSDAs. The left column corresponds to the azamacrocycle OSDA used to obtain the zeolite structures listed in the column in the middle, and the right column corresponds to the N-OSDA typically used to obtain the zeolitic structure

Aza-macrocycle OSDA	Zeolitic structures	N-OSDA	Aza-macrocycle OSDA	Zeolitic structures	N-OSDA
[structure] [154, 159, 160, 162]	ALPO-42 (LTA) [154]	[structure] [84]	[structure] [155, 156, 158, 159]	Mu-5 [155, 156]	No N-OSDA yields Mu-5
	Mu-13 (MSO) [160]	No N-OSDA yields MSO		STA-6 (SAS) [158, 159]	No N-OSDA yields SAS
	ITQ-29 (LTA) [162]	No N-OSDA yields ITQ-29		STA-7 (SAV) [159]	No N-OSDA yields SAV
	MAPO-5 (AFI) [159]	[structure] [140]		MAPO-18 (AEI) [159]	[structure] [90]
	MAPO-42 (LTA) [159]	[structure] [84]	[structure] [159]	MAPO-36 (ATS) [159]	[structure] [164]
[structure] [157]	Zeolites X and Y (FAU) [157]	[structure] [94]		ALPO-21 (AWO) [159]	[structure] [165]
	VPI-5 (VFI) [157]	[structure] [166]	[structure] [159, 161]	STA-6 (SAS) [161]	No N-OSDA yields SAS
[structure] [163]	FAU [163]	No N-OSDA yields all-silica FAU		CHA [161]	[structure] [146]
[structure] [163]	EMT [163]	No N-OSDA yields all-silica EMT		APO-cyclam-1 [159]	No N-OSDA yields APO-cyclam-1

On the other hand, for the synthesis of aluminophosphates where low alkalinity is needed, macrocycle compounds result in very interesting and versatile OSDAs. The possibilities to use aza- and/or oxo-crown macrocycles as OSDAs are almost unlimited due to the large variety of different building blocks available for these

Table 5 Zeolitic structures obtained by using crown ether as OSDAs. The left column corresponds to the crown ether OSDA used to obtain the zeolite structures listed in the column in the middle, and the right column corresponds to the N-OSDA typically used to obtain the zeolitic structure

Crown ether OSDA	Zeolitic structures	N-OSDA	Crown ether OSDA	Zeolitic structures	N-OSDA
[crown ether structure] [172–177]	Hexagonal faujasite (EMT) [172–174]	No N-OSDA yields pure EMT	[crown ether structure] [172–174, 178, 179]	High-silica faujasite (FAU) [172–174]	No N-OSDA yields high-silica FAU
	Rho (RHO) [175]	No N-OSDA yields RHO		All-silica sodalite (SOD) [178, 179]	[ammonium structure] [181]
	KFI [176]	No N-OSDA yields KFI	[crown ether structure, Cs] [180]	Rho (RHO) [180]	No N-OSDA yields RHO
	MCM-61 (MSO) [177]	No N-OSDA yields MSO			

compounds and the additional possibility of using different cations (whether inorganic or organic, though only inorganic have been tested so far) to build completely different supramolecular species. Although several crown ethers and aza-crown macrocycles are commercially available and thus could be used without further modifications, pricing usually makes them hardly possible to be implemented in large-scale synthesis.

4 Proton Sponges as OSDAs

Proton sponges are neutral compounds with extremely high basicity and kinetic activity in proton exchange reactions. The chemical nature of these compounds is varied; they could be aliphatic diamines, aromatic diamines, polycyclic polyamines, amidines, guanidines, phosphazenes, guanidinophosphazenes and Verkade's proazaphosphatranes. They are typically used as base catalysts in a wide range of organic reactions [29, 183, 184].

Up to date, only two proton sponges have been used as OSDAs. The DMAN, which allowed obtaining the extra-large pore novel aluminophosphate ITQ-51 (IFO) [185], and the phosphazene P1, which allowed obtaining the ITQ-47 (BOG), are previously mentioned [58]. Besides that, no more zeolite syntheses have been reported by using these compounds (Table 6).

Table 6 Zeolitic structures obtained by using proton sponge OSDAs. The left column corresponds to the proton sponge OSDA used to obtain the zeolite structures listed in the column in the middle, and the right column corresponds to the N-OSDA typically used to obtain the zeolitic structure

Proton sponge OSDA	Zeolitic structures	N-OSDA	Proton sponge OSDA	Zeolitic structures	N-OSDA
[185]	ITQ-51 (IFO) [185]	No N-OSDA yields IFO	[58]	ITQ-47 (BOG) [58]	No N-OSDA yields BOG

Proton sponges remain a very interesting option for the synthesis of new zeolites because of the large number of possible compounds, their high basicity and their rigid structure, which makes them suitable for the synthesis of large pore zeolites. However, their high basicity is both an advantage and a drawback at the same time, because very high pH hampers the crystallization of the zeolite. Less alkaline synthesis conditions by adding HF (when stable in these media under hydrothermal conditions) could help the formation of crystalline materials with large pore channels, as the synthesis of the ITQ-51 has demonstrated. There are several proton sponges commercially available that can be used without further modifications, and although being considerably more expensive than common N-OSDAs, their pricing is not prohibitive.

5 Metal Complexes as OSDAs

Metal complexes and organometallic compounds have been used for years in the synthesis of zeolites to incorporate metal complexes into zeolites. This strategy has permitted to maintain properties of the original organometallic compounds (or the metal clusters that could be created after decomposition of the organic compounds) in a stable support. The most studied system for the incorporation of several organometallic compounds has been the large pore zeolite Y (FAU) [186–191].

However, the synthesis of all-silica clathrasil nonasil (NON), octadecasil (AST) and dodecasil 1H (DOH) by using $Co(Me_5Cp)_2$ as OSDA [192] and, shortly after, the synthesis of the novel extra-large pore zeolite UTD-1 (DON), also with the Co $(Me_5Cp)_2$ compound, opened a new frontier in the synthesis of zeolites [193–195]. Since then, several zeolites and zeotypes have been synthesized with these compounds. Examples of this strategy for zeolites include the synthesis of the aluminophosphate ALPO-16 (AST) by using $Co(Me_5Cp)_2$ [196]; the synthesis of new aluminophosphates STA-6 (SAS) and STA-7 (SAV) by using nickel aza-macrocycle complexes [197]; the synthesis of the novel gallogermanates GaGeO-

CJ63 (JST) [198] and JU-64 (JSR) [199] by using nickel-diamino complexes; the synthetic equivalent of phillipsite DAF-8 (PHI) by using cobalt-diamino complexes [200]; the synthesis of novel aluminophosphates UCSB-6 (SBS), UCSB-8 (SBE) and UCSB-10 (SBT) [201]; and the synthesis of Rho-like aluminophosphates (RHO) by using cobalt-, magnesium- and manganese-diamino complexes [202].

These compounds have also been applied in the dual OSDA synthesis approach by using metal complexes jointly with alkylammonium cations. By using this methodology, the metal complexes would work as pore space fillers instead of proper OSDAs. This approach has yielded the synthesis of the zeolite Cu-SSZ-13 (CHA) [203, 204] and the silicoaluminophosphates Cu-SAPO-34 (CHA) [205, 206] and Cu-STA-7 (SAV) [207]. This methodology has allowed to homogeneously disperse metal functionalities inside the channels and cavities of small pore zeo-types that would be difficult, if not impossible, by post-synthesis treatments [208, 209]. And additionally, the number of intermediate stages is reduced by several steps that decrease the energy consumption and the waste disposal of metal-containing water streams when compared to traditional cation exchange methods [210] (Table 7).

The potential of these compounds for the synthesis of zeolites lies on the high variety of possible organometallic compounds and their commercial availability (and their low pricing for in situ built complexes), as well as the large range of metallic cations that can be used. However, properties as water solubility, alkalinity and stability under hydrothermal conditions must be tested and studied in detail to be successful OSDAs. Although not tested yet, chiral metallic complexes could be very interesting for the synthesis of chiral zeotypes. These chiral complexes include P-containing chiral ligands, giving the possibility to combine the advantages of the phosphorus-containing OSDAs and metal complexes [215–219].

6 Self-Assembled OSDAs

Self-assembled OSDAs consist in organic compounds with aromatic rings that are able to self-assemble to form macromolecular moieties. This is usually achieved by parallel π–π interactions between aromatic rings. The self-assembled OSDAs are then relatively simple molecules that, when assembled, constitute OSDAs with a suitable size, rigidity, thermal stability and hydrophobicity properties. Under appropriate conditions, these compounds are able to create zeolitic topologies with large pore openings and cavities due to their supramolecular aggregation [220].

The first examples of the use of this kind of OSDA yielded already known zeolitic structures with interesting novel features. Thus, the all-silica ITQ-29 (LTA), never obtained previously and with very interesting properties for separation processes, was obtained using a quinolinium derivative [221]. Also, the alumi-nophosphates ALPO-5 and SAPO-5 (AFI) were obtained using benzylpyrrolidine derivatives as OSDAs. The mechanism of the aggregation and assembling of those molecules filling the channels of ALPO-5 was studied in detail confirming the

Table 7 Zeolitic structures obtained by using metal complex OSDAs. The left column corresponds to the metal complex OSDA used to obtain the zeolite structures listed in the column in the middle, and the right column corresponds to the NOSDA typically used to obtain the zeolitic structure

Metal complex OSDA	Zeolitic structures	N-OSDA	Metal complex OSDA	Zeolitic structures	N-OSDA
[192–196]	Nonasil (NON) [192]	[211]	[199]	JU-64 (JSR) [199]	No N-OSDA yields JSR
	Octadecasil (AST) [192]	[212]	M²⁺: Co, Zn, Mn, Mg [200, 201]	DAF-8 (PHI) [200]	[213]
	Dodecasil 1H (DOH) [192]	[211]		UCSB-6 (SBS) [201]	No N-OSDA yields SBS
				UCSB-8 (SBE) [201]	No N-OSDA yields SBE
	UTD-1 (DON) [193–195]	No N-OSDA yields DON	M²⁺: Co, Zn, Mn, Mg [201]	UCSB-10 (SBT) [201]	No N-OSDA yields SBT
	ALPO-16 (AST) [196]	[212]	M²⁺: Co, Zn, Mn, Mg [202]	Rho-like ALPOs (RHO) [202]	No N-OSDA yields RHO
[197]	STA-6 (SAS) [197]	No N-OSDA yields SAS	[203, 204]	Cu-SSZ-13 (CHA) [203, 204]	[109]

(continued)

Table 7 (continued)

Metal complex OSDA	Zeolitic structures	N-OSDA	Metal complex OSDA	Zeolitic structures	N-OSDA
	STA-7 (SAV) [197]	No N-OSDA yields SAV	[structure] + [structure] [205, 206] [structure] + [structure] [205, 206]	Cu-SAPO-34 (CHA) [205, 206]	CuO + [structure] [214]
[structure] [198]	GaGeO-CJ63 (JST) [198]	No N-OSDA yields JST	[structure] + [structure] [207]	Cu-STA-7 (SAV) [207]	No N-OSDA yields SAV

presence of π–π interactions between aromatic rings that lead to the formation of AFI structure [222–224].

Recently, the synthesis of MTN-type zeolite with a pyridyl-triazine derivative [225], and the synthesis of all-silica zeolite ZSM-12 (MTW) [226] using benzyl-pyrrolidine derivatives as OSDAs have been reported. Also, the synthesis of the extra-large pore zeolite ITQ-51 (IFO) was attained by self-assembled naphthalene-like proton sponge as has been previously mentioned [227]. Other publications report the synthesis of novel extra-large pore germanosilicates NUD-1 and NUD-2 in the presence of an aryl imidazolium derivative [228, 229], the synthesis of nanosheet zeolite ZSM-5 (MFI) with bulky ammonium groups having biphenyl and naphthyl groups [230] and the synthesis of MWW-like zeolite [231] and the extra-large pore zeolite ITQ-37 (-ITV) [232] with imidazolium derivatives.

Several syntheses of zeotypes aided by self-assembled OSDAs have also been described, for example, the synthesis of aluminophosphate STA-1 (SAO) with benzylpyrrolidine derivatives, allowing the isomorphic incorporation of Zn in the SAO structure [233, 234]; the synthesis of AFI-like aluminophosphate with chiral ephedrine derivatives, which shows a chiral distribution of heteroatoms within the AFI structure [235–237]; the synthesis of the novel aluminophosphate ICP-1 using 1,3-diphenylguanidine [238]; the synthesis of three novel layered aluminophos-phates with orto-, meta- and para-fluorinated benzylpyrrolidine derivatives [239]; and the synthesis of the small pore silicoaluminophosphates STA-6 (SAS) [240] and SAPO-42 (LTA) [241] in the presence of quinolinium derivatives (Table 8).

The potential of self-assembled compounds when used as OSDAs of micropo-rous solids has been proved just in few years due to their versatility. They are usually commercially available and low priced (chiral compounds are expensive, like most). Moreover, they could be used as such without further modifications or

Table 8 Zeolitic structures obtained by using self-assembled OSDAs. The left column corresponds to the self-assembled OSDA used to obtain the zeolite structures listed in the column in the middle, and the right column corresponds to the N-OSDA typically used to obtain the zeolitic structure

Self-assembled OSDA	Zeolitic structures	N-OSDA	Self-assembled OSDA	Zeolitic structures	N-OSDA
[221, 240, 241]	ITQ-29 (LTA) [221]	No N-OSDA yields ITQ-29	[227]	ITQ-51 (IFO) [227]	No N-OSDA yields IFO
	STA-6 (SAS) [240]	No N-OSDA yields STA-6	[228, 229]	NUD-1 and NUD-2 [228, 229]	No N-OSDA yields NUD-1 and NUD-2
	SAPO-42 (LTA) [241]	[84]	[230]	Nanosheet ZSM-5 (MFI) [230]	[242]
[223, 233, 234]	ALPO-5 and SAPO-5 (AFI) [223]	[140]	[231]	Nanosheet MWW [231]	[243]
	STA-1 (SAO) [233, 234]	[245]	[232]	ITQ-37 (−ITV) [232]	[244]
			[235–237]	AFI [235–237]	[140]

MTN [225]	[225]	[78]	[238]	ICP-1 [238]	No N-OSDA yields ICP-1
ZSM-12 (MTW) [226]	S,S and R,S diastereo-isomers [226]	[75]	[239]	oFBP, mFBP and pFBP AlPOs [239]	No N-OSDA yields oFBP, mFBP and pFBP AlPOs

conversely be easily modified, yielding bulky, soluble and, especially, offering the possibility to build chiral supramolecular species. Thus, a main focus in this field is to achieve the transfer of chirality from these OSDAs to zeolitic materials with potential separation and catalytic applications, a long-sought objective [246–248].

7 Ionothermal Synthesis

The ionothermal synthesis of materials is characterized by the use of ionic compounds as solvents that replace water in classical hydrothermal synthesis and, at the same time, work as OSDA, alone or with the joint cooperation of another OSDA. The main feature of these conditions is the prevalence of the ionic interaction with the material building blocks during crystallization processes, and thus, little or no water is required. In this review, only syntheses with stoichiometric amounts of water will be taken in account as ionothermal synthesis ($H_2O/SiO_2 \leq 5$–7) [249].

7.1 Ionic Liquids

Ionic liquids (ILs) are a class of organic solvents with high polarity, excellent solvating properties and, usually, high thermal stability. They consist in molten salts with melting points below 100°C with an almost negligible vapour pressure. As molten salts, they consist in a variety of fused anions and cations that could be combined between them, giving rise to an almost unlimited number of salts (as long as their melting point would be <100°C) with tuneable properties depending on the ions forming them. Typical cations include choline, alkylammonium, alkylphosphonium, N-alkylpyridinium and, especially, N,N'-dialkylimidazolium cations, among others, and the anions are also quite varied [250–252]. They have been typically used as alternative solvents to the commonly used volatile organic solvents, which are often more hazardous due to their high vapour pressure. Recently, energy and catalytic applications of IL are being explored [253–258].

The use of ILs as both OSDAs and solvents (known as ionothermal synthesis) arose from the fact that many of the cations forming ILs are chemically very similar to known OSDAs in zeolite syntheses, as well as their polarity also allows using them as solvents, which, together with their low vapour pressure, remove most of the safety issues regarding hydrothermal synthesis [259–261].

The first ionothermal synthesis made use of 1-methyl-3-ethyl imidazolium bromide as IL, which yielded the new aluminophosphate SIZ-1, and other aluminophosphates with already known structures: SIZ-3 (AEL), SIZ-4 (CHA) and SIZ-5 (AFO) [262]. Since then, a large number of ionothermal syntheses have been reported. Most of them relied on the use of imidazolium-based ionic liquids (ImILs) and less frequently pyrrolidinium-based ionic liquids (PyILs), mainly because of pricing and commercial availability. Several aluminophosphates have been obtained with ImIL and its derivatives: the layered aluminophosphate SIZ-6 [263], the novel aluminophosphate SIZ-7 (SIV) and the known SIZ-8 (AEI) and SIZ-9 (SOD) [264], the Fe and Co-substituted aluminophosphates with SOD structure [265], the novel gallophosphate GaPO-a and known CLO and LTA [266], the microcrystalline aluminophosphate MnAlPO-5 (AFI) [267], the pure aluminophosphate AlPO-5 (AFI) [268], the pure aluminophosphates AlPO-5 (AFI) and AlPO-11 (AEL) [269], the zinc aluminophosphate DAF-1 (DFO) [270] and the aluminophosphate SAPO-34 (CHA) [271].

Ionic liquids have been also intensively used in cooperation with amines, as the presence of the IL changes the phase selectivity of the amines with respect to the conventional syntheses. Thus, aluminophosphates AlPO-5 (AFI) and AlPO-25 (ATV) by different amines in ImIL medium [272]; the aluminophosphate AlPO-5 (AFI) with morpholine and ImIL [273]; the aluminophosphates AlPO-42 (LTA), AlPO-5 (AFI) and AlPO-11 (AEL) with ImIL and different amines [274]; the aluminophosphate LTA with ImIL and an amine [275]; and the aluminophosphate CoAPO-34 (CHA) with ImIL and a diamine [276] have been obtained. Importantly, two new structures have been reported, the aluminophosphate JIS-1 by applying a dual-template ImIL-amine approach [277] and the extra-large aluminophosphate DNL-1 with an ImIL and an amine [278].

Finally, just few silica-based zeolites have been successfully obtained with ILs, as only the synthesis of all-silica silicalite-1 (MFI) and theta-1 (TON) zeolites with a mixture of bromide and hydroxide ImIL has been obtained [279] (Table 9).

Additionally, an interesting idea arose after the first synthesis with ILs. ILs, because of their high ionic conductivity, favour a strong interaction with microwave radiation, and their nearly zero vapour pressure allows the synthesis of materials using low pressure-resistant vessels, which is an issue in the microwave-assisted synthesis of materials [280–282]. Because of that, ILs have been also used in the microwave-assisted synthesis of zeolites, although the number of zeotypes synthesized by this way is still few. The first one of this type of synthesis was the use of ImIL to yield AlPO-11 (AEL) [283], and later, the synthesis of all-silica MFI with a dry-gel microwave-assisted using ImILs was reported [284].

7.2   Deep Eutectic Solvents (DESs)

Deep eutectic solvents (DESs) are another kind of ILs. They consist in binary or ternary mixtures of compounds with an extremely low freezing point compared to that of the separate components. The most used DESs comprise mixtures of organic halide salts, such as choline chloride, with hydrogen bond donors, such as amides, amines, alcohols and carboxylic acids [290–292].

The first synthesis using DESs dates back to the first ionothermal synthesis, when choline chloride/urea eutectic mixture IL yielded the novel aluminophosphate SIZ-2 [262]. Next, several materials were obtained, as the novel aluminophosphates SIZ-13 and SIZ-14 (LEV) by using choline chloride-carboxylic acid DESs and

Table 9 Zeolitic structures obtained by using ionic liquid OSDAs. The left column corresponds to the ionic liquid OSDA used to obtain the zeolite structures listed in the column in the middle, and the right column corresponds to the N-OSDA typically used to obtain the zeolitic structure

Ionic liquid OSDA	Zeolitic structures	N-OSDA	Ionic liquid OSDA	Zeolitic structures	N-OSDA
[263, 266, 271]	SIZ-6 [263]	No N-OSDA yields SIZ-6	[272, 274, 284]	AlPO-5 (AFI) [272]	[140]
	GaPO-a [266]	No N-OSDA yields GaPO-a		AlPO-25 (ATV) [272]	—NH₂ [285]
	CLO [266]	[286]		AlPO-11 (AEL) [274]	—NH₂ [143]
	LTA [266]	[84]		ZSM-5 (MFI) [284]	[287]
	SAPO-34 (CHA) [271]	[146]	[273]	AlPO-5 (AFI) [273]	[140]
	AlPO-11 (AEL) [283]	—NH₂ [143]	[277]	JIS-1 [277]	No N-OSDA yields JIS-1
[264]	SIZ-7 (SIV) [264]	No N-OSDA yields SIV	[278]	DNL-1 [278]	No N-OSDA yields DNL-1
	SIZ-8 (AEI) [264]	[90]	[274]	AlPO-42 (LTA) [274]	[84]
	SIZ-9 (SOD) [264]	[181]	[274]	AlPO-5 (AFI) [274]	[140]

(continued)

Table 9 (continued)

Ionic liquid OSDA	Zeolitic structures	N-OSDA	Ionic liquid OSDA	Zeolitic structures	N-OSDA
[265]	SOD [265]	[181]	[275]	LTA [275]	[84]
[267]	MnAPO-5 (AFI) [267, 269]	[140]	[276]	CoAPO-34 (CHA) [276]	[146]
[269]	AlPO-5 (AFI) [269]	[140]	[279]	Silicalite-1 (MFI) [279]	[287]
	AlPO-11 (AEL) [269]	[143]		Theta-1 (TON) [279]	[288]
[270]	DAF-1 (DFO) [270]	[289]			

cyclam or pyridine as auxiliary base [293]; the zinc phosphate ZnPO4-EU1 (DFT) with a choline chloride-imidazolidone DES, which decomposes at high temperature into ethylenediamine [294]; the aluminophosphates UiO-7 (ZON), AlPO-17 (ERI), AlPO-22 (AWW), AlPO-5 (AFI) and SIZ-10 (CHA) with different cations and pentaerythritol [295]; the aluminophosphate FeAlPO-16 (AST) with succinic acid and choline chloride DES and microwave heating [296]; and the aluminophosphate FeAlPO-5 (AFI) with succinic acid and choline chloride DES and microwave heating [297] and the SAPO-5 (AFI) with pentaerythritol and choline chloride DES and microwave heating [298].

Again, publications reporting the synthesis of zeolites using DESs are very scarce. To our best knowledge, only the all-silica SOD with choline chloride/urea DES [299] and the all-silica ZSM-5 (MFI), ZSM-11 (MEL), beta (BEA) and ZSM-39 (MTN) with urea/choline chloride mixture DES and an auxiliary OSDA have been described [300] (Table 10).

The use of ILs and DESs has resulted in the successful synthesis of a relatively large number of zeolitic structures, and future prospects indicate that the number of new porous structures will keep increasing, especially when using them as co-templating agents as the number of possibilities is extremely large. Also, the lowering of water content as main solvent allows for improving safety issues regarding hydrothermal synthesis and results in the removal of waste water streams,

Table 10 Zeolitic structures obtained by using Deep-Eutectic Solvents (DES) OSDAs. The left column corresponds to the DES OSDA used to obtain the zeolite structures listed in the column in the middle, and the right column corresponds to the N-OSDA typically used to obtain the zeolitic structure

DES OSDA	Zeolitic structures	N-OSDA	DES OSDA	Zeolitic structures	N-OSDA
(structure) Cl⁻, N; HO~; H₂N–NH₂ [262, 299]	SIZ-2 [262]	No N-OSDA yields SIZ-2	(structure) Br⁻, N⁺; HO~OH HO~OH [295]	AlPO-17 (ERI) [295]	(structure) N⊕ [103]
	SOD [299]	(structure) N⊕ [181]		AlPO-22 (AWW) [295]	(structure) N [301]
(structure) Cl⁻, N; HO~; HO–O–OH; HO–O–OH [293]	SIZ-13 [293]	No N-OSDA yields SIZ-13	(structure) N⁺ Cl⁻; HO~OH HO~OH [295]	AlPO-5 (AFI) [295]	(structure) N⊕ [140]
(structure) Cl⁻, N; HO~ [293]	SIZ-14 (LEV) [293]	(structure) NH₂ [302]	(structure) N⁺–N 2xBr⁻; HO~OH HO~OH [295]	SIZ-10 (CHA) [295]	(structure) N⊕ [146]
(structure) Cl⁻, N; HO~, H₂N; HN–NH → NH₂ [294]	ZnPO4-EU1 (DFT) [294]	(structure) H₂N, NH₂ [303]	(structure) Cl⁻, N; HO~; HO~OH HO~OH [297]	FeAlPO-16 (AST) [297]	(structure) N [212]
(structure) Cl⁻, N; HO~; HO~OH HO~OH [295, 297]	UiO-7 (ZON) [295]	(structure) N⊕ [304]	(structure) Cl⁻, N; HO~; H₂N–NH₂; N⁺ Br⁻ [300]	ZSM-5 (MFI) [300]	(structure) N⊕ [287]
	FeAlPO-5 (AFI) [297]	(structure) N⊕ [140]	(structure) Cl⁻, N; HO~; H₂N–NH₂ and Cl⁻, N; HO~; H₂N–NH₂; N⁺ I⁻ and N I⁻ [300]	ZSM-11 (MEL) [300]	(structure) N⊕ [145]

(continued)

Table 10 (continued)

DES OSDA	Zeolitic structures	N-OSDA	DES OSDA	Zeolitic structures	N-OSDA
[chemical structure: Cl⁻, HO–…–N⁺, H₂N–(C=O)–NH₂; and structure with Br⁻] [300]	Beta (BEA) [300]	[chemical structure] [74]	[chemical structure: Cl⁻, HO–…–N⁺, H₂N–(C=O)–NH₂; and structure with Br⁻] [300]	ZSM-39 (MTN) [300]	[chemical structure] [78]

favouring both factors of the implementation of large-scale synthesis. In this regard, the use of ILs as solvents is especially interesting as it allows the easy reuse of the most expensive component in the synthesis process, the OSDA. The main drawback of these compounds is their "neutral" nature and, as such, their limitation in the synthesis of aluminosilicate zeolites, which usually requires more alkaline media, and thus, co-templating agents must be used or, alternatively, mixed with hydroxide and non-hydroxide anions in the ILs. Also, only imidazolium-based ILs and DESs have been used for the time being, and many new ILs are available each year. In this regard, the use of phosphonium-based ILs, which possess better thermal and hydrothermal stability than N-based ILs, could be a fruitful alternative for silica-based zeolites.

8 Summary

Although the number of syntheses obtained using alternative OSDAs with respect to the classical ammonium and amine OSDAs is still scarce, this review has shown that the current trend shifts towards using less common OSDAs, as most of them allowed obtaining new compositional ranges and/or structure topologies unaffordable for N-OSDAs. During the last decades, in addition to new structures, numerous isostructural zeotypes have been obtained using different families of OSDAs, which have allowed obtaining chemical compositions beyond those achieved by the simple modification of the conditions of synthesis in the presence of N-based OSDAs (chemical composition, different solvents, temperature, time, ageing, stirring, etc.). The current trend shows that the number of zeolitic structures will continuously increase.

In addition, most of alternative OSDAs are further superior to conventional molecular N-containing OSDAs with regard to their hydrothermal stability, synthesis efficiency and selectivity, OSDA reusability, other specific properties (introduction of probe atoms or functionalities, removal of hydrothermal safety issues) and, last but not the least, an improved safety and environmental impact during crystallization of zeolites. However, significant disadvantages still work against alternative OSDAs, especially regarding availability and price. However, the same issues were found for N-OSDAs when first used compared with alkaline inorganic SDAs. Then, the industrial demands of cheap and, especially, reliable and active materials pushed on for the development of these materials and their synthesis. Nowadays, industrial demands remain playing a great role on the amelioration of the properties of known materials and, overall, the development of new materials with novel properties for new processes demanding very specific properties, as chirality (in active sites and/or channels), channel topologies (very large pores, multisized pores, specific intra-channel cage sizes, channels shape, etc.), crystal topologies (crystal shape and size, nanosheets, hierarchical, etc.) and specific functionalities (incorporation of metal complexes and clusters, isomorphic substitution of tetrahedral atoms).

Finally, it should be reminded that the synthesis of materials, especially under hydrothermal conditions, essentially remains being an experimental science. Although the knowledge and application of computational chemistry are rapidly advancing, there is still a lot of ground to be laid. Therefore, the research in the synthesis of materials does not have to be exclusively focused on the industrial application, because achieving the same end (zeotype materials) through multiple and different paths (alternative OSDAs) will undoubtedly be beneficial to deepen the fundamental knowledge about the synthesis pathways that favour specific features in the final materials, which will be very beneficial in the future for the rational design of porous materials for catalytic and adsorption applications.

References

1. Barrer RM (1948). J Chem Soc 127:2158–2163
2. Vollhardt P, Schore N (2011) Organic chemistry. 6th edn
3. Baerlocher C (2017) Database of zeolite structures. http://www.iza-structure.org/databases/
4. Li J, Corma A, Yu J (2015). Chem Soc Rev 44(20):7112–7127
5. Li Y, Yu J (2014). Chem Rev 114(14):7268–7316
6. Moliner M, Rey F, Corma A (2013). Angew Chem Int Ed 52(52):13880–13889
7. Moliner M, Martínez C, Corma A (2013). Chem Mater 26(1):246–258
8. Meng X, Xiao F-S (2013). Chem Rev 114(2):1521–1543
9. Davis ME (2013). Chem Mater 26(1):239–245
10. Cundy CS, Cox PA (2005). Microporous Mesoporous Mater 82(1–2):1–78
11. Davis ME, Lobo RF (1992). Chem Mater 4(4):756–768
12. Olah GA (1998) Onium ions. Wiley, Hoboken
13. Corbridge DE (2013) Phosphorus: chemistry, biochemistry and technology. CRC press, Boca Raton
14. Bradaric CJ, Downard A, Kennedy C, Robertson AJ, Zhou Y (2003). Green Chem 5 (2):143–152
15. Bradaric-Baus CJ, Zhou Y (2014) US 8901338 B2
16. Atefi F, Garcia MT, Singer RD, Scammells PJ (2009). Green Chem 11(10):1595–1604
17. Fraser KJ, MacFarlane DR (2009). Aust J Chem 62(4):309–321
18. Selva M, Perosa A, Noè M (2016). Org Chem 45:132

19. Bryant D (2004). Phosp Environ Technol 1
20. Quin LD (2000) A guide to organophosphorus chemistry. Wiley, Hoboken
21. Hudson H, Hartley F (1990) Primary, secondary and tertiary phosphates and heterocyclic organophosphorus (III) compounds, p 1
22. Stewart B, Harriman A, Higham LJ (2011). Organometallics 30(20):5338–5343
23. Xie W, Xie R, Pan W-P, Hunter D, Koene B, Tan L-S, Vaia R (2002). Chem Mater 14 (11):4837–4845
24. Zanger M, Vander Werf CA, McEwen WE (1959). J Am Chem Soc 81(14):3806–3807
25. Van Kruchten EMGA (2000) US 6124508 A
26. Ates A, Hardacre C (2012). J Colloid Interface Sci 372(1):130–140
27. Gopalakrishnan J (2009). Appl Organomet Chem 23(8):291–318
28. Schwesinger R, Schlemper H (1987). Angew Chem Int Ed Engl 26(11):1167–1169
29. Ishikawa T (2009) Superbases for organic synthesis: guanidines, amidines, phosphazenes and related organocatalysts. Wiley, Hoboken
30. Stringfellow G (1989). San Diego:109
31. Edmundson R (1987) Dictionary of organophosphorus compounds. CRC Press, Boca Raton
32. Whittam TV (1976) DE 2548695
33. Butter S, Jurewicz A, Kaeding W, Chang CD, Silvestri AJ, Smith RL (1976) US Patent 3:483
34. Araya AL (1983) Barrie Milner EP 108486
35. Rieck HPD, Litterer HD (1985) US 4528172 A
36. Tuel A, Taarit YB (1993). Zeolites 13(5):357–364
37. Tuel A, Taarit YB (1994). Zeolites 14(4):272–281
38. Lingappan N, Krishnasamy V (1996). Bull Chem Soc Jpn 69(4):1125–1128
39. Marler B, Daniels P, i Muné JS (2003). Microporous Mesoporous Mater 64(1):185–201
40. Dorset DL, Kennedy GJ, Strohmaier KG, Diaz-Cabañas MJ, Rey F, Corma A (2006). J Am Chem Soc 128(27):8862–8867
41. Dorset DL, Strohmaier KG, Kliewer CE, Corma A, Díaz-Cabañas MJ, Rey F, Gilmore CJ (2008). Chem Mater 20(16):5325–5331
42. Corma A, Diaz-Cabanas MJ, Jorda JL, Rey F, Sastre G, Strohmaier KG (2008). J Am Chem Soc 130(49):16482–16483
43. Corma A, Díaz-Cabañas M, Jiang J, Afeworki M, Dorset D, Soled S, Strohmaier K (2010). Proc Natl Acad Sci 107(32):13997–14002
44. Corma A, Rey García F, Navarro Villalba MT, Simancas R, Velamazan N, Cantín Á, Jordá Moret JL (2012) WO 2012/049344
45. García FR, Rodriguez MH, Moret JLJ (2014) US 20150038756 A1
46. Simancas R, Jordá JL, Rey F, Corma A, Cantín A, Peral I, Popescu C (2014). J Am Chem Soc 136(9):3342–3345
47. Yun Y, Hernandez M, Wan W, Zou X, Jorda JL, Cantin A, Rey F, Corma A (2015). Chem Commun 51(36):7602–7605
48. Simancas J, Simancas R, Bereciartua PJ, Jorda JL, Rey F, Corma A, Nicolopoulos S, Pratim Das P, Gemmi M, Mugnaioli E (2016). J Am Chem Soc 138(32):10116–10119
49. Cantín Á, Corma A, Díaz-Cabañas MJ, Jordá JL, Moliner M, Rey F (2006). Angew Chem Int Ed 45(47):8013–8015
50. Schmidt JE, Chen C-Y, Brand SK, Zones SI, Davis ME (2016) Chem A Eur J 22(12):4022–4029
51. Dou TL, Xiaofeng L, Xu J, Gui P, Wang L (2007) CN101054183
52. Calvert RB, Chang CD, Rubin MK, Valyocsik EW (1987) US 4642226 A
53. Burton A (2007) US 7682600 B2
54. Valyocsik EW (1986) US 4568654 A
55. Hernández-Rodríguez M, Jordá JL, Rey F, Corma A (2012). J Am Chem Soc 134 (32):13232–13235
56. Smeets S, McCusker LB, Baerlocher C, Xie D, Chen C-Y, Zones SI (2015). J Am Chem Soc 137(5):2015–2020

57. Itabashi K, Okubo T, Elangovan SP (2010) JP 2010260777
58. Simancas R, Dari D, Velamazán N, Navarro MT, Cantín A, Jordá JL, Sastre G, Corma A, Rey F (2010). Science 330(6008):1219–1222
59. Canós AC, García FR, Villalba MTN, Coloma RS, Cirujeda NV, Sanz ÁC, Moret JLJ (2012) US 20130046123 A1
60. Kun Q, Xiao-Wei S, Da X, Ji-Yang L (2012). Chem J Chin Univ 33(10):2141–2145
61. Zhang X, Liu D, Xu D, Asahina S, Cychosz KA, Agrawal KV, Al Wahedi Y, Bhan A, Al Hashimi S, Terasaki O (2012). Science 336(6089):1684–1687
62. Wang Y, Li ZD, Jian, Wang Q, He M, Liu Q, Gao X, Yan L, Pang X, Li F (2014) CN 104098111
63. Sonoda T, Maruo T, Yamasaki Y, Tsunoji N, Takamitsu Y, Sadakane M, Sano T (2015). J Mater Chem A 3(2):857–865
64. Tsapatsis M, Zhang X (2015) US 9180413 B2
65. Yamasaki Y, Tsunoji N, Takamitsu Y, Sadakane M, Sano T (2016). Microporous Mesoporous Mater 223:129–139
66. Lemishko T, Simancas J, Hernández-Rodríguez M, Jiménez-Ruiz M, Sastre G, Rey F (2016). Phys Chem Chem Phys 18:17244–17252
67. Martín N, Li Z, Martínez-Triguero J, Yu J, Moliner M, Corma A (2016). Chem Commun 52 (36):6072–6075
68. Kokotailo G, Lawton S, Olson D (1978). Nature 272:437–438
69. Chou YH, Cundy CS, Garforth AA, Zholobenko VL (2006). Microporous Mesoporous Mater 89(1–3):78–87
70. Chauhan NL, Das J, Jasra RV, Parikh PA, Murthy ZVP (2012). Mater Lett 74:115–117
71. Gies H, Marker B (1992). Zeolites 12(1):42–49
72. Franklin KR, Lowe BM (1989). Stud Surf Sci Catal 49:179–188
73. Tuel A, Taärit YB, Naccache C (1993). Zeolites 13(6):454–461
74. Wadlinger RL, Kerr GT, Rosinski EJ (1967) US 3308069 A
75. Rosinski E, Rubin M (1974) US 3832449 A
76. Ciric J (1977) US 4021331 A
77. Rollmann LD, Valyocsik EW (1980) US 4205052 A
78. Schlenker J, Dwyer F, Jenkins E, Rohrbaugh W, Kokotailo G, Meier W 1981
79. Vaughan DE (1985) US 4554146 A
80. Vaughan DEW, Strohmaier KG (1987) US 4661332 A
81. Marcus BK, Lok BM (1989) US 4857288 A
82. Franco M, Perez-Pariente J, Fornes V (1991). Zeolites 11(4):349–355
83. Annen MJ, Davis ME, Higgins JB, Schlenker JL (1991). J Chem Soc Chem Commun (17):1175–1176
84. Lok BM, Messina CA, Patton RL, Gajek RT, Cannan TR, Flanigen EM (1984). J Am Chem Soc 106(20):6092–6093
85. Wilson ST, Lok BM, Messina CA, Cannan TR, Flanigen EM (1983). J Am Chem Soc 218:79–106
86. Akporiaye D, Fjellvag H, Halvorsen E, Haug T, Karlsson A, Lillerud K (1996). Chem Commun 13:1553–1554
87. Vaughan DE, Strohmaier KG (1995) US 5455020 A
88. Yoshikawa M, Zones SI, Davis ME (1997). Microporous Mater 11(3–4):127–136
89. Miller SJ (1998) US 5716593 A
90. Simmen A, McCusker L, Baerlocher C, Meier W (1991). Zeolites 11(7):654–661
91. Bibby D, Milestone N, Aldridge L (1979). Nature 280(5724):664–665
92. Klotz MR (1981) US 4269813 A
93. Davis M, Montes C, Hathaway P, Garces J (1989). Stud Surf Sci Catal 49:199–214
94. Vaughan DE, Strohmaier KG (1990) US 4931267 A
95. Kowalak S, Jankowska A, Mikołajska E (2010). Microporous Mesoporous Mater 127 (1):126–132

96. Bai R, Sun Q, Wang N, Zou Y, Guo G, Iborra S, Corma A, Yu J (2016). Chem Mater 28 (18):6455–6458
97. Corma A, Díaz-Cabañas MJ, Jordá JL, Martinez C, Moliner M (2006). Nature 443 (7113):842–845
98. Nair S, Jeong H-K, Chandrasekaran A, Braunbarth CM, Tsapatsis M, Kuznicki SM (2001). Chem Mater 13(11):4247–4254
99. Anderson M, Terasaki O, Ohsuna T, Malley P, Philippou A, MacKay S, Ferreira A, Rocha J (1995) Lidin, S. Philos Mag B 71(5):813–841
100. Kustova MY, Hasselriis P, Christensen CH (2004). Catal Lett 96(3-4):205–211
101. Briscoe N, Johnson D, Shannon M, Kokotailo G, McCusker L (1988). Zeolites 8(1):74–76
102. Barrer R, Denny P (1961). J Chem Soc:971–982
103. Aiello R, Barrer R (1970). J Chem Soc A:1470–1475
104. Hopkins P (1989) Role of the tetramethylammonium cation in the synthesis of zeolites ZK-4, Y, and HS. ACS Publications, Washington, DC
105. Patarin J, Caullet P, Marler B, Faust A, Guth J (1994). Zeolites 14(8):675–681
106. Schoeman BJ, Sterte J, Otterstedt J-E (1995). J Colloid Interface Sci 170(2):449–456
107. Valtchev VP (1994). J Chem Soc Chem Commun (3):261–262
108. Wagner P, Nakagawa Y, Lee GS, Davis ME, Elomari S, Medrud RC, Zones S (2000). J Am Chem Soc 122(2):263–273
109. Zones SI (1985) US 4544538 A
110. Vortmann S, Marler B, Gies H, Daniels P (1995). Microporous Mater 4(2):111–121
111. Butter SA, Kaeding WW (1976) US 3972832 A
112. Blasco T, Corma A, Martínez-Triguero J (2006). J Catal 237(2):267–277
113. Caeiro G, Magnoux P, Lopes J, Ribeiro FR, Menezes S, Costa A, Cerqueira H (2006). Appl Catal A Gen 314(2):160–171
114. Xue N, Olindo R, Lercher JA (2010). J Phys Chem C 114(37):15763–15770
115. van der Bij HE, Weckhuysen BM (2015). Chem Soc Rev 44(20):7406–7428
116. Rahimi N, Karimzadeh R (2011). Appl Catal A Gen 398(1):1–17
117. Damodaran K, Wiench J, de Menezes SC, Lam Y, Trébosc J, Amoureux J-P, Pruski M (2006). Microporous Mesoporous Mater 95(1):296–305
118. Cabral de Menezes SM, Lam YL, Damodaran K, Pruski M (2006). Microporous Mesoporous Mater 95(1–3):286–295
119. Corma A, Mengual J, Miguel PJ (2012). Appl Catal A Gen 421–422:121–134
120. Galadima A, Muraza O (2017) Microporous and Mesoporous Mater 249:42–54
121. Védrine JC, Auroux A, Dejaifve P, Ducarme V, Hoser H, Zhou S (1982). J Catal 73 (1):147–160
122. Vinek H, Rumplmayr G, Lercher JA (1989). J Catal 115(2):291–300
123. Xue N, Chen X, Nie L, Guo X, Ding W, Chen Y, Gu M, Xie Z (2007). J Catal 248(1):20–28
124. Lee H, Zones SI, Davis ME (2003). Nature 425(6956):385–388
125. Lee H, Zones SI, Davis ME (2005). J Phys Chem B 109(6):2187–2191
126. Margarit VJ, Martínez-Armero ME, Navarro MT, Martínez C, Corma A (2015). Angew Chem 127(46):13928–13932
127. Kloetstra KR, Zandbergen HW, Jansen JC, van Bekkum H (1996). Microporous Mater 6 (5):287–293
128. Na K, Choi M, Ryoo R (2013). Microporous Mesoporous Mater 166:3–19
129. Kakiuchi Y, Yamasaki Y, Tsunoji N, Takamitsu Y, Sadakane M, Sano T (2016) Chem Lett (0)
130. Matsumoto H, Matsuda T, Miyazaki Y (2000). Chem Lett 29(12):1430–1431
131. Fang S, Yang L, Wei C, Peng C, Tachibana K, Kamijima K (2007). Electrochem Commun 9 (11):2696–2702
132. Paulsson H, Hagfeldt A, Kloo L (2003). J Phys Chem B 107(49):13665–13670
133. Butte W, Eilers J, Kirsch M (1982). Anal Lett 15(10):841–850
134. Large GB (1982) US 4315765 A

135. Paulsson H, Berggrund M, Svantesson E, Hagfeldt A, Kloo L (2004). Sol Energy Mater Sol Cells 82(3):345–360
136. Gerhard D, Alpaslan SC, Gores HJ, Uerdingen M, Wasserscheid P (2005). Chem Commun (40):5080–5082
137. Lee C-P, Peng J-D, Velayutham D, Chang J, Chen P-W, Suryanarayanan V, Ho K-C (2013). Electrochim Acta 114:303–308
138. Evans ST (1996) US 5552132 A
139. Jo C, Lee S, Cho SJ, Ryoo R (2015). Angew Chem Int Ed 54(43):12805–12808
140. Bennett J, Cohen J, FLANIGEN EM, Pluth J, Smith J (1983) Crystal structure of tetrapropylammonium hydroxide—aluminum phosphate number 5. ACS Publications, Washington, DC
141. Kirchner RM, Bennett JM (1994). Zeolites 14(7):523–528
142. Villaescusa LA, Barrett PA, Camblor MA (1999). Angew Chem Int Ed 38(13-14):1997–2000
143. Bennett J, Richardson J, Pluth J, Smith J (1987). Zeolites 7(2):160–162
144. Parise JB (1986). Acta Crystallogr Sect C: Cryst Struct Commun 42(6):670–673
145. Kokotailo G, Chu P, Lawton S, Meier W (1978). Nature 275(5676):119–120
146. Wilson ST, Flanigen EM (1989) Synthesis and characterization of metal aluminophosphate molecular sieves. ACS Publications, Washington, DC
147. Dickins R, Parker D, Gloe K (2005) Macrocyclic chemistry: current trends and future perspectives. Springer, Dordrecht
148. Timmons JC, Hubin TJ (2010). Coord Chem Rev 254(15–16):1661–1685
149. Curtis N (1960) J Chem Soc 11:4409–4413
150. Bradshaw JS, Krakowiak KE, Izatt RM (2009) The chemistry of heterocyclic compounds, aza-crown macrocycles, vol 51. Wiley, Hoboken
151. Wragg DS, Morris R, Burton AW, Zones SI, Ong K, Lee G (2007). Chem Mater 19 (16):3924–3932
152. Wright PA, Morris RE, Wheatley PS (2007). Dalton Trans (46):5359–5368
153. Millini R, Carluccio L, Frigerio F, O'Neil Parker Jr W, Bellussi G (1998). Microporous Mesoporous Mater 24(4–6):199–211
154. Schreyeck L, D'Agosto F, Stumbe J, Caullet P, Mougenel J (1997). Chem Commun 13:1241–1242
155. Wessels T, McCusker L, Baerlocher C, Reinert P, Patarin J (1998). Microporous Mesoporous Mater 23(1):67–77
156. Wragg DS, Hix GB, Morris RE (1998). J Am Chem Soc 120(27):6822–6823
157. Khan TA, Hriljac JA (1999). Inorg Chim Acta 294(2):179–182
158. Patinec V, Wright PA, Lightfoot P, Alan Aitken R, Cox PA (1999). J Chem Soc Dalton Trans 22:3909–3911
159. Maple MJ, Philp EF, Slawin AMZ, Lightfoot P, Cox PA, Wright PA (2001). J Mater Chem 11 (1):98–104
160. Paillaud JL, Caullet P, Schreyeck L, Marler B (2001). Microporous Mesoporous Mater 42 (2–3):177–189
161. Wheatley PS, Morris RE (2002). J Solid State Chem 167(2):267–273
162. Huang A, Weidenthaler C, Caro J (2010). Microporous Mesoporous Mater 130(1):352–356
163. Dhainaut J, Daou TJ, Chappaz A, Bats N, Harbuzaru B, Lapisardi G, Chaumeil H, Defoin A, Rouleau L, Patarin J (2013). Microporous Mesoporous Mater 174:117–125
164. Smith JV, Pluth JJ, Andries KJ (1993). Zeolites 13(3):166–169
165. Bennett JM, Cohen JM, Artioli G, Pluth JJ, Smith JV (1985). Inorg Chem 24(2):188–193
166. Davis ME, Saldarriaga C, Montes C, Garces J, Crowdert C (1988). Nature 331:698–699
167. Pedersen CJ (1967). J Am Chem Soc 89(26):7017–7036
168. Yoshio M, Noguchi H (1982). Anal Lett 15(15):1197–1276
169. Alexander V (1995). Chem Rev 95(2):273–342
170. Gokel GW, Leevy WM, Weber ME (2004). Chem Rev 104(5):2723–2750
171. Hiraoka M (2016) Crown ethers and analogous compounds, vol 45. Elsevier, Amsterdam

172. Delprato F, Delmotte L, Guth J, Huve L (1990). Zeolites 10(6):546–552
173. Dougnier F, Patarin J, Guth J, Anglerot D (1992). Zeolites 12(2):160–166
174. Burkett SL, Davis ME (1993). Microporous Mater 1(4):265–282
175. Chatelain T, Patarin J, Fousson E, Soulard M, Guth J, Schulz P (1995). Microporous Mater 4 (2-3):231–238
176. Chatelain T, Patarin J, Farre R, Petigny O, Schulz P (1996). Zeolites 17(4):328–333
177. Shantz DF, Burton A, Lobo RF (1999). Microporous Mesoporous Mater 31(1):61–73
178. van de Goor G, Behrens P, Felsche J (1994). Microporous Mater 2(6):501–514
179. Keijsper J, Den Ouden C, Post M (1989). Stud Surf Sci Catal 49:237–247
180. Ke Q, Sun T, Cheng H, Chen H, Liu X, Wei X, Wang S (2017) Chem Asian J 12(10):1043–1047
181. Baerlocher C, Meier W (1969). Helv Chim Acta 52(7):1853–1860
182. Feijen E, Devadder K, Bosschaerts MH, Lievens JL, Martens JA, Grobet PJ, Jacobs PA (1994). J Am Chem Soc 116(7):2950–2957
183. Llamas-Saiz AL, Foces-Foces C, Elguero J (1994). J Mol Struct 328:297–323
184. Staab HA, Saupe T (1988). Angew Chem Int Ed Engl 27(7):865–879
185. Martínez-Franco R, Moliner M, Yun Y, Sun J, Wan W, Zou X, Corma A (2013). Proc Natl Acad Sci 110(10):3749–3754
186. Möller K, Borvornwattananont A, Bein T (1989). J Phys Chem 11:4562–4571
187. DeWilde W, Peeters G, Lunsford JH (1980). J Phys Chem (United States) 84 (18): 2306–2310
188. Kawi S, Chang J, Gates B (1993). J Am Chem Soc 115
189. Recchia S, Dossi C, Fusi A, Sordelli L, Psaro R (1999). Appl Catal A Gen 182(1):41–51
190. Dams M, Drijkoningen L, Pauwels B, Van Tendeloo G, De Vos D, Jacobs P (2002). J Catal 209(1):225–236
191. Basset J, Choplin A (1983). J Mol Catal 21(1-3):95–108
192. van de Goor G, Freyhardt CC, Behrens P (1995). Z Anorg Allg Chem 621(2):311–322
193. Balkus K, Gabrielov A, Zones S (1995). Stud Surf Sci Catal 97:519–525
194. Lobo RF, Tsapatsis M, Freyhardt CC, Khodabandeh S, Wagner P, Chen C-Y, Balkus KJ, Zones SI, Davis ME (1997). J Am Chem Soc 119(36):8474–8484
195. Freyhardt C, Tsapatsis M, Lobo R, Balkus K, Davis M (1996). Nature 381(6580):295–298
196. Balkus KJ, Gabrielov AG, Shepelev S (1995). Microporous Mater 3(4):489–495
197. Garcia R, Philp EF, Slawin AM, Wright PA, Cox PA (2001). J Mater Chem 11(5):1421–1427
198. Han Y, Li Y, Yu J, Xu R (2011). Angew Chem Int Ed 50(13):3003–3005
199. Xu Y, Li Y, Han Y, Song X, Yu J (2013). Angew Chem Int Ed 52(21):5501–5503
200. Barrett PA, Sankar G, Stephenson R, Catlow CRA, Thomas JM, Jones RH, Teat SJ (2006). Solid State Sci 8(3):337–341
201. Bu X, Feng P, Stucky GD (1997). Science 278(5346):2080–2085
202. Feng P, Bu X, Stucky GD (1998). Microporous Mesoporous Mater 23(5):315–322
203. Martínez Franco R, Moliner M, Thogersen JR, Corma A (2013). ChemCatChem 5 (11):3316–3323
204. Ren L, Zhu L, Yang C, Chen Y, Sun Q, Zhang H, Li C, Nawaz F, Meng X, Xiao F-S (2011). Chem Commun 47(35):9789–9791
205. Deka U, Lezcano-Gonzalez I, Warrender SJ, Lorena Picone A, Wright PA, Weckhuysen BM, Beale AM (2013). Microporous Mesoporous Mater 166:144–152
206. Martínez-Franco R, Moliner M, Franch C, Kustov A, Corma A (2012). Appl Catal B Environ 127:273–280
207. Lorena Picone A, Warrender SJ, Slawin AMZ, Dawson DM, Ashbrook SE, Wright PA, Thompson SP, Gaberova L, Llewellyn PL, Moulin B, Vimont A, Daturi M, Park MB, Sung SK, Nam I-S, Hong SB (2011). Microporous Mesoporous Mater 146(1–3):36–47
208. Maes A, Cremers A (1973). J Am Chem Soc 121:230–239
209. Guczi L, Kiricsi I (1999). Appl Catal A Gen 186(1–2):375–394
210. Moliner M (2012). ISRN Mater Sci 2012:1–24

211. Marler B, Dehnbostel N, Eulert H-H, Gies H, Liebau F (1986). J Incl Phenom Macrocycl Chem 4(4):339–349
212. Schott-Darie C, Patarin J, Le Goff P, Kessler H, Benazzi E (1994). Microporous Mater 3 (1-2):123–132
213. Feng P, Bu X, Stucky GD (1997). Nature 388(6644):735–741
214. Palella B, Cadoni M, Frache A, Pastore H, Pirone R, Russo G, Coluccia S, Marchese L (2003). J Catal 217(1):100–106
215. Gennari C, Piarulli U (2003). Chem Rev 103(8):3071–3100
216. Franciò G, Faraone F, Leitner W (2000). Angew Chem Int Ed 39(8):1428–1430
217. Fernández-Pérez H, Etayo P, Panossian A, Vidal-Ferran A (2011). Chem Rev 111 (3):2119–2176
218. Pfaltz A, Drury WJ (2004). Proc Natl Acad Sci U S A 101(16):5723–5726
219. Dolhem F, Johansson MJ, Antonsson T, Kann N (2007). J Comb Chem 9(3):477–486
220. Gómez-Hortigüela L, López-Arbeloa F, Corà F, Pérez-Pariente J (2008). J Am Chem Soc 130 (40):13274–13284
221. Corma A, Rey F, Rius J, Sabater MJ, Valencia S (2004). Nature 431(7006):287–290
222. Gómez-Hortigüela L, Corà F, Catlow CRA, Pérez-Pariente J (2004). J Am Chem Soc 126 (38):12097–12102
223. Gómez-Hortigüela L, Pérez-Pariente J, Corà F, Catlow CRA, Blasco T (2005). J Phys Chem B 109(46):21539–21548
224. Gómez-Hortigüela L, López-Arbeloa F, Pérez-Pariente J (2009). Microporous Mesoporous Mater 119(1–3):299–305
225. Huang A, Caro J (2009). J Cryst Growth 311(21):4570–4574
226. García R, Gómez-Hortigüela L, Sánchez F, Pérez-Pariente J (2010). Chem Mater 22 (7):2276–2286
227. Martínez-Franco R, Sun J, Sastre G, Yun Y, Zou X, Moliner M, Corma A (2014). Proc R Soc A 470:20140107
228. Chen FJ, Xu Y, Du HB (2014). Angew Chem Int Ed 53(36):9592–9596
229. Gao Z-H, Chen F-J, Xu L, Sun L, Xu Y, Du H-B (2016). Chem Eur J 22(40):14367–14372
230. Xu D, Ma Y, Jing Z, Han L, Singh B, Feng J, Shen X, Cao F, Oleynikov P, Sun H (2014). Nat Commun 5:4262
231. Xu L, Ji X, Li S, Zhou Z, Du X, Sun J, Deng F, Che S, Wu P (2016). Chem Mater 28 (12):4512–4521
232. Chen F-J, Gao Z-H, Liang L-L, Zhang J, Du H-B (2016). CrystEngComm 18(15):2735–2741
233. Gómez-Hortigüela L, Pinar AB, Pérez-Pariente J, Corà F (2009). Chem Mater 21 (14):3447–3457
234. Pinar AB, Gomez-Hortiguela L, McCusker LB, Perez-Pariente J (2011). Dalton Trans 40 (32):8125–8131
235. Alvaro-Munoz T, Lopez-Arbeloa F, Perez-Pariente J, Gómez-Hortigüela L (2014). J Phys Chem C 118(6):3069–3077
236. Gomez-Hortiguela L, Alvaro-Munoz T, Bernardo-Maestro B, Perez-Pariente J (2015). Phys Chem Chem Phys 17(1):348–357
237. Bernardo-Maestro B, Vos E, López-Arbeloa F, Pérez-Pariente J, Gómez-Hortigüela L (2017). Microporous Mesoporous Mater 239:432–443
238. Álvaro-Muñoz T, Pinar AB, Šišak D, Pérez-Pariente J, Gómez-Hortigüela L (2014). J Phys Chem C 118(9):4835–4845
239. Gómez-Hortigüela L, Sanz A, Álvaro-Muñoz T, López-Arbeloa F, Pérez-Pariente J (2014). Microporous Mesoporous Mater 183:99–107
240. Martínez-Franco R, Cantín Á, Moliner M, Corma A (2014). Chem Mater 26(15):4346–4353
241. Martínez-Franco R, Cantín Á, Vidal-Moya A, Moliner M, Corma A (2015). Chem Mater 27 (8):2981–2989
242. Choi M, Na K, Kim J, Sakamoto Y, Terasaki O, Ryoo R (2009). Nature 461(7261):246–249

243. Luo HY, Michaelis VK, Hodges S, Griffin RG, Román-Leshkov Y (2015). Chem Sci 6 (11):6320–6324
244. Sun J, Bonneau C, Cantín Á, Corma A, Díaz-Cabañas MJ, Moliner M, Zhang D, Li M, Zou X (2009). Nature 458(7242):1154–1157
245. Noble GW, Wright PA, Lightfoot P, Morris RE, Hudson KJ, Kvick Å, Graafsma H (1997). Angew Chem Int Ed Engl 36(1-2):81–83
246. Zaera F (2012). Catal Lett 142(5):501–516
247. Coronas J (2010). Chem Eng J 156(2):236–242
248. Bellussi G, Carati A, Rizzo C, Millini R (2013). Cat Sci Technol 3(4):833–857
249. Ma H, Tian Z, Xu R, Wang B, Wei Y, Wang L, Xu Y, Zhang W, Lin L (2008). J Am Chem Soc 130(26):8120–8121
250. Welton T (1999). Chem Rev 99(8):2071–2084
251. Wasserscheid P, Welton T (2008) Ionic liquids in synthesis. Wiley, Hoboken
252. Ventura SP, Gonçalves AM, Sintra T, Pereira JL, Gonçalves F, Coutinho JA (2013). Ecotoxicology 22(1):1–12
253. Seddon KR (1997). J Chem Technol Biotechnol 68(4):351–356
254. Plechkova NV, Seddon KR (2008). Chem Soc Rev 37(1):123–150
255. Smiglak M, Pringle J, Lu X, Han L, Zhang S, Gao H, MacFarlane D, Rogers R (2014). Chem Commun 50(66):9228–9250
256. MacFarlane DR, Tachikawa N, Forsyth M, Pringle JM, Howlett PC, Elliott GD, Davis JH, Watanabe M, Simon P, Angell CA (2014). Energy Environ Sci 7(1):232–250
257. Ho TD, Zhang C, Hantao LW, Anderson JL (2013). Anal Chem 86(1):262–285
258. Steinrueck H-P, Wasserscheid P (2015). Catal Lett 145(1):380–397
259. Parnham ER, Morris RE (2007). Acc Chem Res 40(10):1005–1013
260. Morris RE (2009). Chem Commun (21):2990–2998
261. Tian Z-J, Liu H (2016) Zeolites in sustainable chemistry. Springer, Berlin, Heidelberg, pp 37–76
262. Cooper ER, Andrews CD, Wheatley PS, Webb PB, Wormald P, Morris RE (2004). Nature 430(7003):1012–1016
263. Parnham ER, Wheatley PS, Morris RE (2006). Chem Commun 4:380–382
264. Parnham ER, Morris RE (2006). J Am Chem Soc 128(7):2204–2205
265. Han L, Wang Y, Li C, Zhang S, Lu X, Cao M (2008). AICHE J 54(1):280–288
266. Ma H, Xu R, You W, Wen G, Wang S, Xu Y, Wang B, Wang L, Wei Y, Xu Y, Zhang W, Tian Z, Lin L (2009). Microporous Mesoporous Mater 120(3):278–284
267. Ng E-P, Sekhon SS, Mintova S (2009). Chem Commun (13):1661–1663
268. Khoo DY, Kok W-M, Mukti RR, Mintova S, Ng E-P (2013). Solid State Sci 25:63–69
269. Shi Y, Liu G, Wang L, Zhang X (2014). Microporous Mesoporous Mater 193:1–6
270. Pinar AB, McCusker LB, Baerlocher C, Hwang S-J, Xie D, Benin AI, Zones SI (2016). New J Chem 40(5):4160–4166
271. Sánchez-Sánchez M, Romero ÁA, Pinilla-Herrero I, Sastre E. Catal Today 296:239–246
272. Wang L, Xu Y, Wei Y, Duan J, Chen A, Wang B, Ma H, Tian Z, Lin L (2006). J Am Chem Soc 128(23):7432–7433
273. Xu R, Zhang W, Guan J, Xu Y, Wang L, Ma H, Tian Z, Han X, Lin L, Bao X (2009). Chem Eur J 15(21):5348–5354
274. Pei R, Tian Z, Wei Y, Li K, Xu Y, Wang L, Ma H (2010). Mater Lett 64(19):2118–2121
275. Fayad EJ, Bats N, Kirschhock CEA, Rebours B, Quoineaud A-A, Martens JA (2010). Angew Chem 122(27):4689–4692
276. Musa M, Dawson DM, Ashbrook SE, Morris RE (2017). Microporous Mesoporous Mater 239:336–341
277. Xing H, Li J, Yan W, Chen P, Jin Z, Yu J, Dai S, Xu R (2008). Chem Mater 20 (13):4179–4181
278. Wei Y, Tian Z, Gies H, Xu R, Ma H, Pei R, Zhang W, Xu Y, Wang L, Li K, Wang B, Wen G, Lin L (2010). Angew Chem 122(31):5495–5498

279. Wheatley PS, Allan PK, Teat SJ, Ashbrook SE, Morris RE (2010). Chem Sci 1(4):483–487
280. Cundy CS (1998). Collect Czechoslov Chem Commun 63(11):1699–1723
281. Morris RE (2008). Angew Chem Int Ed 47(3):442–444
282. Martínez-Palou R (2010). Mol Divers 14(1):3–25
283. Wang L, Xu Y-P, Wang B-C, Wang S-J, Yu J-Y, Tian Z-J, Lin L-W (2008). Chem Eur J 14 (34):10551–10555
284. Cai R, Liu Y, Gu S, Yan Y (2010). J Am Chem Soc 132(37):12776–12777
285. Parise JB (1985). Chem Commun 9:606–607
286. Estermann M, McCusker L (1991). Nature 352(6333):320
287. Argauer RJ, Landolt GR (1972) US 3702886 A
288. Barri S, Smith G, White D, Young D (1984) Nature 312(5994):533–534
289. Robert G, MeurigáThomas J (1993). Chem Commun 7:633–635
290. Abbott AP, Boothby D, Capper G, Davies DL, Rasheed RK (2004). J Am Chem Soc 126 (29):9142–9147
291. Zhang Q, Vigier KDO, Royer S, Jérôme F (2012). Chem Soc Rev 41(21):7108–7146
292. Zhao H, Baker GA (2013). J Chem Technol Biotechnol 88(1):3–12
293. Drylie EA, Wragg DS, Parnham ER, Wheatley PS, Slawin AMZ, Warren JE, Morris RE (2007). Angew Chem Int Ed 46(41):7839–7843
294. Liu L, Kong Y, Xu H, Li JP, Dong JX, Lin Z (2008). Microporous Mesoporous Mater 115 (3):624–628
295. Liu L, Li X, Xu H, Li J, Lin Z, Dong J (2009). Dalton Trans (47):10418–10421
296. Zhao X, Kang C, Wang H, Luo C, Li G, Wang X (2011). J Porous Mater 18(5):615–621
297. Zhao X, Wang H, Dong B, Sun Z, Li G, Wang X (2012). Microporous Mesoporous Mater 151:56–63
298. Zhao X, Wang H, Kang C, Sun Z, Li G, Wang X (2012). Microporous Mesoporous Mater 151:501–505
299. Zhou XT, Liu QF, Liu Y (2012) Ionothermal synthesis of sodalite from metakaolin, advanced materials research. Trans Tech Publications, Switzerland, pp 789–792
300. Lin ZS, Huang Y (2016). Can J Chem 94(6):533–540
301. Richardson J, Pluth J, Smith J (1989). Naturwissenschaften 76(10):467–469
302. Barrett PA, Jones RH (2000). Phys Chem Chem Phys 2(3):407–412
303. Chen J, Natarajan S, Thomas JM, Jones RH, Hursthouse MB (1994). Angew Chem Int Ed Engl 33(6):639–640
304. Marler B, Patarin J, Sierra L (1995). Microporous Mater 5(3):151–159

Struct Bond (2018) 175: 139–178
DOI: 10.1007/430_2017_11
© Springer International Publishing AG 2017
Published online: 13 October 2017

Role of Supramolecular Chemistry During Templating Phenomenon in Zeolite Synthesis

Cecilia Paris and Manuel Moliner

Abstract In the last years, there is an increasing interest in the use of organic molecules with the appropriate functionalities to interact with other organic molecules and/or inorganic cations through non-covalent supramolecular interactions, as very specific organic structure-directing agents (OSDAs) for zeolite synthesis. These assembled molecular subunits allow directing the crystallization of zeolite structures with particular physico-chemical properties, such as novel framework topologies, crystal size, chemical compositions, acid-base properties, or metal incorporation, which otherwise would not be achieved using "classical" amine or ammonium-based OSDA molecules. Along the present chapter, different zeolite synthesis strategies employing assembled molecular subunits will be presented, including the use of crown ether-based supramolecular templates, metal-organic complexes, aromatic molecules able to interact through π–π interactions, or supramolecular assembled amphiphilic molecules, among others. The most relevant results described in the literature using these supramolecular-based templating routes will be discussed, together with the current challenges and perspectives.

Keywords Advanced functional materials • Heterogeneous catalysis • Organic structure-directing agents (OSDAs) • Supramolecular chemistry • Zeolite synthesis

Contents

C. Paris and M. Moliner (✉)
Instituto de Tecnología Química, Universitat Politècnica de València-Consejo Superior de Investigaciones Científicas, Avenida de los Naranjos s/n, 46022 Valencia, Spain
e-mail: mmoliner@itq.upv.es

1   Introduction

The introduction of organic molecules in the synthesis of zeolites could be considered as one of the most outstanding contributions in the field [1, 2], since it has allowed not only controlling their physico-chemical properties (i.e. hydrothermal stability and acidity, among others) but also the discovery of most of the 232 zeolites accepted today by the International Zeolite Association (IZA) (http://www.iza-structure.org/index.htm). In this sense, since the early zeolite synthesis descriptions using organic molecules by Barrer and Denny in the 1960s [3], zeolites with very diverse framework structures, presenting from small-pore openings (below 3.5 Å) to extra-large pores (above 8 Å), have been reported, thanks to the use of specific organic molecules acting as organic structure-directing agents (OSDAs) [1, 2].

In general, the most employed OSDA molecules in zeolite synthesis are amines and ammonium cations [4], but in the last years, different new zeolite structures have been achieved by using novel phosphorous-derived organic molecules as OSDAs [5, 6]. Although the exact templating roles of the organic molecules in zeolite synthesis are not completely understood, it has been suggested that their presence in the synthesis gels favours the assembly and organization of the silicon-containing species around the OSDAs through mainly van der Waals forces [7]. Amine- and ammonium-based OSDA molecules have been intensively prepared with a variety of sizes and shapes [4], and it has been generally observed that, particularly for high-silica zeolites, there is a relationship between the structural properties of the organic molecules and the resulting pores/cavities of the crystalized zeolites [1, 2]. This fact could be attributed to the unique stabilization energy that the OSDA molecules provide to favour the crystallization of particular zeolite structures that otherwise would not be achieved.

Many studies have shown that the selectivity of a particular OSDA towards a specific zeolite structure remarkably increases with the size and rigidity of the OSDA, approaching a "tight lock-and-key fit", where the organic molecule could be considered as a real "template" of the crystallized zeolite [1]. According to this, in the last years, many efforts have been devoted to the design of bulky and rigid OSDA molecules to synthesize zeolite structures with particular large or extra-large pores and/or large cavities [4, 8, 9]. However, the design of bulky OSDA molecules presents different drawbacks. On the one hand, more synthesis and purification steps will be required in their preparation, considerably increasing the costs associated with their manufacture [10]. On the other hand, the bulky amine- or ammonium-based OSDAs tend to show high C/N molar ratios, resulting in highly hydrophobic molecules, which could present solubility problems within the aqueous synthesis gels, mostly precluding the zeolite crystallization [1].

In order to overcome these problems, a very interesting and rational alternative is to take advantage of the fact that some organic molecules are able to interact with other organic molecules and/or inorganic cations through weaker non-covalent supramolecular interactions, to form bulkier assembled molecular subunits that will be used as OSDAs for zeolite synthesis [11–13]. In the last 20 years, the use of assembled molecular subunits has allowed preparing different new zeolite structures, particularly those with large or extra-large pores, or improving the physico-chemical properties of other known zeolites, such as controlling their crystal sizes or their chemical compositions by simple one-pot methods. Some of the synthesis strategies include the use of crown ether-based supramolecular templates [14, 15], metal-organic complexes [16–19], supramolecular π–π aromatic interactions [12, 20], or supramolecular interactions of amphiphilic molecules [21, 22], among others.

Along the present chapter, we will overview the different "supramolecular OSDA" approaches described in the literature for the synthesis of zeolites, highlighting the most relevant results achieved, with the principal aim to show the great challenges that can be faced in zeolite synthesis by using "non-classical" assembled molecular subunits as OSDAs.

2 Crown Ether-Based Supramolecular Templates

Crown ethers are well-known molecules that have been extensively employed in classic supramolecular host-guest chemistry because they can act as very efficient hosts, being capable of binding guest ions [23]. Taking advantage of these supramolecular interactions of crown ether molecules with several ions, the synthesis of different cage-based zeolites has been accomplished, particularly large-pore FAU-related zeolites and small-pore zeolites.

2.1 Synthesis of FAU-Related Zeolite Structures

The crystalline structure of the FAU zeolite is formed by a three-dimensional large-pore system with large supercages of ~1.3 nm of diameter (see Fig. 1a) [25], and this unique crystalline structure allows its application as catalyst in many industrial chemical processes [26]. However, one of the main problems related to the FAU

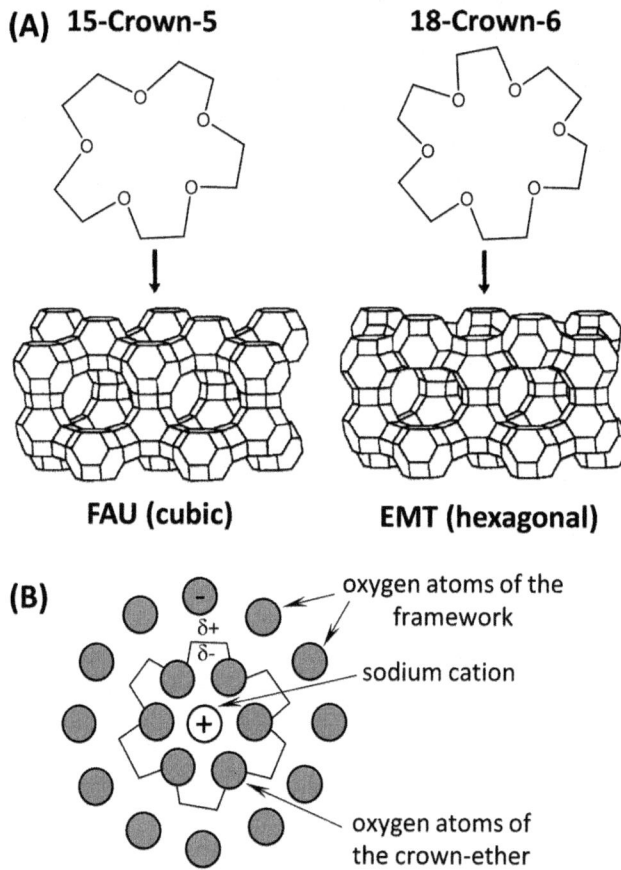

Fig. 1 (a) Crown ethers employed for the synthesis of the FAU-related zeolites and (b) proposed supramolecular assembling of crown ether molecules. Reproduced from [15, 24]

synthesis is that this material can hardly be achieved with Si/Al molar ratios higher than 2.6, and thus, different post-synthetic treatments are required to finally obtain a high-silica FAU material with the appropriate acidity and stability to be used as acid catalyst [26]. Many descriptions can be found in the literature attempting to crystallize the high-silica FAU by direct synthesis methods, but most of them have not allowed increasing significantly the framework Si/Al molar ratio of the resulting FAU materials [27]. In general, the main requirement for synthesizing a high-silica FAU material is to find an appropriate large and bulky ODSA molecule able to fit and stabilize the large supercages present within the FAU structure, allowing the increase of the Si/Al molar ratio by, at least, partial substitution of the sodium cations.

Having that in mind, Delprato et al. rationalized that the organic molecules to attempt the synthesis of FAU should present the following criteria: (1) to be soluble and stable under synthesis conditions, (2) to have steric compatibility with synthesis conditions, and (3) to have stabilizing interactions with framework elements and alkali cations [15]. Among other large cyclic OSDA molecules proposed, crown ethers are able to accomplish all the above criteria.

Delprato et al. showed that the use of the 15-crown-5 and 18-crown-6 molecules combined with sodium in the synthesis media allowed the crystallization of the high-silica cubic FAU and hexagonal FAU (named EMT), respectively (see Fig. 1a), with Si/Al molar ratios approaching the value of ~5 [15]. The use of crown ethers has also allowed the synthesis of different FAU-type intergrowths between the cubic and hexagonal FAU polymorphs [28–31]. As a first approximation, Delprato et al. proposed that the particular supramolecular [(crown ether, Na^+), (OH^-)] complexes would be real species acting as OSDAs, being finally integrated within the FAU pores during the crystallization (see Fig. 1b).

In order to further study the conformation and distribution of the sodium/crown ether complexes, the hydrothermal synthesis of FAU and EMT using crown ethers and the crystalline structure of the FAU-related materials in their as-prepared forms have been evaluated by using different characterization techniques [24, 32, 33]. Davis et al. have observed by Raman spectroscopy and solid-state ^1H-^{13}C CP NMR that the sodium/crown ether complexes facilitate the arrangement of sodium-templated aluminosilicate networks into FAU and EMT [24]. Indeed, these authors propose a mechanism involving the assembly of some structure subunits with the sodium/crown ether complexes through sandwich-like interactions between the sodium/crown ether complex and the six T-atom oxide rings of the supercage [24]. Jacobs et al. have proposed a more detailed crystallization mechanism based on the results obtained from different characterization techniques, including ^{13}C CP MAS NMR, TGA, and nitrogen adsorption [32]. They suggest that the sodium/crown ether complexes are initially placed in the hypoholes of preformed FAU sheets, allowing the formation of the 3-D FAU-related structures by a self-assembly mechanism of these organic-inorganic macromolecules [32]. Finally, Baerlocher et al. show the high specificity of the sodium/crown ether complex towards the FAU cavities, precisely locating these complexes within the as-prepared EMT zeolite by Rietveld refinement [33].

Interestingly, the method involving inorganic cation/crown ether complexes for synthesizing FAU-related zeolites can be extended to Gd complexes of crown ethers [34]. In this case, Balkus et al. have proposed the use of Gd(III) complex of 18-crown-6 as OSDA for the preparation of the Gd-containing hexagonal FAU, observing the final encapsulation of the Gd^{3+}-18-crown-6 complex within the cavities of the hexagonal FAU zeolite. The good stability of this Gd-based FAU zeolite at low pHs implies its potential applicability as oral magnetic resonance imaging (MRI) agent.

2.2 Synthesis of Small-Pore Zeolites with Large Cavities (RHO, KFI, and MCM-61)

Besides the synthesis of the FAU-related zeolites, different small-pore zeolites presenting large cavities have been synthesized by using cation/crown ethers supramolecular complexes as OSDA molecules, such as RHO [35–37], KFI [38], or MCM-61 [39].

RHO zeolite is a three-dimensional small-pore zeolite containing large cavities, named α-cages, within its structure (see Fig. 2a). This material was traditionally synthesized under OSDA-free conditions with Si/Al ratios below 3 [40]. Chatelain et al. have described the synthesis of the high-silica RHO, with Si/Al ratios of ~4.5 by using the 18-crown-6 ether and Na^+ and Cs^+ cations in the synthesis gel. This was the first synthesis description allowing the crystallization of the RHO zeolite as pure phase, but in this early manuscript, the potential role of the Cs^+-crown ether complex as OSDA in the synthesis of the RHO zeolite was not studied.

More recently, Wang et al. have described the preparation of the RHO zeolite with Si/Al molar ratios as high as 10, by systematically adjusting the ratio of the alkali metal ions, particularly Na^+ and Cs^+, and the crown ether, 18-crown-6, in the preparative gels [37]. These authors have found that two-crown ether molecules interact with Cs^+ cations to form a bulky supramolecular sandwich complex, which is preferentially placed within the α-cages of the RHO structure (see Fig. 2a). Interestingly, this high-silica RHO zeolite can be Cu-exchanged, and the resulting Cu-RHO zeolite shows good catalytic activity and hydrothermal stability for the selective catalytic reduction (SCR) of NOx [37].

Under similar synthesis conditions but combining the 18-crown-6 ether with potassium and strontium cations, Chatelain et al. described the synthesis of the high-silica KFI zeolite [38]. The KFI zeolite structure is very similar to the above-described RHO zeolite, presenting also a three-dimensional small-pore system with the presence of the large α-cages/α-cavities. Up to that moment, the KFI zeolite was mainly prepared under OSDA-free synthesis conditions with Si/Al molar ratios below 3 [41, 42]. The use of the 18-crown-6 ether with potassium and strontium

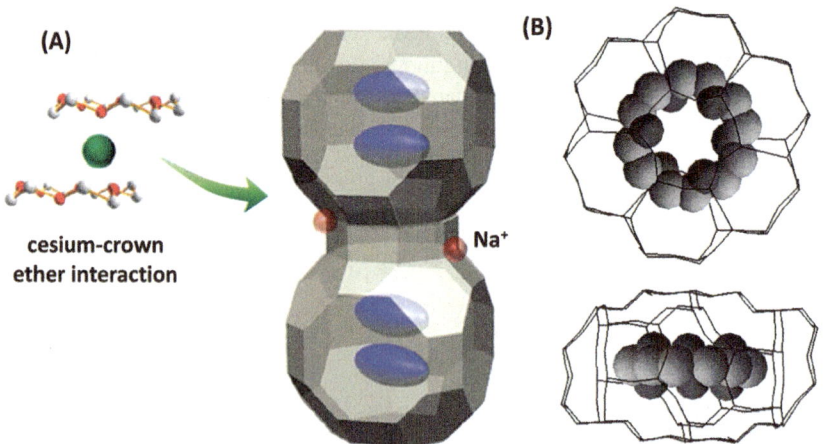

Fig. 2 (a) Supramolecular assembly of crown ether molecules with Cs^+ cations to synthesize the high-silica RHO zeolite and (**b**) large $[6^{20}4^6]$ cavity present within the MCM-61 structure with the entrapped K^+/crown ether complex. Reproduced from [37, 39], respectively

cations allowed the increase of the crystallinity of the achieved KFI zeolites, and the framework Si/Al molar ratio was increased up to 4 [38]. This fact could be explained by the unique directing effects of the cation/crown ether complexes towards the α-cages present within the KFI framework, as approximately one crown ether molecule was observed to be present within each unit cell [38].

Finally, another example describing the synthesis of a new zeolite by combining 18-crown-6 molecules with alkali cations is the case of the high-silica MCM-61 zeolite [39, 43]. MCM-61 cannot be strictly considered a small-pore zeolite, since the MCM-61 framework is only accessible through six-membered rings (see Fig. 2b) [39]. However, the synthesis of the MCM-61 is a very interesting example describing a real "template" interaction between the OSDA and the zeolite framework, since the potassium-18-crown-6 complexes are perfectly bounded by the $[6^{20}4^6]$ cages present within the MCM-61 structure (see Fig. 2b). Moreover, these large cavities contain 18-membered rings, suggesting that the use of potassium-18-crown-6 complexes could be proper OSDA molecules to attempt the synthesis of extra-large-pore zeolites under the adequate synthesis conditions.

3 Transition Metal Organometallic Complexes

The particular ability of different organic ligands to interact with transition metals has allowed the supramolecular design of different bulky and stable organometallic complexes to attempt the preparation of some new zeolite structures or the one-pot preparation of efficient and functional metal-containing zeolites for particular catalytic reactions.

3.1 Cyclopentadienyl Ligands

One of the former transition metal organometallic complexes employed as OSDAs for zeolite synthesis was those using cyclopentadienyl ligands [16]. These early complexes were designed with two main objectives: (1) to provide a direct method for efficiently encapsulated metal complex within the zeolite pores and (2) to influence the nucleation and crystallization processes favouring the crystallization of new zeolites [16].

In this sense, Valyocsik and Balkus et al. described the use of bis(cyclopentadienyl) cobalt(III) cation (see Fig. 3a) for the synthesis of the clathrasil ZSM-51, presenting the framework topology of nonasil (NON) structure [16, 44]. Despite the fact that the void volume of ZSM-51 is not accessible, the NON structure contains large $[5^8 6^{12}]$ cavities that can accommodate the organometallic complex (see Fig. 3b). This methodology was extended to other non-accessible clathrasil structures, such as octadecasil (AST) or dodecasil-1H (DOH) using different metal-based cyclopentadienyl complexes, such as 1,1′-dimethylcobaltocenium, $[Co(C_5H_4Me)_2]^+$, or benzene(cyclopentadienyl)iron(II),

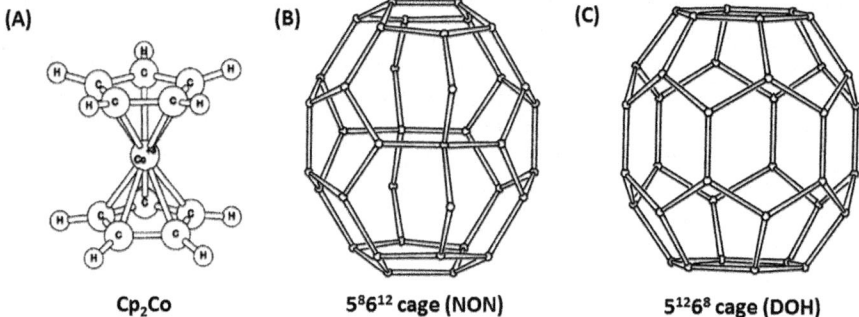

Fig. 3 (**a**) Supramolecular assembly of bis(cyclopentadienyl)cobalt(III), (**b**) large [$5^8 6^{12}$] cavity present in the nonasil (NON) structure, and (**c**) large [$5^{12} 6^8$] cavity present in the dodecasil-1H (DOH) structure

[Fe(C$_6$H$_6$)(C$_5$H$_5$)]$^+$ [45]. As occurred above, the achieved clathrasils contain large cavities within their structures to properly encapsulate the organometallic complexes (see, for instance, the [$5^{12} 6^8$] cavity present within the DOH structure in Fig. 3c).

These preliminary results on clathrasils paved the way for the synthesis of the first high-silica zeolite presenting 14-membered ring pore openings (~7.5 × 10 Å), UTD-1, using bis(cyclopentadienyl)cobalt(III) as single OSDA (see Fig. 3a) [17]. The structural analysis of the as-prepared UTD-1 zeolite reveals that there are two Co complexes per unit cell of the UTD-1 material, highlighting the unique organic directing role of the supramolecular assembly between two cyclopentadienyl molecules and cobalt. Interestingly, the hydrothermal stability of the organometallic-free UTD-1 is very high, with excellent Brønsted acidity for its application as acid catalyst in different hydrocarbon reactions requiring the use of large reactants or products [46].

Similar cyclopentadienyl-based complexes have also been employed for the synthesis of large-pore AlPO-5 (AFI) and clathrate AlPO-16 (AST) [47]. Particularly, the bis(cyclopentadienyl)cobalt(III) complex has been shown to be a very efficient complex for synthesizing these two aluminophosphate-based materials, where the metal complex has been included intact within the molecular sieves.

3.2 Polyamine Ligands

3.2.1 Metal-Chelate Complexes

The preparation of ship-in-a-bottle metal complexes within zeolite structures has been intensively studied in the literature [48]. In those cases, the cationic metal species are first introduced within the crystalline zeolite by cationic exchange, and later the metal-exchanged zeolite is reacted with a flexible chelate (i.e. salen) that can diffuse into the zeolite, achieving the desired organometallic complex encapsulated within the large zeolitic cavities [48].

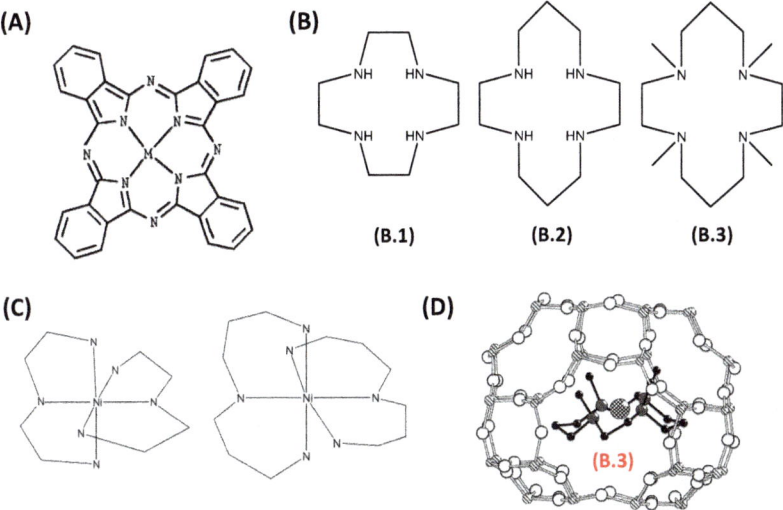

Fig. 4 (**a**) Metallophthalocyanine complex, (**b**) azamacrocycles ligands, (**c**) Ni^{2+} complexes of deta and dpta, and (**d**) azamacrocycle-based Ni complex (B.2) within the supercage of STA-7. Partially reproduced from [18]

In order to do the metal encapsulation of the metal-chelate complexes in a more efficient manner, the direct crystallization of the zeolite structure around the metal complex has been intensively studied using the metal-chelate complexes as single OSDAs [49]. This methodology to encapsulate the metal-chelate complex has been named by some authors as "build-bottle-around-ship" method [50]. Following this strategy, different metallophthalocyanine (see Fig. 4a) or metalloporphyrin complexes, among others, have been encapsulated within X and Y zeolites (FAU structure), and these materials have been employed for selective oxidation reactions, such as cyclohexene oxidation [49, 50]. However, it is important to remark that both X and Y zeolites can be easily prepared under OSDA-free conditions at low-synthesis temperatures and short crystallization times, which can be considered as "mild conditions" in zeolite synthesis and, thus, favour the stability of these bulky metal-chelate complexes in the synthesis media. The direct hydrothermal synthesis of other zeolite structures using these bulky metal-chelate complexes has been rarely described in the literature [49].

3.2.2 Ni- and Cu-Polyamine Complexes

In order to overcome the problems that the above-mentioned bulky metal-chelate complexes can present in directing the crystallization of zeolites, Wright et al. have nicely rationalized the preparation of supramolecular Ni-containing complexes using more simple polyamine-based ligands, including different azamacrocycles (see

Fig. 4b), or linear polyamines, as diethylenetriamine (DETA) and dipropylenetriamine (DPTA) (see Ni-DETA and Ni-DPTA complexes in Fig. 4c) [18, 51, 52].

Wright et al. have described the preparation of some small-pore-based AlPO and SAPO materials, such as STA-6 (SAS), STA-7 (SAV), SAPO-34 (CHA), or AlPO-18 (AEI), using different supramolecular assembled Ni-polyamine complexes as single OSDA molecules [18, 51, 52]. All these AlPO-based materials contain large cavities, which can accommodate the Ni-polyamine complexes, including those prepared with azamacrocycles or linear polyamines (see as example the azamacrocycle-based Ni complex within the supercage of STA-7 in Fig. 4d). The authors have demonstrated by crystallographic and structural studies that the Ni complexes are intact within the large cavities of the above-described small-pore AlPO-based molecular sieves [18, 51, 52]. The regular calcination in air of the Ni complex-containing AlPO-related materials results in the formation of extra-framework Ni(II) species [52]. Despite the fact that the catalytic activity of these Ni-containing AlPO-based zeolites has not been intensively explored, preliminary results on the n-butane conversion at high reaction temperatures (375–575°C) indicate that the Ni-SAPO-34 catalysts show higher catalytic activity than SAPO-34 [52].

Very recently, the synthesis of the Ni-containing aluminosilicate form of CHA, Ni-SSZ-13, has been reported using also the nickel-diethylenetriamine complex (see Ni-DETA in Fig. 4c) as OSDA [53]. The resultant Ni-SSZ-13 materials show Si/Al molar ratios of ~2.6–4.2, which are remarkably lower than the initial Si/Al molar ratios in the synthesis gels (Si/Al ~5–20), suggesting limited zeolite solid yields during the synthesis. However, the methodology presented in this report is very interesting since it describes, for the first time, the synthesis of a Ni-containing small-pore aluminosilicate material following a one-pot method and using a simple and inexpensive Ni-polyamine complex. This novel method would offer new opportunities for attempting the direct synthesis of other Ni-containing zeolites under alkaline conditions and their posterior catalytic application in industrially relevant chemical processes, such as olefin oligomerization [54].

In a similar way, different authors have described the one-pot synthesis of different Cu-containing small-pore zeolites by using organometallic Cu-polyamine complexes as single OSDAs [55, 56] or in a cooperative manner with other organic molecules [19, 57, 58]. Xiao et al. first reported the synthesis of the Cu-containing aluminosilicate form of CHA, Cu-SSZ-13, using copper-tetraethylenepentamine (Cu-TEPA) as single OSDA (see Fig. 5a) [55, 56]. These fresh Cu-SSZ-13 materials show good catalytic performance for the selective catalytic reduction (SCR) of NOx, but, unfortunately, they present limited hydrothermal stability when aged under steaming at temperatures above 700°C (treatment required to envisage the stability under operative conditions). The reason of that is the relatively low Si/Al molar ratios (Si/Al ~4–7) and high Cu contents (~10%) in the final Cu-SSZ-13 materials [55], both affecting the hydrothermal stability of the catalyst. In order to overcome the above problems, Martinez-Franco et al. have proposed the cooperative use of Cu-TEPA with N,N,N-trimethyladamantammonium (TMAda), which is a well-stablished organic molecule for the synthesis of the high-silica SSZ-13 zeolite, as co-OSDAs

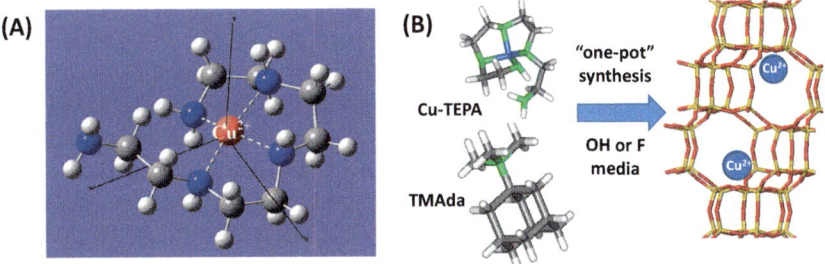

Fig. 5 (**a**) Structure of the Cu-tetraethylenepentamine (Cu-TEPA) complex and (**b**) cooperative OSDAs, Cu-TEPA, and TMAda, for the one-pot synthesis of Cu-SSZ-13. Reproduced from [19, 55], respectively

for the synthesis of the Cu-SSZ-13 material (see Fig. 5b) [19]. This particular combination allows controlling the Si/Al molar ratios and the Cu content in the final Cu-SSZ-13 materials in the ranges between 10–30 and 1–7 %wt., respectively, resulting in SCR-NOx catalysts with excellent catalytic activities and improved hydrothermal stabilities [19]. The costs associated with the manufacture of the one-pot Cu-SSZ-13 zeolite can be further improved by substituting the above TMAda cation by the commercially available and remarkably cheaper tetraethyl ammonium (TEA) cation while maintaining excellent catalytic activities and hydrothermal stabilities for the SCR-NOx [59].

Similar OSDA cooperative approaches have been described for the synthesis of Cu-containing SAPO-related materials by combining a supramolecular assembled Cu-polyamine complex with an additional organic molecule [57, 58, 60, 61]. Indeed, Picone et al. efficiently prepared the Cu-STA-7 material by combining copper-cyclam and TEA as cooperative OSDAs [57], whereas Martinez-Franco et al. reported the effective synthesis of the Cu-SAPO-34 and Cu-SAPO-18 materials by combining Cu-TEPA and TEA and Cu-TEPA and N,N-dimethyl-3,5-dimethylpiperidinium cation, respectively [58, 61]. In general, it has been observed that if the copper content is controlled between 2 and 4%wt. and the silicon species are preferentially placed in isolated form within the small-pore zeotypes, the resultant Cu-SAPO materials show not only excellent catalytic activities for the SCR of NOx reaction but also extremely high hydrothermal stabilities when steamed at high reaction temperatures [58, 61].

3.3 Thiol Complexes

The ability of thiol groups to interact with late transition metals to form highly stable complexes is well-known [62, 63]. Taking advantage of the high stability of the complexes formed by thiol groups and late transition metals, which can resist the severe hydrothermal synthesis conditions required for zeolite synthesis, Iglesia et al. proposed the one-pot synthesis of different metal-containing LTA zeolites by combining the use of (3-mercaptopropyl)trimethoxysilane and different metal precursors, including Pt, Pd, Rh, or Ir (see Fig. 6a) [64]. This methodology is relevant for designing small-pore zeolites containing late transition metals within their structure, because the post-synthetic introduction of metal precursors presenting large ionic radius can be severely restricted by the pore dimensions (~3.5 Å). Following this approach, well-encapsulated metal nanoparticles of ~1 nm can be achieved within the pores of the LTA zeolite, and the resulting materials present good shape selectivity for diverse oxidation and hydrogenation reactions, confirming the overall metal encapsulation [64, 66, 67].

However, it is worth noting that the above syntheses have been carried out using only the supramolecular water-soluble metal complex (see Fig. 6a) as the single OSDA combined with sodium cations [64]. This results in metal-containing LTA zeolites with very low Si/Al molar ratios (below 3), presenting limited applicability as heterogeneous catalysts due to their low hydrothermal stability and Brønsted acidity.

As an alternative synthesis method to increase the Si/Al molar ratio, Moliner et al. have proposed the combination of TMAda and Pt-mercapto complexes as OSDAs to direct the crystallization of the Pt-containing high-silica CHA zeolite [65]. Following this methodology, the Si/Al ratio of the small-pore zeolite can be increased up to ~9, and the Pt content can be controlled between 0.2 and 0.4 %wt. Advanced characterization techniques, such as X-ray absorption spectroscopy and aberration-corrected HAADF scanning transmission electron microscopy, reveal a fine control of the metal structures from ~1 nm nanoparticles to site-isolated single Pt atoms via reversible interconversion of one species into another (see Fig. 6b). Moreover, the resultant metal-containing high-silica CHA materials show high hydrothermal stability, even when treated with steam at high temperatures (above 600°C). These facts open great opportunities in many catalytic applications where well-defined and controlled metallic active sites are required [68].

4 Supramolecular Aromatic Interactions

The supramolecular aggregation of organic molecules containing aromatic rings within their structures has been proposed in the literature to design bulkier and rigid OSDAs, allowing the crystallization of different molecular sieves with particular pore topologies and chemical compositions. Indeed, depending on the size, shape, and nature of the organic molecules, different bulky self-assembled dimers can be achieved in the synthesis media, undoubtedly influencing the nucleation and crystallization processes.

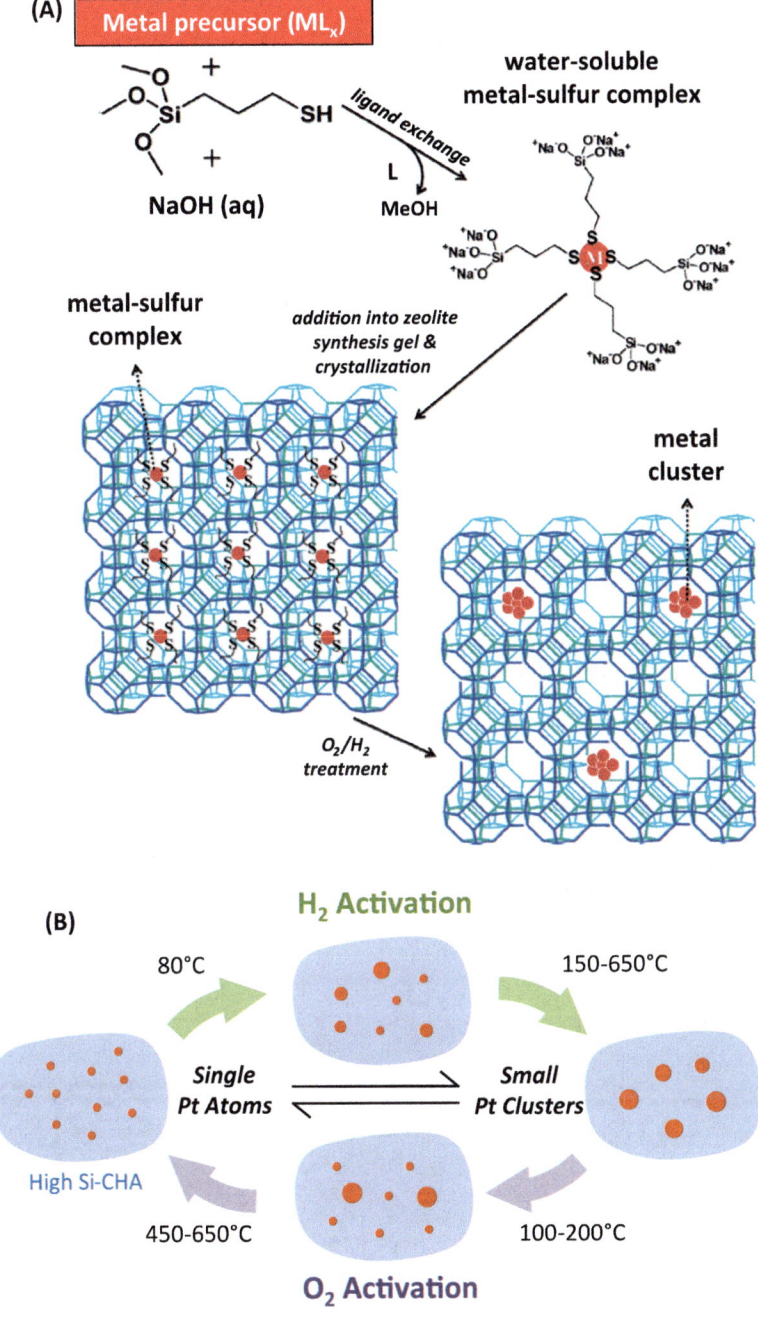

Fig. 6 (**a**) Scheme for thiol-assisted metal encapsulation and (**b**) scheme for the reversible Pt transformation from single atoms to ~1 nm nanoparticles within CHA crystals. Reproduced from [64, 65], respectively

4.1 π–π Stacking

4.1.1 Synthesis of Small-Pore Zeolites with Controlled Chemical Compositions

The use of self-assembled aromatic molecules through supramolecular π–π interactions for zeolite synthesis was first described by Corma et al. for the rational preparation of the high-silica form of LTA [12]. Up to that moment, LTA zeolite was only achieved with very low Si/Al molar ratios (below 3), mainly because it was synthesized under OSDA-free conditions in presence of alkali cations (i.e. Na) [69]. The structure of LTA is a small-pore zeolite containing large spherical cages (sodalite or β-cages), and its preparation as high-silica or even pure silica forms would offer this zeolite excellent properties for gas separations and catalytic applications [12]. Thus, it would be required to find a bulky OSDA molecule able to fit within the sodalite cages, allowing the partial or total replacement of the alkali cations in the final LTA zeolite. However, the design of a bulky single organic molecule presenting the adequate size and shape to stabilize the sodalite cages is not a simple task, and it could be expected that this potential single OSDA would be highly hydrophobic, maybe limiting the organic-inorganic interactions in the synthesis media.

As an alternative, Corma et al. proposed the self-assembling of a simple organic molecule into a bulky dimer, and to do this, the 4-methyl-2,3,6,7-tetrahydro-1H,5H-pyrido [3.2.1-ij] quinolinium molecule was selected (MTPQ; see Fig. 7a) [12]. This MTPQ molecule is able to form a soluble dimer in the synthesis media through a supramolecular π–π interaction between the aromatic rings of two MTPQ molecules, and the resultant bulky and rigid assembled dimer perfectly fits the sodalite cages present in the LTA (see Fig. 7b). Following this self-assembled

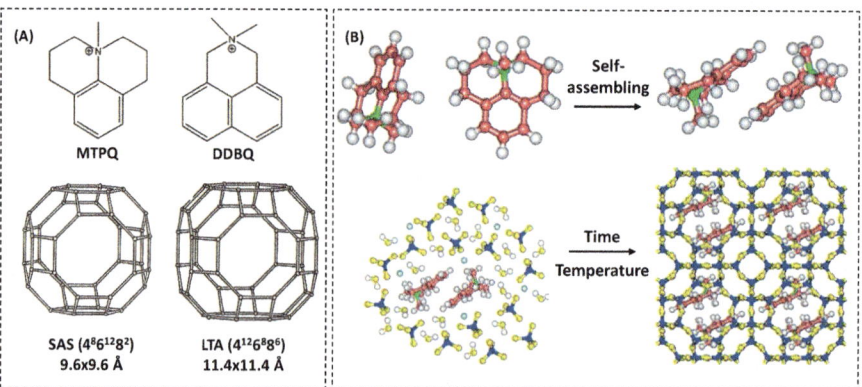

Fig. 7 (a) Aromatic molecules employed for the supramolecular π–π stacking: MTPQ and DDBQ and (b) scheme proposed for the synthesis of ITQ-29 using supramolecular self-assembled OSDAs. Reproduced from [12]

methodology, the LTA zeolite with controlled Si/Al molar ratios (from 20 to infinity) can be achieved, which has been named ITQ-29 [12]. The Al-ITQ-29 shows good catalytic properties for the methanol-to-olefins (MTO) process [12], whereas the pure silica ITQ-29 has been proposed as an efficient material to perform gas separations, particularly linear from branched olefins [12], and for low-dielectric constant (low-k) applications [70, 71].

A similar self-assembling approach with related aromatic molecules has been proposed to control the Si distribution within small-pore silicoaluminophosphate (SAPO) materials [72, 73]. In the last years, small-pore SAPO materials have been applied as efficient catalysts in different industrial processes, such as MTO [74] or the selective catalytic reduction (SCR) of NOx [75]. In both processes, it has been observed that the selective location of silicon as isolated species within the crystalline structure of the SAPO materials is a critical parameter, improving the catalytic activity and the hydrothermal stability of these SAPO-related catalysts [58, 61, 76]. These facts could be explained by a better control of the acid properties for the MTO process and a better distribution of the metal-exchangeable positions for the required extra-framework copper cations in the case of SCR-NOx applications [58, 61]. It is important to note that the isomorphic substitution of P^{5+} by Si^{4+} in a SAPO material would induce the formation of a negative charge in the crystalline framework that could be balanced by a proton or by an extra-framework cation.

Having that in mind, Martinez-Franco et al. suggested that the use of bulky self-assembled aromatic molecules through π–π interactions could be a good method not only to control the crystallization of small-pore SAPO materials with large cavities but also to preferentially distribute the silicon species in isolated environments along the crystalline framework [72]. They based the above hypotheses on the maximization of the volume/charge ratio achieved by these self-assembled molecules allowing the stabilization of large cavities (see Fig. 7b) and, in addition, on the steady allocation of the positive charges within the rigid paired OSDA that would force the silicon species to stay in isolated form to properly balance these positive charges [72, 73]. Indeed, Martinez-Franco et al. observed that depending on the size of the former aromatic molecule, the size of the self-assembled paired OSDA could also be controlled, resulting in the crystallization of different cage-based small-pore SAPOs [72, 73]. The dimer formed by the π-stacked MTPQ molecules (see Fig. 7a) favours the crystallization of the silicoaluminophosphate STA-6 (SAS), whereas the dimer formed by π-stacked DDBQ molecules (2,2-dimethyl-2,3-dihydro-1H-benzo[de]-isoquinoline-2-ium, DDBQ; see Fig. 7a) favours the crystallization of the silicoaluminophosphate SAPO-42 (LTA), which has a slightly larger cavity than the SAS structure (Fig. 7a). In both materials, the silicon distribution could be controlled, observing the presence of two isolated silicon species per each cavity, in good balance with the two positive charges introduced by the self-assembled paired OSDA. These materials are now being tested in gas separation processes, where different gas adsorption capacities can be observed depending on their silicon distribution.

4.1.2   Towards Chiral Distributions Within Microporous Solids

Gómez-Hortigüela et al. proposed the use of different aromatic molecules, containing one or two benzyl rings, as OSDA molecules for the synthesis of AlPO-related structures [20]. They initially proposed five different aromatic molecules, including amines and ammonium cations: benzylpyrrolidine, benzylpiperidine, benzyl-hexamethylenimine, dibenzylpiperazine, and dibenzyldimethylammonium [20]. All these molecules allowed the crystallization of the same crystalline structure, which is the AFI (AlPO-5, 12-ring pores), but they observed different organic packing within the AFI pores depending on if the former OSDA molecules present one or two benzyl rings (see Fig. 8a). For instance, the benzylpyrrolidine (BP) molecule shows the formation of self-assembled dimers with the benzyl groups parallel to each other, whereas dibenzyldimethylammonium (DBDM) assembles into longer chains in which the benzyl ring of one molecule faces the ring of the subsequent molecule (see Fig. 8a) [20]. Subsequently, the authors carried out more fundamental studies in order to understand the supramolecular chemistry involving the occlusion of these aromatic molecules [79, 80]. From both fluorescence and computational studies, the authors observed that the OSDA/H_2O ratio selected to undergo the zeolite synthesis clearly influences the formation of the dimeric species in the synthesis gel and, consequently, the crystallization rates and the assembling nature of the aromatic molecules [79, 80].

Based on the excellent directing roles of the self-assembled rigid dimer formed by the benzylpyrrolidine (BP) molecule (see Fig. 8a), Gomez-Hortigüela et al. proposed the use of the related (S)-(-)-N-benzylpyrrolidine-2-methanol (BPM), taking into account that this is a chiral molecule (see Fig. 8b) [77]. The authors hypothesized that the chiral and rigid self-assembled dimer could be considered as a potential OSDA molecule able to influence the spatial distribution of metallic dopants along the tetrahedral network, then promoting chirality into achiral crys-talline structures to induce enantioselectivity for catalytic applications (see Fig. 8b) [77]. Using BPM as OSDA, the crystallization of the AFI material containing diverse active metals incorporated in the AFI framework, such as magnesium, silicon, cobalt, zinc, or vanadium, has been obtained [77]. The authors performed different theoretical studies to evaluate the arrangement of the BPM molecules within the pores of the AFI structure and their potential ability to create the long-range chirality [77, 81].

Gomez-Hortigüela et al. have extended the chirality transfer approach to another aromatic molecule, (1R,2S)-(-)-ephedrine (EPH), whose structure contains two chiral centres (see Fig. 8c), and thus, the potential chirality transfer to inorganic zeolite-based frameworks can be enhanced [82–84]. Self-assembling of EPH mol-ecules would be favoured not only by the π–π interactions between the aromatic rings but also by intermolecular H-bonds between NH_2 and OH groups (see Fig. 8c). As occurred above with the benzylpyrrolidine (BP) molecule, EPH is also capable to direct the crystallization of the AFI zeotype with different metal dopants within its structure, such as Mg, Si, Co, or Zn [84]. A molecular-mechanics

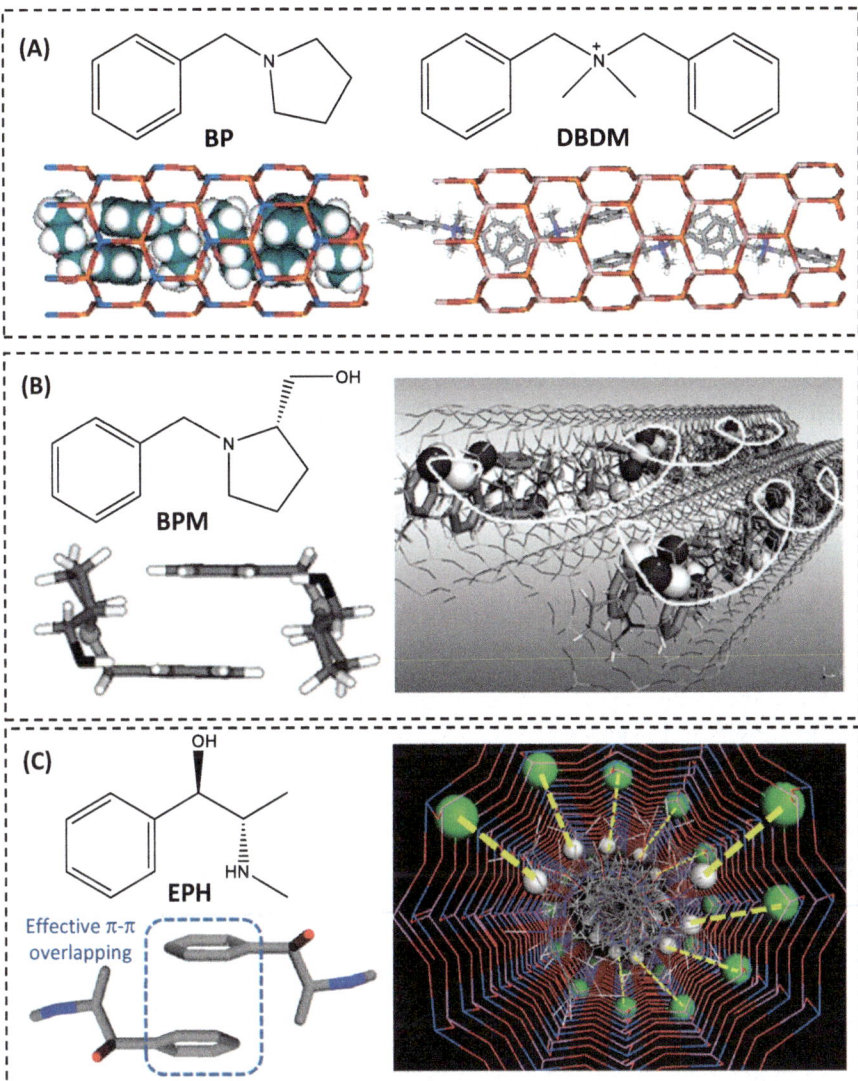

Fig. 8 (a) Benzylpyrrolidine (BP) and dibenzyldimethylammonium (DBDM) molecules and their corresponding self-assembled species within the AFI structure; (b) (S)-(-)-*N*-benzylpyrrolidine-2-methanol (BPM) molecule, its corresponding self-assembled dimer, and the potential chirality transfer within AFI structure; and (c) (1*R*,2*S*)-ephedrine (EPH) molecule, its corresponding self-assembled dimer and the potential chirality transfer within AFI structure. Reproduced from [20, 77, 78], respectively

computational study indicates that an H-bonded chain-like arrangement of π-stacked self-assembled EPH dimers is present within the AFI structure, where the asymmetric nature of EPH together with the H-bond interactions between NH_2

and OH groups forces a rotation of ~30° each two consecutive paired EPH molecules, favouring the helicoidal distribution of dopants (see Fig. 8c) [78]. Now, this
novel and interesting concept comprising new type of chirality in microporous
materials should be experimentally exploited in enantiomeric separation or asymmetric catalysis.

4.1.3 Synthesis of Extra-large-Pore Zeolites

Many new extra-large-pore zeolites, presenting pore openings larger than 7 Å, have
been accomplished in the last years, thanks to the introduction of germanium in the
synthesis media [85–89]. The presence of germanium allows the formation of some
small cages, such as double-3-rings (D3R) or double-4-rings (D4R) [8], which
favour the crystallization of zeolites with low framework densities [90], as it is
the case of most of the achieved extra-large-pore zeolites [8]. Some of the reported
extra-large-pore germanosilicates require the use of large and rigid OSDA molecules [8]. However, as it has been described above, the design of bulky OSDAs
would involve different multistep organic synthesis, increasing the costs associated
with their preparation and maybe also limiting the overall organic product yields. In
addition, if the OSDA presents high hydrophobicity, its directing effects could be
clearly affected by a limited solubility in the synthesis media.

   According to this, Du et al. have recently proposed the use of diverse simple
aromatic molecules as precursors of bulky self-assembled OSDA dimers, which, in
combination with germanium, have allowed the crystallization of different extra-
large-pore germanosilicates (see Fig. 9) [91–93]. In this sense, the supramolecular
assembly of the 1-methyl-3-(4-methylbenzyl)imidazolium directs the crystallization of the extra-large-pore NUD-1 germanosilicate, whose structure is formed by
intersecting 18-, 12-, and 10-ring pores (see Fig. 9a) [91]. The structure of NUD-1 is
highly related to the previously described ITQ-33 and ITQ-44 zeolites [85], all of
them presenting identical sheets with different packing along the c axis [91]. On the

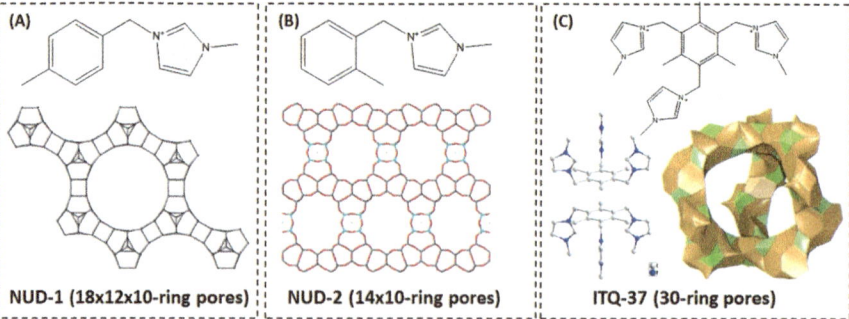

Fig. 9 Aromatic-based OSDA molecules employed for the synthesis of extra-large-pore
germanosilicates: (**a**) NUD-1, (**b**) NUD-2, and (**c**) ITQ-37. Reproduced from [86, 91–93],
respectively

other hand, the use of 1-methyl-3-(2′-methylbenzyl)imidazolium has allowed the crystallization of the new NUD-2 zeolite, which is a germanosilicate presenting interconnected 14- and 10-ring pores (see Fig. 9b) [92]. The layers of the NUD-2 zeolite are FER-type layers connected by D4Rs, suggesting that most of the Ge atoms would be placed in the interlayer region, and thus, this material could be an excellent candidate to be post-synthetically treated, for instance, through ADOR methods, to achieve other related highly stable zeolites [94]. Finally, the combination of germanium and a semi-rigid imidazolium, such as 1,1′,1″-(2,4,6-trimethylbenzene-1,3,5-triyl)-tris(methylenetris(3-methyl-1H-imidazol-3-ium) [TMBI], has permitted the crystallization of the previously described ITQ-37 zeolite (see Fig. 9c) [86, 93]. For this particular case, the authors show that the supramolecular assembly of the aromatic TMBI molecule through π–π stacking produces an extremely bulky and rigid OSDA, which permits the formation of the very large 30-ring pores present within the ITQ-37 zeolite (see Fig. 9c) [93].

The use of supramolecular aromatic assembled OSDA molecules combined with other inorganic directing agents, such as germanium, could be considered a general and profitable methodology to direct the crystallization of other novel large and extra-large-pore structures.

4.2   Proton Sponges

Proton sponges are aromatic diamines with the amine groups in close proximity (see Fig. 10a), providing high basicity to these molecules ($pK_a > 12.1$) [96]. This high pK_a value allows them to remain in their protonated form even at high pHs, which are mostly required in zeolite synthesis, favouring their potential use as OSDAs [97]. Moreover, the presence of the aromatic rings allows envisioning that the proton sponges could ensemble into bulky and rigid paired molecules, maybe providing the adequate templating role to direct the formation of extra-large-pore zeolites.

According to the above premises, Martinez-Franco et al. have reported the synthesis of the new extra-large-pore ITQ-51 zeotype by using the commercially available proton sponge, 1,8-bis(dimethylamino)naphthalene (DMAN, see Fig. 10a) [97]. The ITQ-51 structure has a monodirectional 16-ring pore system, presenting oval pore openings of 9.9 × 7.7 Å, and in addition, the crystalline structure of the ITQ-51 remains stable after calcination treatments, being one of the very few examples of hydrothermally stable extra-large-pore zeolites [97]. Different characterization techniques have demonstrated the self-aggregation of the DMAN molecules into dimeric species, both in the synthesis gels and within the as-prepared ITQ-51 material (see Fig. 10b, c) [95]. The geometry of the dimeric species reveals that the aromatic rings are parallel to each other, with the methyl groups on the opposite side of each DMAN molecule (see Fig. 10b, c). However, in this particular case, a π–π stacking interaction between the aromatic rings is not observed in the experimental and theoretical studies, being the OSDA dimer interactions mainly driven by electrostatic and van der Waals

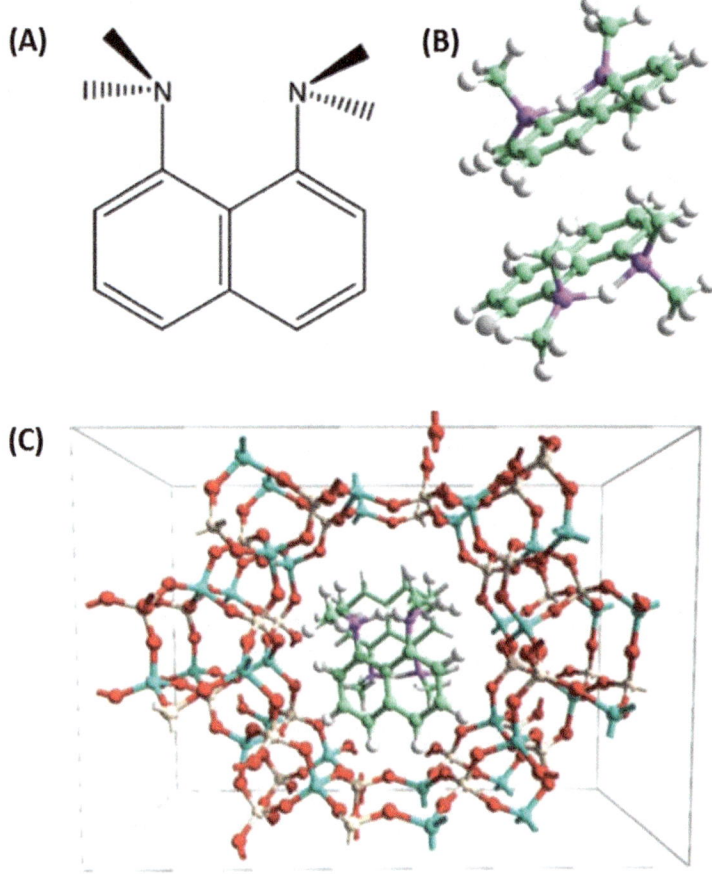

Fig. 10 (a) Structure of the proton sponge, (b) supramolecular assembly of proton sponges, and (c) self-assembled proton sponges molecules within the pores of the ITQ-51 zeotype. Reproduced from [95]

interactions [95]. Two different metals, Mg and Co, have been introduced within the crystalline structure of the ITQ-51, resulting in metal-containing extra-large-pore zeotypes with high hydrothermal stability and high Brønsted acidity [98]. These metal-ITQ-51 materials show good properties to be applied as heterogeneous acid catalysts in chemical processes involving bulky large reactants and/or products.

4.3   Fluorinated Molecules

Along the years, the influence of the size, shape, and rigidity of the organic molecules employed as OSDAs in zeolite synthesis has been intensively studied [1, 2], but the chemical character of these organic molecules has not been

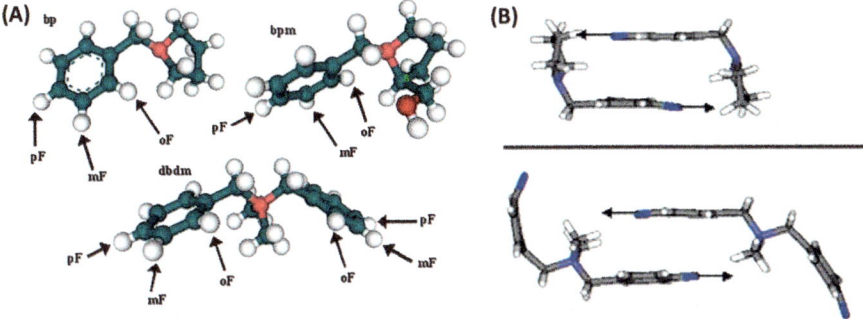

Fig. 11 (a) Different proposed fluorinated aromatic molecules with the fluorine moiety in different positions of the aromatic ring and (b) electrostatic repulsion observed for the fluorinated molecules in para-positions. Reproduced from [101]

sufficiently considered. Indeed, the organic-inorganic interactions in zeolite synthesis preferentially take place through the hydrogen atoms present in the organic molecules. Thus, these nonbonded interactions could be substantially modified by replacing some of the hydrogen atoms by other elements, as, for instance, fluorine atoms, affecting the electronic, chemical, and surface properties of the organic molecule [99]. In addition, the fluorinated molecules would be more hydrophobic than the all-hydrogen organic counterparts [100], property that, as it is well-known, clearly influences the crystallization of the microporous crystalline materials.

In a series of papers, Gomez-Hortigüela et al. have studied different fluorinated aromatic organic molecules for the synthesis of microporous materials, including amines and ammonium cations, where the fluorine moiety has been introduced selectively in different positions within the organic molecules (see Fig. 11a) [101–104]. In general, these authors have observed that depending on the fluorine position in the aromatic ring, the interactions between the organic molecules in the synthesis media would be different and, consequently, their ability to act as structure-directing agents [101]. When the fluorine is in *meta*-position in those organic molecules, it increases the electrostatic contribution of the aromatic ring, improving in general the structure-directing ability of the fluorinated-organic compound. But, in addition, the fluorine position in the benzyl ring can favour or prevent the supramolecular interaction of the organic molecules in the synthesis media. For instance, the fluorine in *para*-positions mostly precludes the dimerization of the aromatic molecules [101], and in these cases, the fluorinated molecules are not able to direct the crystallization of any zeolite structure. By theoretical calculations, the authors conclude that the fluorine atom in *para*-positions of the aromatic ring faces the other potential molecule present in the stacked dimer, inducing an electrostatic repulsion that avoids the dimer formation (see Fig. 11b).

These results suggest that the rational modification of the chemical nature of the organic molecules that will be used as OSDA in zeolite synthesis would offer novel and interesting supramolecular interactions between the organic molecules, which could induce unique organic directing effects towards new or improved molecular sieves.

5 Amphiphilic Surfactants

In the 1990s, researchers at Mobil described the use of alkyltrimethylammonium surfactants $[C_nH_{2n+1}(CH_3)_3N^+]$, as OSDAs for the formation of the ordered mesoporous MCM-41 material when the length of the alkyl chain was preferentially comprised between 8 and 16 [105, 106]. The long alkyl chain favours the surfactant aggregation in the form of micelles under the reaction conditions, acting as supramolecular templates for directing the synthesis of the siliceous MCM-41, which present pores with openings above 20 Å after removing the entrapped organic surfactant molecules by regular calcination treatments with air (see Fig. 12a) [106]. Depending on the surfactant concentration and the synthesis

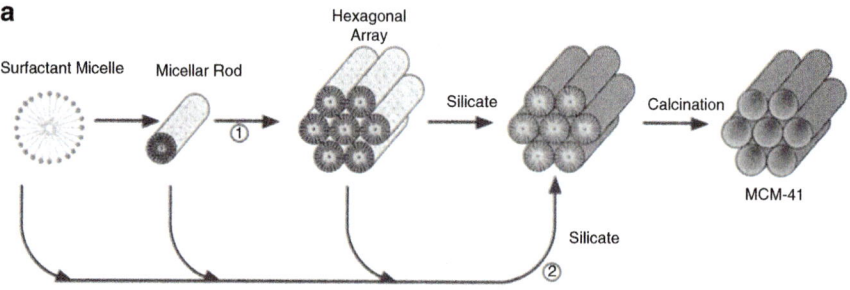

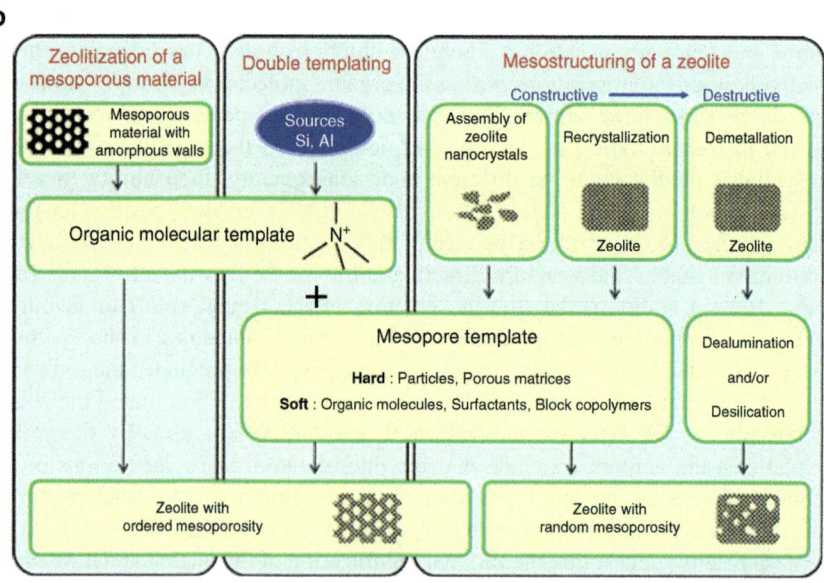

Fig. 12 (**a**) Scheme proposed for the formation of the supramolecular-templated MCM-41 material and (**b**) synthesis routes for designing hierarchical zeolites. Reproduced from [106, 107], respectively

conditions, different supramolecular surfactant aggregation can be achieved, allowing the formation of other materials, such as MCM-48 and MCM-50, which are cubic or lamellar materials, respectively [108]. Despite the discovery of the supramolecular-templated ordered mesoporous materials permitted expanding notoriously the number of applications where siliceous materials could be employed, particularly those involving large organic molecules [108], the amorphous nature of their walls resulted in materials with limited acidity and hydrothermal stability, avoiding their use in many catalytic applications.

5.1   Synthesis of Hierarchical Zeolites by Using Amphiphilic Surfactants

Thus, since the discovery of the mesoporous MCM-type materials by Mobil, there has been an intensification of the research attempting the design of novel materials combining micropores with mesopores, in order to fine-tune the pore accessibility, acidity, and hydrothermal stability.

Different strategies have been proposed to control the mesoporosity within well-crystallized zeolites, including "destructive" post-synthetic treatments or "constructive" synthesis methods (see Fig. 12b) [107]. The "destructive" methods comprise post-synthetic acid or base treatments to remove tetrahedrally coordinated atoms from the zeolite structure, but in general, the mesoporosity created within the zeolite crystals is not well controlled, and, in addition, relatively low solid yields are achieved after these treatments. On the other hand, the "constructive" synthetic methods mostly involve the use of amphiphilic surfactants in the synthesis media as supramolecular templates to generate the mesopores within the zeolite crystals.

5.1.1   Non-covalently Bonded Surfactants Combined with Micropore-Directing Agents

Many attempts have been described in the literature using non-covalently bonded surfactants to favour the formation of a controlled mesoporosity within the zeolite crystals during the zeolite crystallization. In general, the preferred strategy involves the combination of zeolite templates and surfactant-based mesopore templates in the synthesis media, but following this methodology, the formation of separate phases is mostly observed. Some examples describing the combination of micro- and meso-templates are next described.

One of the former works was described by van Bekkum et al., where they combined sodium and cetyltrimethylammonium cations to favour the formation of a zeolite Y with controlled mesopores [109]. The authors were able to obtain FAU/MCM-41 composites, where thin MCM-41 layers were preferentially placed on the external surface of the FAU zeolite crystals, but the presence of separate MCM-41 particles was also observed within the solids.

The mixture of two alkyl ammonium cations with different alkyl length chains (C_6 and C_{14}) was employed by Karlsson et al. to favour the crystallization of a zeolitic phase containing a mesoporous structure [110]. Characterization techniques revealed the formation of fairly complex aggregates of MFI- and MCM-41-type materials, where a physical mixture of segregated MFI and MCM-41 particles was also observed.

An improved method to achieve MCM-41/ZSM-5 composites was reported by Li et al. following a two-step crystallization process [111]. The mesoporous MCM-41 was first prepared by using the supramolecular assembling of cetyltrimethyl-ammonium (CTMA), and later, the amorphous MCM-41 was recrystallized using tetrapropylammonium (TPA) to achieve a MCM-41/ZSM-5 composite containing interconnected mesopores and micropores. The authors show that the MCM-41/ ZSM-5 composites present improved acidity and catalytic activity for cracking reactions compared to physical mixtures containing MCM-41 and ZSM-5 materials.

Another example using the combination of micro- and meso-templates has been described by Vernimmen et al. for attempting the synthesis of the hierarchical Ti-silicalite zeolite (TS-1) [112]. For this purpose, the authors have partially modified the "classical" synthesis recipe for the TS-1 zeolite, replacing part of the microtemplate (TPA) by a meso-template (CTMA) and systematically studying many synthesis conditions, such as temperature and time of crystallization or microtemplate/meso-template molar ratios. From this extensive study, the authors concluded that the crystallization of true hierarchical zeolites cannot be claimed, observing instead the formation of zeolitic/mesoporous composites.

5.1.2 Covalently Bonded Surfactants

In the previous section, it has been shown that the direct combination of micro- and meso-templates is not a very efficient method to allow the direct synthesis of mesoporous zeolites, since the two different templates act in a competitive manner. At that point, Ryoo et al. speculated that the mixture of phases could be explained by the preferential exclusion of the surfactants from the aluminosilicate domain during the zeolite crystallization [113]. In order to avoid this, Ryoo et al. proposed the use of amphiphilic surfactant molecules containing a hydrolysable methoxysilyl group in the molecule, which will form covalent bonds with the other SiO_2 and Al_2O_3 sources, preventing their segregation from the aluminosilicate domains and, then, favouring the formation of intracrystalline mesoporosity during the zeolite crystallization (see Fig. 13a) [113]. Following this rationalized methodology, Ryoo et al. were able to synthesize different hierarchical zeolites, as LTA and MFI [113, 114], and AlPO-based materials, as SAPO-5 and SAPO-11 [115], all of them with controlled mesopore diameters, typically in the range of 2–20 nm, depending on the molecular structure of the mesopore template. For the particular case of the hierarchical MFI zeolite, its crystal morphology (see Fig. 13b) enhances

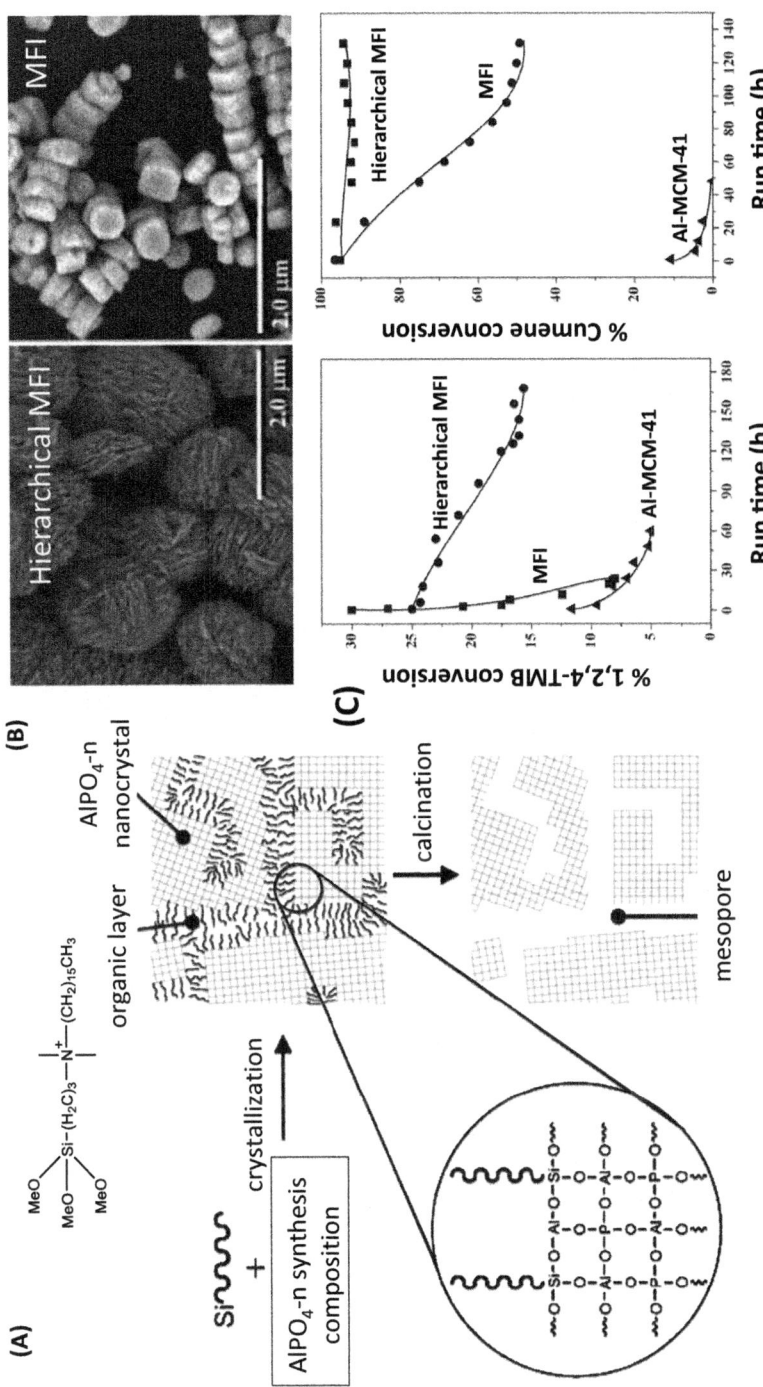

Fig. 13 (a) Scheme proposed for the formation of hierarchical zeolites using the supramolecular assembly of covalently attached amphiphilic surfactants, (b) SEM images of hierarchical MFI and classical MFI, and (c) comparison of the catalytic activity observed for 1,2,4-TMB isomerization and cumene cracking using hierarchical MFI and classical MFI. Reproduced from [114, 115], respectively

the stability against deactivation for various catalytic reactions, such as 1,2,4-trimethylbenzene isomerization and cumene cracking (see Fig. 13c), thanks to the improved diffusion pathways of the large molecule trough the intracrystalline mesopores [114].

More recently, Inayat et al. have described the synthesis of the mesoporous FAU-type zeolite nanosheets by also using an organosilane amphiphilic surfactant, as 3-(trimethoxysilyl)propyl hexadecyl dimethyl ammonium (TPHA) [116]. The achieved FAU-type crystals are mainly constructed of zeolitic nanosheets in a house-of-cards-like assembly, presenting wide macropores between the nanosheet stacks.

5.1.3   Amphiphilic Surfactants Combined with Pre-crystallized Zeolites

A different approach attempting the formation of mesoporous zeolites combines the use of pre-crystallized zeolites with surfactant molecules (see Fig. 14) [117]. This methodology, reported by García-Martínez et al., permits the creation of mesopores within pre-crystallized high-silica USY zeolites by introducing cationic surfactants, such as cetyltrimethylammonium (CTMA) in diluted basic solutions (i.e. NaOH) [117–119]. The authors proposed a synthesis mechanism based on the diffusion of single cationic surfactants within the zeolite crystals by electrostatic driving forces, and once the surfactant concentration is relatively high within zeolite crystals, it could self-assemble into supramolecular ordered micelles, resulting in the formation of mesopores (see Fig. 14) [118].

The authors claim that the sizes of the mesopores can be controlled by modifying the alkyl chain of the surfactants used, achieving narrower pore-size distributions than those observed by desilication treatments. The mesostructured USY zeolite obtained by this treatment shows high crystallinity, acidity, and hydrothermal stability, suggesting that this modified USY zeolite can perform as good acid catalyst for reactions involving large molecules. In this sense, the authors have reported improved catalytic properties for the fluid catalytic cracking (FCC) process when using the mesostructured USY zeolite compared to classical FAU catalysts, including higher selectivities to gasoline and less coke production, among others [118].

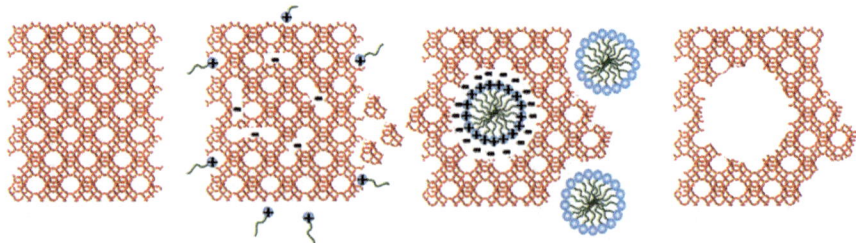

Fig. 14 Proposed scheme for the synthesis of mesostructured zeolites combining pre-crystallized zeolites and surfactant-based templates. Reproduced from [117]

5.2 Synthesis of Zeolite Nanosheets

In the previous section, the creation of controlled mesoporosity within the zeolite crystals has been attempted to improve the diffusion limitations that the restricted pores of well-crystallized zeolites can induce when large molecules are involved in the chemical process. However, a different strategy to overcome the diffusion limitations could be the reduction of the thickness of the zeolite crystals, which would also favour the molecular diffusion of large organic molecules [120].

Several years ago, Corma et al. showed a new concept to design ultrathin zeolites, such as ITQ-2 and ITQ-6, based on the delamination of lamellar zeolitic precursors [121, 122]. For this purpose, Corma et al. proposed a two-step-based methodology, where once the layered precursors were prepared following classical hydrothermal synthesis procedures, these layered precursors were swelled by using supramolecular assembled amphiphilic surfactants, resulting in single-crystalline layers of the zeolite precursor (see Fig. 15a). The ultrathin nature of these delaminated materials not only has allowed its application in many catalytic reactions presenting bulky molecules, both in petrochemical and fine chemistry-based industrially relevant processes, but also their outstanding external surface area has permitted their use as efficient inorganic supports of functional molecules, i.e. enzymes or particular organocatalysts [124].

More recently, Ryoo et al. have extended the design of zeolite nanosheets to other zeolite structures by following a rationalized one-pot synthesis methodology using dual amphiphilic templates as single OSDAs (see Fig. 15b) [22]. These dual templates contain a long-chain alkyl group (above C_{20}) combined with two ammonium groups spaced by a C_6 alkyl chain (see Fig. 15b), where the long hydrophobic tail forms a supramolecular micellar structure that avoids the zeolite growth and the diammonium group allows crystallization of the zeolite layers (see Fig. 15c) [22]. The former syntheses using these bifunctional templates allowed the preparation of nanosheets of zeolite MFI under different chemical compositions, including aluminosilicate, titanosilicate, or stannosilicate [22, 125–128]. These materials have been tested in diverse catalytic applications requiring Brønsted or Lewis acid sites, such as the gas-phase Beckmann rearrangement [125], the selective oxidation of large alkenes to epoxides [126, 127], or the Baeyer-Villiger oxidation of cyclic ketones [128].

The modification of the structure of the bifunctional OSDA molecules and/or the synthesis conditions has allowed the preparation of other zeolite-related nanosheets, including aluminosilicates (MTW, Beta, MRE, and MWW structures) [129–131] and aluminophosphates (AEL, AEI, and ATO framework structures) [132]. For instance, for the synthesis of the MWW nanosheets, named MIT-1 [130], Roman-Leshkov et al. have designed a bifunctional amphiphilic template containing a head segment that resembles the adamantammonium-type OSDA employed to synthesize the high-silica MCM-22P (see Fig. 16a) [134]. The bulky head could stabilize the formation of the external cups present in MWW-layers, and the long alkyl chain would preclude the crystal growth along the perpendicular axes

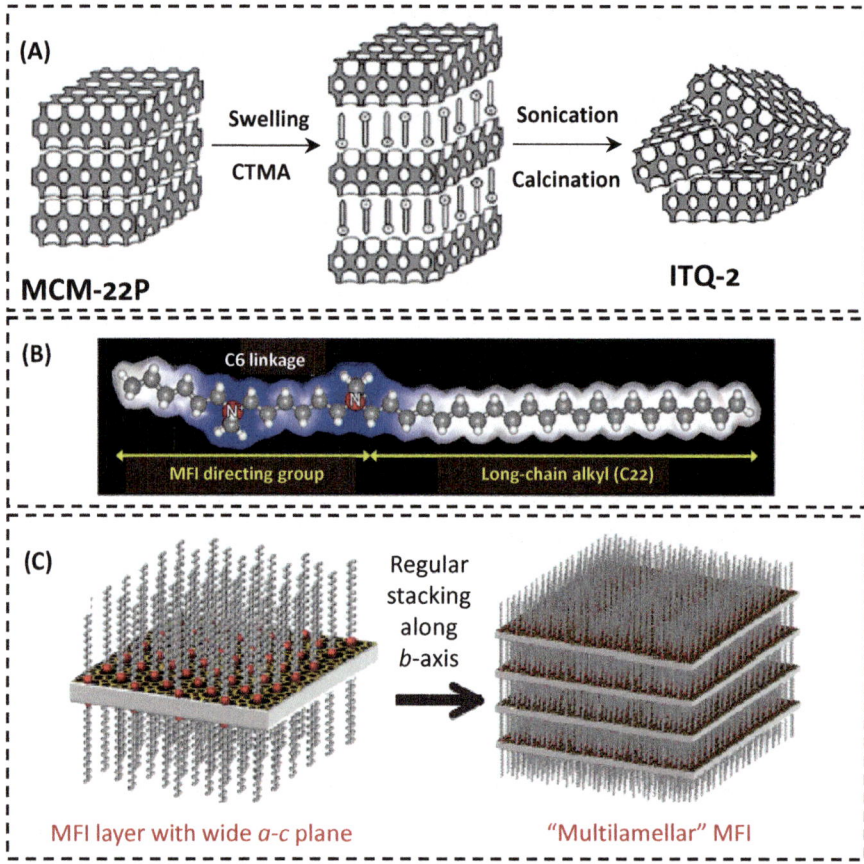

Fig. 15 (a) Two-step methodology for the preparation of ultrathin zeolites, (b) amphiphilic dual template for the direct synthesis of nanosheets, and (c) multilamellar stacking of the as-prepared nanosheets. Reproduced from [22, 123], respectively

to the MWW layer (see Fig. 16a). The resultant MIT-1 has been formerly employed by Roman-Leshkov et al. for the Friedel-Crafts alkylation of benzene with benzyl alcohol [130]. They showed a threefold enhancement in the catalytic activity compared to other MWW-related catalysts, thanks to the remarkable increase of the external surface area and, consequently, higher accessibility to the external MWW cups, which have been proposed as the sites where the alkylation reaction preferentially takes place [135].

Very recently, Gallego et al. have proposed a new concept for synthesizing active and selective zeolites by using OSDAs that mimic the transition state (TS) of pre-established reactions to be catalysed (see Fig. 16b) [133]. The principal aim is the design of zeolites with the specific cavity to allocate the TS of the reaction to be studied, allowing the increase of the activity and/or selectivity towards the desired

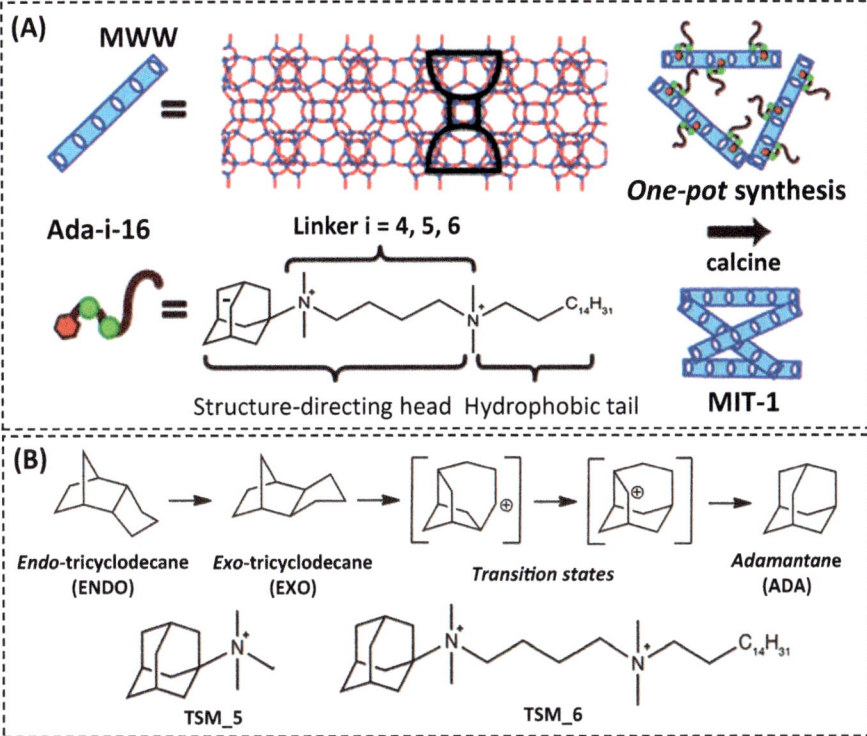

Fig. 16 (**a**) One-pot synthesis of the delaminated MWW material and (**b**) reaction mechanism for the adamantane synthesis and proposed transition state mimics as OSDAs for zeolite synthesis. Reproduced from [130, 133], respectively

product [133]. An important factor to take into account in this "ab initio" design of zeolite catalysts is if the size of the TS is larger or similar to the size of the desired products, because in the latter case, the products can suffer diffusion limitations when formed within the zeolite crystals. As seen in Fig. 16b, this can occur for the tricyclodecane isomerization reaction, where the proposed transition states of the reaction show similar size to the adamantane product. However, the above-presented MIT-1 material could show excellent catalytic properties for the adamantane production by the following two reasons: (1) it has been synthesized with a bifunctional OSDA containing an adamantane-like head, which can be considered as the TS of the tricyclodecane isomerization reaction (see Fig. 16a, b), and (2) MIT-1 shows high accessibility, thanks to its lamellar nature, which has been achieved by the supramolecular assembly of the long alkyl chains present in the bifunctional OSDA (see Fig. 16a, b). According to these hypotheses, MIT-1 has been tested for the tricyclodecane isomerization reaction, resulting in a remarkable increase of the catalytic activity and adamantane selectivity compared to other commercial zeolite catalysts [133].

Finally, Che et al. have also described the synthesis of nanosheet zeolite layers, particularly for the MFI structure, proposing the use of novel single-head quaternary ammonium surfactants (see Fig. 17a) instead of multiple quaternary ammonium head groups, as reported previously by Ryoo et al. (see Fig. 15b) [136, 137]. Up to that moment, only amorphous mesoporous materials were obtained in the literature using single-head quaternary ammonium surfactants [22, 105]. To favour the formation of the lamellar structure, Che et al. proposed the introduction of rigid aromatic fragments, such as biphenyl or naphthyl groups, within the long alkyl chain (see Fig. 17a), since the rigid aromatic fragments form strong self-assembling

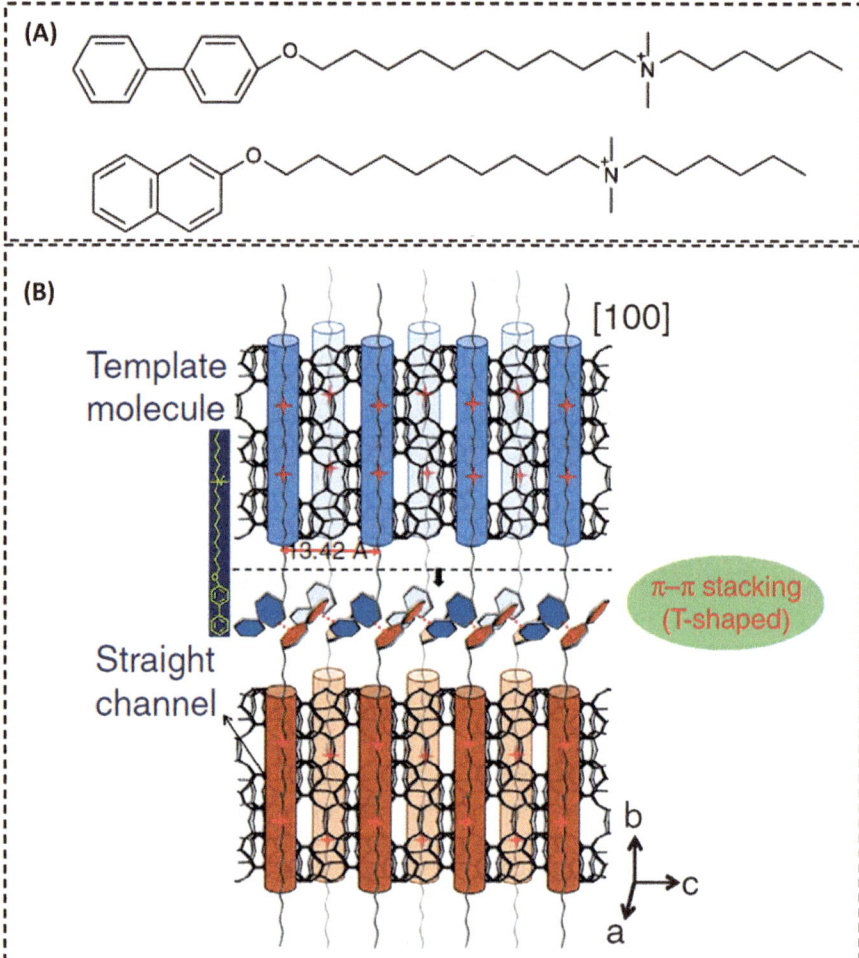

Fig. 17 (**a**) Single-ammonium head surfactants proposed for the synthesis of nanosheet zeolites and (**b**) supramolecular assembly of the single-ammonium cations allowing the formation of nanosheet-based zeolites. Reproduced from [136]

π–π interactions between two organic molecules, stabilizing the overall supramo-lecular lamellar assembly (see Fig. 17b) [136, 137]. This methodology could be employed in the preparation of other ultrathin zeolite structures.

6 Conclusions and Perspectives

It has been shown that the supramolecular assembly of organic molecules with other organic molecules and/or inorganic cations has allowed the rational prepara-tion of many zeolite structures with controlled physico-chemical properties, such as acidity, pore accessibility, and metal distribution, among others, resulting in mate-rials with excellent applicability in catalysis and gas separations.

In general, the use of supramolecular assembled crown ethers or π–π stacked aromatic molecules preferentially directs the crystallization of small-pore or large-pore zeolites, both presenting large cavities within their structures. Despite the interesting results achieved using these molecules, there is still room for improve-ment, in particular in the design of high-silica FAU-related zeolites, whose direct synthesis under higher Si/Al molar ratios would have a remarkable impact in the catalysis field, especially in petrochemistry. It seems reasonable that the use of bulky supramolecular assembled OSDAs would be the proper methodology to stabilize the large cavities present within the FAU-related zeolites. Thus, more efforts should be done in the near future designing new bulky assembled OSDAs capable to fit better the FAU cavities, and in this sense, the use of ab initio theoretical calculations could help in defining the best organic structures to be used as OSDAs [138, 139].

In a similar way, different extra-large-pore zeolites have been reported combin-ing the use of bulky π–π stacked aromatic molecules with germanium atoms in the synthesis media. However, the current challenge in synthesizing extra-large-pore zeolites would be their preparation under Ge-free conditions in order to not only improve their hydrothermal stability and acidity but also reduce their synthesis costs. In this sense, the use of supramolecular assembled aromatic molecules could favour the formation and stabilization of the extra-large pores, but the presence of other inorganic directing agents would be maybe required, as a replacement for germanium, being also able to stabilize the formation of small cages (i.e. D4R), which are mainly present in extra-large-pore zeolites. The potential use of pre-crystallized zeolites, containing these small cages in their former structure, as initial silicon source to attempt the synthesis of extra-large-pore zeolite through inter-zeolite transformations could be an interesting procedure.

Another important point in the field is the design of enantioselective zeolite-based catalysts. In the present chapter, it has been shown that the use of supramo-lecular assembled chiral aromatic molecules could be used as potential OSDAs to direct the chiral distribution of metallic dopants within an achiral zeolite structure. In order to extend the chirality to the inorganic framework, very recently, Davis et al. have designed different rigid dicationic chiral molecules by connecting the ammonium cations through rigid cyclopropane or cyclobutane rings, allowing the crystallization of highly enantiomerically enriched STW zeolite [140]. Considering these recent results, it could be envisioned that the design of bulky dicationic molecules combining the presence of aromatic rings, able to form supramolecular assemblies through π–π interactions, and rigid cycloalkane groups would enforce the rigidity of the organic molecules to transfer their chirality to the inorganic frameworks.

The use of amphiphilic surfactants has allowed the design of zeolite crystals with improved diffusion pathways, both including hierarchical and ultrathin zeolites. The achieved zeolites show remarkable improvements in industrially relevant catalytic applications, offering unique opportunities to these materials for their commercial implementation. Nevertheless, a challenging point would be the design of similar zeolites with low diffusion pathways but using organic cations with shorter alkyl chains (i.e. below C_6). This fact would reduce considerably the costs associated with the preparation of both hierarchical and ultrathin zeolites, making them more competitive for industrial applications.

Recently, the supramolecular assembled organometallic complexes have been intensively employed in the design of metal-containing zeolites, particularly within small-pore zeolites for their applications as efficient and inexpensive SCR catalysts. Moreover, the new descriptions on the fine-tuning reversible control of the metallic active sites within small-pore zeolites open new possibilities for the design of multimetallic zeolites and their potential implications in novel heterogeneous catalytic processes [65, 68].

Finally, most of the materials synthesized following the supramolecular templating methodology have been applied as active and selective catalysts after the removal of the organic moieties entrapped within the pores of the zeolite structures by calcination treatments with air. However, in the last years, there is an increasing interest in the design of advanced functional hybrid organic-inorganic materials for their implementation in novel applications, such as biomedicine, chemosensors, or light-emitting, among others [141]. In most of these cases, the particular interaction between the guest organic molecules, or organometallic complexes, and the inorganic host structures is crucial for improving their applicability, for instance, for high energy transfer capabilities in solar energy harvesting, information processing, or nanodiagnostics. In this sense, the post-synthetic insertion of different aromatic dyes within crystalline zeolites for their use as optical devices, forming self-assembled close-packed die architectures within the pores of the zeolite, has been recently described [142]. This work permits envisioning that the direct synthesis of zeolites could be attempted using the proper aromatic dyes as potential bulky self-assembled OSDAs, maximizing the packing and the interactions within the final hybrid organic-inorganic material. In this sense, there is a recent synthesis description by Sola-Llano et al. attempting the one-pot design of an optical host-guest hybrid material using a hemicyanine dye as OSDA [143]. More efforts should be done on this topic for the rational preparation of sophisticated functional materials.

Acknowledgements This work has been supported by the Spanish Government-MINECO through "Severo Ochoa" (SEV-2016-0683) and MAT2015-71261-R and by the Fundación Ramón Areces through a research contract of the "Life and Materials Science" program.

References

1. Lobo RF, Zones SI, Davis ME (1995) Structure-direction in zeolite synthesis. J Incl Phenom Mol Recognit Chem 21:47–78
2. Moliner M, Rey F, Corma A (2013) Towards the rational design of efficient organic structure-directing agents for zeolite synthesis. Angew Chem Int Ed 52:13880–13889
3. Barrer RM, Denny PJ (1961) Hydrothermal chemistry of the silicates. Part IX. Nitrogenous aluminosilicates. J Chem Soc 971–982. https://doi.org/10.1039/JR9610000971
4. Burton AW, Zones SI (2007) Organic molecules in zeolite synthesis: their preparation and structure-directing effects. Stud Surf Sci Catal 168:137–179
5. Dorset DL, Kennedy GJ, Strohmaier KG, Diaz-Cabañas MJ, Rey F, Corma A (2006) P-derived organic cations as structure-directing agents: synthesis of a high-silica zeolite (ITQ-27) with a two-dimensional 12-ring channel system. J Am Chem Soc 128:8862–8867
6. Simancas R, Dari D, Velamazan N, Navarro MT, Cantin A, Jorda JL, Sastre G, Corma A, Rey F (2010) Modular organic structure-directing agents for the synthesis of zeolites. Science 330:1219–1222
7. Burkett SL, Davis ME (1994) Mechanism of structure direction in the synthesis of Si-ZSM-5: an investigation by intermolecular 1H-29Si CP MAS NMR. J Phys Chem 98:4647–4653
8. Jiang J, Yu J, Corma A (2010) Extra-large-pore zeolites: bridging the gap between micro and mesoporous structures. Angew Chem Int Ed 49:3120–3145
9. Moliner M, Martinez C, Corma A (2014) Synthesis strategies for preparing useful small pore zeolites and zeotypes for gas separations and catalysis. Chem Mater 26:246–258
10. Zones SI (2011) Translating new materials discoveries in zeolite research to commercial manufacture. Microporous Mater 144:1–8
11. De Vos DE, Jacobs PA (2001) Zeolite-based supramolecular assemblies. Stud Surf Sci Catal 137:957–985
12. Corma A, Rey F, Rius J, Sabater MJ, Valencia S (2004) Supramolecular self-assembled molecules as organic directing agent for synthesis of zeolites. Nature 431:287–290
13. Moliner M (2015) Design of zeolites with specific architectures using self-assembled aromatic organic structure directing agents. Top Catal 58:502–512
14. Ozin GA, Steele MR (1994) The zeolite ligand; zeolite encapsulated semiconductor nanomaterials. Macromol Symp 80:45–61
15. Delprato F, Delmotte L, Guth JL, Huve L (1990) Synthesis of new silica-rich cubic and hexagonal faujasites using crown-ether based supramolecules as templates. Zeolites 10:546–552
16. Balkus Jr KJ, Shepelev S (1993) Synthesis of nonasil molecular sieves in the presence of cobalticinium hydroxide. Microporous Mater 1:383–391
17. Freyhardt CC, Tsapatsis M, Lobo RF, Balkus Jr KJ, Davis ME (1996) A high-silica zeolite with a 14-tetrahedral-atom pore opening. Nature 381:295–298
18. Garcia R, Philp EF, Slawin AMZ, Wright PA, Cox PA (2001) Nickel complexed within an azamacrocycle as a structure directing agent in the crystallization of the framework metalloaluminophosphates STA-6 and STA-7. J Mater Chem 11:1421–1427
19. Martinez-Franco R, Moliner M, Thogersen JR, Corma A (2013) Efficient one-pot preparation of Cu-SSZ-13 materials using cooperative OSDAs for their catalytic application in the SCR of NOx. ChemCatChem 5:3316–3323
20. Gomez-Hortigüela L, Perez-Pariente J, Cora F, Catlow RA, Blasco T (2005) Structure-directing role of molecules containing benzyl rings in the synthesis of a large-pore

aluminophosphate molecular sieve: an experimental and computational study. J Phys Chem B 109:21539–21548

21. Beck JS, Vartuli JC, Kennedy GJ, Kresge CT, Roth WJ, Schrammt SE (1994) Molecular or supramolecular templating: defining the role of surfactant chemistry in the formation of microporous and mesoporous molecular sieves. Chem Mater 6:1816–1821

22. Choi M, Na K, Kim J, Sakamoyo Y, Terasaki O, Ryoo R (2009) Stable single-unit-cell nanosheets of zeolite MFI as active and long-lived catalysts. Nature 461:246–249

23. Sherman J (2003) Molecules that can't resist templation. Chem Commun 14:1617–1623

24. Burkett SL, Davis ME (1993) Structure-directing effects in the crown ether-mediated syntheses of FAU and EMT zeolites. Microporous Mater 1:265–282

25. Parise JB, Corbin DR, Abrams L, Cox DE (1984) Structure of dealuminated Linde Y-zeolite: Si139.7Al52.3O384 and Si173.1Al18.9O384: presence of non-framework Al species. Acta Cryst C40:1493–1497

26. Vogt ETC, Weckhuysen BM (2015) Fluid catalytic cracking: recent developments on the grand old lady of zeolite catalysis. Chem Soc Rev 44:7342–7370

27. Woltermann GM, Magee JS, Griffith SD (1993) Commercial preparation and characterization of FCC catalysts. Stud Surf Sci Catal 76:105–144

28. Anderson MW, Pachis KS, Prebin F, Carr SW, Terasaki O, Ohsuna T, Alfreddson V (1991) Intergrowths of cubic and hexagonal polytypes of faujasitic zeolites. J Chem Soc Chem Commun 23:1660–1664

29. Dougnier F, Patarin J, Guth JL, Anglerot D (1992) Synthesis, characterization, and catalytic properties of silica-rich faujasite-type zeolite (FAU) and its hexagonal analog (EMT) prepared by using crown-ethers as templates. Zeolites 12:160–166

30. Dwyer J, Karim K (1992) Synthesis of ZSM-20 using crown ethers. Zeolites 12:412–414

31. Gonzalez G, Soraya-Gonzalez C, Stracke W, Reichelt R, Garcia L (2007) New zeolite topologies based on intergrowths of FAU/EMT systems. Microporous Mesoporous Mater 101:30–42

32. Feijen EJP, De Vadder K, Bosschaerts MH, Lievens JL, Martens JA, Crobet PJ, Jacobs PA (1994) Role of 18-crown-6 and 15-crown-5 ethers in the crystallization of polytype faujasite zeolites. J Am Chem Soc 116:2950–2957

33. Baerlocher C, McCusker L, Chiappetta R (1994) Location of the 18crown-6 template in EMC-2 (EMT) Rietveld refinement of the calcined and as-synthesized forms. Microporous Mater 2:269–280

34. Balkus Jr KJ, Shi J (1997) Synthesis of hexagonal Y type zeolites in the presence of Gd(III) complexes of 18-crown-6. Microporous Mater 11:325–333

35. Chatelain T, Patarin J, Fousson E, Soulard M, Guth JL, Schulz P (1995) Synthesis and characterization of high-silica zeolite RHO prepared in the presence of 18-crown-6 ether as organic template. Microporous Mater 4:231–238

36. Araki S, Kiyohara Y, Tanaka S, Miyake Y (2012) Crystallization process of zeolite rho prepared by hydrothermal synthesis using 18-crown-6 ether as organic template. J Colloid Interface Sci 376:28–33

37. Ke Q, Sun T, Cheng H, Chen H, Liu X, Wei X, Wang S (2017) Targeted synthesis of ultrastable high-silica RHO zeolite through alkali metal-crown ether interaction. Chem Asian J 12(10):1043–1047. https://doi.org/10.1002/asia.201700303

38. Chatelain T, Patarin J, Farre R, Petigny O, Schulz P (1996) Synthesis and characterization of 18-crown-6 ether-containing KFI-type zeolite. Zeolites 17:328–333

39. Shantz DF, Burton A, Lobo RF (1999) Synthesis, structure solution, and characterization of the aluminosilicate MCM-61: the first aluminosilicate clathrate with 18-membered rings. Microporous Mesoporous Mater 31:61–73

40. Robson HE, Shoemaker DP, Ogilvie RA, Manor PC (1973) Synthesis and crystal structure of zeolite rho – a new zeolite related to linde type A. Adv Chem Ser 121:106–115

41. Barrer RM (1948) Synthesis of a zeolitic mineral with chabazite-like sorptive properties. J Chem Soc 2:127–132

42. Barrer RM, Hinds L, White EA (1953) The hydrothermal chemistry of silicates. Part III. Reactions of analcite and leucite. J Chem Soc 2:1466–1475

43. Valyocsik EW (1997) US Patent 5,670,131

44. Valyocsik EW (1986) US Patent 4,568,654

45. van de Goor G, Lindlar B, Felsche J, Behrens P (1995) Solvent-free synthesis of clathrasils using metal-organic complexes as structure-directing agents. J Chem Soc Chem Commun 24:2559–2561

46. Lobo RF, Tsapatsis M, Freyhardt CC, Khodabandeh S, Wagner P, Chen CY, Balkus Jr KJ, Zones SI, Davis ME (1997) Characterization of the extra-large-pore zeolite UTD-1. J Am Chem Soc 119:8474–8484

47. Balkus Jr KJ, Gabrielov AG, Shepelev S (1995) Synthesis of aluminum phosphate molecular sieves using cobalticinium hydroxide. Microporous Mater 3:489–495

48. Corma A, Garcia H (2004) Supramolecular host-guest systems in zeolites prepared by ship-in-a-bottle synthesis. Eur J Inorg Chem 6:1143–1164

49. Balkus Jr KJ, Kowalak S, Ly KT, Hargis DC (1991) Zeolite synthesis with metal chelate complexes. Stud Surf Sci Catal 69:93–99

50. Zhan BZ, Li XY (1998) A novel 'build-bottle-around-ship' method to encapsulate metalloporphyrins in zeolite-Y. An efficient biomimetic catalyst. Chem Commun 3:349–350

51. Garcia R, Shannon IJ, Slawin AMZ, Zhou W, Cox PA, Wright PA (2003) Synthesis, structure and thermal transformations of aluminophosphates containing the nickel complex [Ni (diethylenetriamine)2]2+ as a structure directing agent. Microporous Mesoporous Mater 58:91–104

52. Garcia R, Coombsa TD, Shannon IJ, Wright PA, Cox PA (2003) Nickel amine complexes as structure-directing agents for aluminophosphate molecular sieves: a new route to supported nickel catalysts. Top Catal 24:115–124

53. Cui Y, Tong X, Li Y, Chen M, Zhou W, Ren S, Li L, Yan Z, Zhu L (2017) One-pot synthesis of Ni-SSZ-13 zeolite using a nickel-amine complex as an efficient organic template. J Mater Sci 52(17):10156–10162. https://doi.org/10.1007/s10853-017-1171-x

54. Finiels A, Fajula F, Hulea V (2014) Nickel-based solid catalysts for ethylene oligomerization – a review. Cat Sci Technol 4:2412–2426

55. Ren L, Zhu L, Yang C, Chen Y, Sun Q, Zhang H, Li C, Nawaz F, Meng X, Xiao FS (2011) Designed copper-amine complex as an efficient template for one-pot synthesis of Cu-SSZ-13 zeolite with excellent activity for selective catalytic reduction of NOx by NH3. Chem Commun 47:9789–9791

56. Xie L, Liu F, Ren L, Shi X, Xiao FS, He H (2014) Excellent performance of one-pot synthesized Cu-SSZ-13 catalyst for the selective catalytic reduction of NOx with NH3. Environ Sci Technol 48:566–572

57. Picone AL, Warrender SJ, Slawin AMZ, Dawson DM, Ashbrook SE, Wright PA, Thompson SP, Gaberova L, Llewellyn PL, Moulin B, Vimont A, Daturi M, Park MB, Soung SK, Name IS, Hong SB (2011) A co-templating route to the synthesis of Cu SAPO STA-7, giving an active catalyst for the selective catalytic reduction of NO. Microporous Mesoporous Mater 146:36–47

58. Martinez-Franco R, Moliner M, Concepcion P, Thogersen JR, Corma A (2014) Synthesis, characterization and reactivity of high hydrothermally stable Cu-SAPO-34 materials prepared by "one-pot" processes. J Catal 314:73–82

59. Martin N, Moliner M, Corma A (2015) High yield synthesis of high-silica chabazite by combining the role of zeolite precursors and tetraethylammonium: SCR of NOx. Chem Commun 51:9965–9968

60. Martinez-Franco R, Moliner M, Franch C, Kustov A, Corma A (2012) Rational direct synthesis methodology of very active and hydrothermally stable CuSAPO34 molecular sieves for the SCR of NOx. Appl Catal B 127:273–280

61. Martinez-Franco R, Moliner M, Corma A (2014) Direct synthesis design of Cu-SAPO-18, a very efficient catalyst for the SCR of NOx. J Catal 319:36–43

62. Zheng N, Stucky GD (2006) A general synthetic strategy for oxide-supported metal nanoparticle catalysts. J Am Chem Soc 128:14278–14280
63. Brust M, Fink J, Bethell D, Schiffrin DJ, Kiely C (1995) Synthesis and reactions of functionalised gold nanoparticles. J Chem Soc Chem Commun 16:1655–1656
64. Choi M, Wu Z, Iglesia E (2010) Mercaptosilane-assisted synthesis of metal clusters within zeolites and catalytic consequences of encapsulation. J Am Chem Soc 132:9129–9137
65. Moliner M, Gabay JE, Kliewer CE, Carr RT, Guzman J, Casty GL, Serna P, Corma A (2016) Reversible transformation of Pt nanoparticles into single atoms inside high-silica chabazite zeolite. J Am Chem Soc 138:15743–15750
66. Otto T, Zones SI, Iglesia E (2016) Challenges and strategies in the encapsulation and stabilization of monodisperse Au clusters within zeolites. J Catal 339:195–208
67. Wu Z, Goel S, Choi M, Iglesia E (2014) Hydrothermal synthesis of LTA-encapsulated metal clusters and consequences for catalyst stability, reactivity, and selectivity. J Catal 311:458–468
68. Abate S, Barbera K, Centi G, Lanzafame P, Perathoner S (2016) Disruptive catalysis by zeolites. Cat Sci Technol 6:2485–2501
69. Kerr GT (1966) Chemistry of crystalline aluminosilicates II. The synthesis and properties of zeolite ZK-4. Inorg Chem 5:1537–1539
70. Hunt HK, Lew CM, Sun M, Yan Y, Davis ME (2010) Pure-silica zeolite thin films by vapor phase transport of fluoride for low-k applications. Microporous Mesoporous Mater 128:12–18
71. Hunt HK, Lew CM, Sun M, Yan Y, Davis ME (2010) Pure-silica LTA, CHA, STT, ITW and SVR thin films and powders for low-k applications. Microporous Mesoporous Mater 130:49–55
72. Martinez-Franco R, Cantin A, Moliner M, Corma A (2014) Synthesis of the small pore silicoaluminophosphate STA-6 by using supramolecular self-assembled organic structure directing agents. Chem Mater 26:4346–4353
73. Martinez-Franco R, Cantin A, Vidal-Moya A, Moliner M, Corma A (2015) Self-assembled aromatic molecules as efficient organic structure directing agents to synthesize the silicoaluminophosphate SAPO-42 with isolated Si species. Chem Mater 27:2981–2989
74. Tian P, Wei Y, Ye M, Liu Z (2015) Methanol to olefins (MTO): from fundamentals to commercialization. ACS Catal 5:1922–1938
75. Beale AM, Gao F, Lezcano-Gonzalez I, Peden CH, Szanyi J (2015) Recent advances in automotive catalysis for NOx emission control by small-pore microporous materials. Chem Soc Rev 44:7371–7405
76. Martinez-Franco R, Li Z, Martinez-Triguero J, Moliner M, Corma A (2016) Improving the catalytic performance of SAPO-18 for the methanol-to-olefins (MTO) reaction by controlling the Si distribution and crystal size. Cat Sci Technol 6:2796–2806
77. Gomez-Hortigüela L, Blasco T, Perez-Pariente J (2007) (S)-( )-N-benzylpyrrolidine-2-methanol: a new and efficient structure directing agent for the synthesis of crystalline microporous aluminophosphates with AFI-type structure. Microporous Mesoporous Mater 100:55–62
78. Gomez-Hortigüela L, Alvaro-Muñoz T, Bernardo-Maestro B, Perez-Pariente J (2015) Towards chiral distributions of dopants in microporous frameworks: helicoidal supramolecular arrangement of (1R,2S)-ephedrine and transfer of chirality. Phys Chem Chem Phys 17:348–357
79. Gomez-Hortigüela L, Lopez-Arbeloa F, Cora F, Perez-Pariente J (2008) Supramolecular chemistry in the structure direction of microporous materials from aromatic structure-directing agents. J Am Chem Soc 130:13274–13284
80. Gomez-Hortigüela L, Hamad S, Lopez-Arbeloa F, Pinar AB, Perez-Pariente J, Cora F (2009) Molecular insights into the self-aggregation of aromatic molecules in the synthesis of nanoporous aluminophosphates: a multilevel approach. J Am Chem Soc 131:16509–16524
81. Gomez-Hortigüela L, Garcia R, Lopez-Arbeloa F, Cora F, Perez-Pariente J (2010) Structure directing effect of (1S,2S)-2-hydroxymethyl-1-benzyl-1-methylpyrrolidinium in the synthesis of AlPO-5. J Phys Chem C 114:8320–8327

82. Bernardo-Maestro B, Lopez-Arbeloa F, Perez-Pariente J, Gomez-Hortigüela L (2015) Supramolecular chemistry controlled by conformational space during structure direction of nanoporous materials: self-assembly of ephedrine and pseudoephedrine. J Phys Chem C 119:28214–28225

83. Bernardo-Maestro B, Roca-MOreno MD, Lopez-Arbeloa F, Perez-Pariente J, Gomez-Hortigüela L (2016) Supramolecular chemistry of chiral (1R,2S)-ephedrine confined within the AFI framework as a function of the synthesis conditions. Catal Today 277:9–20

84. Alvaro-Muñoz T, Lopez-Arbeloa F, Perez-Pariente J, Gomez-Hortigüela L (2014) (1R,2S)-ephedrine: a new self-assembling chiral template for the synthesis of aluminophosphate frameworks. J Phys Chem C 118:3069–3077

85. Corma A, Díaz-Cabañas MJ, Jorda JL, Martínez C, Moliner M (2006) High-throughput synthesis and catalytic properties of a molecular sieve with 18- and 10-member rings. Nature 443:842–845

86. Sun J, Bonneau C, Cantin A, Corma A, Diaz-Cabañas MJ, Moliner M, Zhang D, Li M, Zou X (2009) The ITQ-37 mesoporous chiral zeolite. Nature 458:1154–1157

87. Corma A, Diaz-Cabañas MJ, Jiang J, Afeworki M, Dorset DL, Soled SL, Strohmaier KG (2010) Extra-large pore zeolite (ITQ-40) with the lowest framework density containing double four- and double three-rings. Proc Natl Acad Sci U S A 107:13997–14002

88. Jiang J, Jorda JL, Yu J, Baumes LA, Mugnaioli E, Diaz-Cabañas MJ, Kolb U, Corma A (2011) Synthesis and structure determination of the hierarchical meso-microporous zeolite ITQ-43. Science 333:1131–1134

89. Jiang J, Yun Y, Zou X, Jorda JL, Corma A (2015) ITQ-54: a multi-dimensional extra-large pore zeolite with 20 × 14 × 12-ring channels. Chem Sci 6:480–485

90. Brunner GO, Meier WM (1989) Framework density distribution of zeolite-type tetrahedral nets. Nature 337:146–147

91. Chen FJ, Xu Y, Du HB (2014) An extra-large-pore zeolite with intersecting 18-, 12-, and 10-membered ring channels. Angew Chem Int Ed 53:9592–9596

92. Gao ZH, Chen FJ, Xu L, Sun L, Xu Y, Du HB (2016) A stable extra-large-pore zeolite with intersecting 14- and 10-membered-ring channels. Chem Eur J 22:14367–14372

93. Chen FJ, Gao ZH, Liang LL, Zhang J, Du HB (2016) Facile preparation of extra-large pore zeolite ITQ-37 based on supramolecular assemblies as structure-directing agents. CrystEngComm 18:2735–2741

94. Eliášová P, Opanasenko M, Wheatley PS, Shamzhy M, Mazur M, Nachtigall P, Roth WJ, Morris RE, Cejka J (2015) The ADOR mechanism for the synthesis of new zeolites. Chem Soc Rev 44:7177–7206

95. Martinez-Franco R, Sun J, Sastre G, Yun Y, Zou X, Moliner M, Corma A (2014) Supramolecular assembly of aromatic proton sponges to direct the crystallization of extra-large-pore zeotypes. Proc Math Phys Eng Sci 470(2166):20140107

96. Staab HA, Saupe T (1988) "Proton sponges" and the geometry of hydrogen bonds: aromatic nitrogen bases with exceptional basicities. Angew Chem Int Ed 27:865–879

97. Martinez-Franco R, Moliner M, Yun Y, Sun J, Wan W, Zou X, Corma A (2013) Synthesis of an extra-large molecular sieve using proton sponges as organic structure-directing agents. Proc Natl Acad Sci U S A 110:3749–3754

98. Martínez-Franco R, Paris C, Moliner M, Corma A (2016) Synthesis of highly stable metal-containing extra-large-pore molecular sieves. Phil Trans R Soc A 374:20150075

99. Shinoda K, Hato M, Hayashi T (1972) Physicochemical properties of aqueous solutions of fluorinated surfactants. J Phys Chem 72:909–914

100. Milioto S, Crisantino R, De Lisi R, Inglese A (1995) Apparent molar volumes of some hydrogenated and fluorinated alcohols in sodium dodecanoate and sodium perfluorooctanoate aqueous solutions. Langmuir 11:718–724

101. Gomez-Hortigüela L, Cora F, Perez-Pariente J (2008) Supramolecular assemblies of fluoro-aromatic organic molecules as structure directing agents of microporous materials: different effects of fluorine. Microporous Mesoporous Mater 109:494–504

102. Gomez-Hortigüela L, Perez-Pariente J, Blasco T (2005) Fluorine-containing organic molecules as structure directing agents in the synthesis of crystalline microporous materials. Part I: Synthesis of AlPO4-5 and SAPO-5 from fluorobenzyl-pyrrolidine. Microporous Mesoporous Mater 78:189–197

103. Gomez-Hortigüela L, Marquez-Alvarez C, Cora F, Lopez-Arbeloa F, Perez-Pariente J (2008) Cooperative effect of hydroxide and fluorinated organic ions as structure directing agent in the synthesis of crystalline microporous aluminophosphates. Chem Mater 20:987–995

104. Gomez-Hortigüela L, Marquez-Alvarez C, Sastre E, Cora F, Perez-Pariente J (2006) Effect of fluorine-containing chiral templates on Mg distribution in the structure of MgAPO-5 and its influence on catalytic activity. Catal Today 114:174–182

105. Kresge CT, Leonowicz ME, Roth WJ, Vartuli JC, Beck JS (1992) Ordered mesoporous molecular sieves synthesized by a liquid-crystal template mechanism. Nature 359:710–712

106. Beck JS, Vartuli JC, Roth WJ, Leonowicz ME, Kresge CT, Schmitt KD, Chu CTW, Olson DH, Sheppard EW, McCullen SB, Higgins JB, Schlenker JL (1992) A new family of mesoporous molecular sieves prepared with liquid crystal templates. J Am Chem Soc 114:10834–10843

107. Chal R, Gerardin C, Bulut M, van Donk S (2011) Overview and industrial assessment of synthesis strategies towards zeolites with mesopores. ChemCatChem 3:67–81

108. Vartuli JC, Shih SS, Kresge CT, Beck JS (1998) Potential applications for M41S type mesoporous molecular sieves. Stud Surf Sci Catal 117:13–21

109. Kloetstra KR, Zandbergen HW, Jansen JC, van Bekkum H (1996) Overgrowth of mesoporous MCM-41 on faujasite. Microporous Mater 6:287–293

110. Karlsson A, Stöcker M, Schmidt R (1999) Composites of micro- and mesoporous materials: simultaneous syntheses of MFI/MCM-41 like phases by a mixed template approach. Microporous Mesoporous Mater 27:181–192

111. Huang L, Guo W, Deng P, Xue Z, Li Q (2000) Investigation of synthesizing MCM-41/ZSM-5 composites. J Phys Chem B 104:2817–2823

112. Vernimmen J, Meynen V, Herregods SJF, Mertens M, Lebedev OI, Van Tendeloo G, Cool P (2011) New insights in the formation of combined zeolitic/mesoporous materials by using a one-pot templating synthesis. Eur J Inorg Chem 27:4234–4240

113. Choi M, Cho HS, Srivastava R, Venkatesan C, Choi DH, Ryoo R (2006) Amphiphilic organosilane-directed synthesis of crystalline zeolite with tunable mesoporosity. Nat Mater 5:718–723

114. Srivastava R, Choi M, Ryoo R (2006) Mesoporous materials with zeolite framework: remarkable effect of the hierarchical structure for retardation of catalyst deactivation. Chem Commun 43:4489–4491

115. Choi M, Srivastava R, Ryoo R (2006) Organosilane surfactant-directed synthesis of mesoporous aluminophosphates constructed with crystalline microporous frameworks. Chem Commun 42:4380–4382

116. Inayat A, Knoke I, Spiecker E, Schwieger W (2012) Assemblies of mesoporous FAU-type zeolite nanosheets. Angew Chem Int Ed 51:1962–1965

117. Garcia-Martinez J, Johnson M, Valla J, Li K, Ying JY (2012) Mesostructured zeolite Y – high hydrothermal stability and superior FCC catalytic performance. Cat Sci Technol 2:987–994

118. Prasomsri T, Jiao W, Weng SZ, Garcia-Martinez J (2015) Mesostructured zeolites: bridging the gap between zeolites and MCM-41. Chem Commun 51:8900–8911

119. Garcia-Martinez J, Li K, Krishnaiah G (2012) A mesostructured Y zeolite as a superior FCC catalyst – from lab to refinery. Chem Commun 48:11841–11843

120. Mintova S, Grand J, Valtchev V (2016) Nanosized zeolites: quo vadis? C R Chim 19:183–191

121. Corma A, Fornes V, Pergher SB, Maesen TLM, Buglass JG (1998) Delaminated zeolite precursors as selective acidic catalysts. Nature 396:353–356

122. Corma A, Diaz U, Domine ME, Fornes V (2000) AlITQ-6 and TiITQ-6: synthesis, characterization, and catalytic activity. Angew Chem Int Ed 39:1499–1501

123. Rothw WJ, Cejka J (2011) Two-dimensional zeolites: dream or reality? Cat Sci Technol 1:43–53
124. Diaz U (2012) Layered materials with catalytic applications: pillared and delaminated zeolites from MWW precursors. ISRN Chem Eng 2012:35. Article ID 537164
125. Kim J, Park W, Ryoo R (2011) Surfactant-directed zeolite nanosheets: a high-performance catalyst for gas-phase Beckmann rearrangement. ACS Catal 1:337–341
126. Na K, Jo C, Kim J, Ahn WS, Ryoo R (2011) MFI titanosilicate nanosheets with single-unit-cell thickness as an oxidation catalyst using peroxides. ACS Catal 1:901–907
127. Wang J, Xu L, Peng H, Wu H, Jiang J, Liu Y, Wu P (2012) Multilayer structured MFI-type titanosilicate: synthesis and catalytic properties in selective epoxidation of bulky molecules. J Catal 288:16–23
128. Luo HY, Bui L, Gunther WR, Min E, Román-Leshkov Y (2012) Synthesis and catalytic activity of Sn-MFI nanosheets for the Baeyer-Villiger oxidation of cyclic ketones. ACS Catal 2:2695–2699
129. Kim W, Kim JC, Kim J, Seo Y, Ryoo R (2013) External surface catalytic sites of surfactant-tailored nanomorphic zeolites for benzene isopropylation to cumene. ACS Catal 3:192–195
130. Luo HY, Michaelis VK, Hodges S, Griffin RG, Román-Leshkov Y (2015) One-pot synthesis of MWW zeolite nanosheets using a rationally designed organic structure-directing agent. Chem Sci 6:6320–6324
131. Margarit VJ, Martinez-Armero M, Navarro MT, Martinez C, Corma A (2015) Direct dual-template synthesis of MWW zeolite monolayers. Angew Chem Int Ed 54:13724–13728
132. Seo Y, Lee S, Jo C, Ryoo R (2013) Microporous aluminophosphate nanosheets and their nanomorphic zeolite analogues tailored by hierarchical structure-directing amines. J Am Chem Soc 135:8806–8809
133. Gallego EM, Portilla MT, Paris C, León-Escamilla A, Boronat M, Moliner M, Corma A (2017) "Ab initio" synthesis of zeolites for preestablished catalytic reactions. Science 355:1051–1054
134. Diaz-Cabañas MJ, Camblor MA, Corell C, Corma A (2000) US Patent 6,077,498
135. Sastre G, Catlow RA, Corma A (1999) Diffusion of benzene and propylene in MCM-22 zeolite. A molecular dynamics study. J Phys Chem B 103:5187–5196
136. Xu D, Ma Y, Jing Z, Han L, Singh B, Feng J, Shen X, Cao F, Oleynikov P, Sun H, Terasaki O, Che S (2014) π–π interaction of aromatic groups in amphiphilic molecules directing for single-crystalline mesostructured zeolite nanosheets. Nat Commun 5:4262
137. Xu D, Jing Z, Cao F, Sun H, Che S (2014) Surfactants with aromatic-group tail and single quaternary ammonium head for directing single-crystalline mesostructured zeolite nanosheets. Chem Mater 26:4612–4619
138. Schmidt JE, Deem MW, Davis ME (2014) Synthesis of a specified, silica molecular sieve by using computationally predicted organic structure-directing agents. Angew Chem Int Ed 53:8372–8374
139. Moliner M, Serna P, Cantín A, Sastre G, Díaz-Cabañas MJ, Corma A (2008) Synthesis of the Ti-silicate form of BEC polymorph of B-zeolite assisted by molecular modeling. J Phys Chem C 112:19547–19554
140. Brand SK, Schmidt JE, Deem MW, Daeyaert F, Ma Y, Terasaki O, Orazov M, Davis ME (2017) Enantiomerically enriched, polycrystalline molecular sieves. Proc Natl Acad Sci U S A 114:5101–5106
141. Li D, Yu J (2016) AIEgens-functionalized inorganic-organic hybrid materials: fabrications and applications. Small 12:6478–6494
142. Gigli L, Arletti R, Tabacchi G, Fois E, Vitillo JG, Martra G, Agostini G, Quartieri S, Vezzalini G (2014) Close-packed dye molecules in zeolite channels self-assemble into supramolecular nanoladders. J Phys Chem C 118:15732–15743
143. Sola-Llano R, Martínez-Martínez V, Fujita Y, Gomez-Hortigüela L, Alfayate A, Uji-i H, Perez-Pariente J, Lopez-Arbeloa I (2016) Formation of a nonlinear optical host-guest hybrid material by tight confinement of LDS 722 into aluminophosphate 1D nanochannels. Chem Eur J 22:15700–15711

Struct Bond (2018) 175: 179–200
DOI: 10.1007/430_2017_12
© Springer International Publishing AG 2017
Published online: 27 October 2017

Metal Complexes as Structure-Directing Agents for Zeolites and Related Microporous Materials

Abigail E. Watts, Alessandro Turrina, and Paul A. Wright

Abstract Metal complexes can act as structure-directing agents (SDAs) for zeolites and zeotypes, either alone or together with additional SDAs in dual-templating approaches. Such complexes include organometallic cobaltocenium ions, alkali metal crown ether complexes, first-row transition-metal (Fe, Co, Ni, Cu) polyamines and thiol-complexed second- and third-row transition metals (Pd, Pt). Their inclusion has been demonstrated in some cases by crystallographic methods but more commonly by spectroscopy (UV-visible, X-ray absorption, Mössbauer). The unique feature of this class of template is that they can not only direct crystallisation but also give solids with homogeneously distributed metal cations or metal oxide species upon calcination, precluding the need for an additional post-synthesis modification step. Materials prepared via this 'one-pot' synthetic route have been shown to give shape-selective catalysts for reactions such as the selective catalytic reduction of NO_x with ammonia and the hydrogenation, dehydration and oxidative dehydrogenation of small hydrocarbons and oxygenates.

Keywords Catalysts · Metal complexes · One-pot synthesis · Structure-directing agents · Templates · Zeolites · Zeotypes

Contents

A.E. Watts and P.A. Wright (✉)
EaStCHEM School of Chemistry, University of St Andrews, Purdie Building, North Haugh,
St Andrews KY16 9ST, UK
e-mail: paw2@st-andrews.ac.uk

A. Turrina
Johnson Matthey Technology Centre, PO BOX 1, Belasis Avenue, Chilton, Billingham TS23
1LB, UK

1   Introduction

One of the defining aspects of the structural chemistry of aluminosilicate zeolites and zeotypes (zeotypes have tetrahedrally connected 'zeolitic' frameworks but different framework compositions) is the variety of different framework types that can be prepared via their hydrothermal synthesis [1]. These are, in their open, porous forms, metastable with respect to denser phases with the same framework composition but are synthetically accessible when species from the synthesis mixture occupy the pores of the crystallising zeolites and stabilise them. Such guests include hydrated metal cations (and water itself), organic molecules and, the subject of this chapter, metal complexes. While the interactions in all cases act to increase the overall thermodynamic stability of the framework, it is convenient to define structure-directing agents (SDAs) as those that strongly favour the crystallisation of one framework over another, whereas in other cases, guest species can play a less specific framework-stabilising role, sometimes described as void filling. Alkali and alkaline earth metal cations have been investigated extensively and can demonstrate specificity (e.g. potassium cations are observed to direct the cancrinite cage in zeolites L and offretite), but it has been the successful exploitation of organics (especially alkylammonium cations, but also amines and even phosphonium ions) that has resulted in the steady growth in the number of new zeolitic silicate and aluminophosphate structures over the last decades. By contrast, relatively little attention has been paid to metal complexes as SDAs.

Although surprising at first glance, given the wealth of information available on metal complexes in the field of coordination and organometallic chemistry, this reflects the relatively low stability of many complexes under the conditions of zeolite synthesis. There is more literature on the post-synthetic assembly of metal complexes within zeolite pores, exemplified by the metal-templated synthesis of phthalocyanines [2] and the preparation of manganese complexes of trimethyltriazacyclononane in zeolite Y [3]. The aim of such preparations is to generate catalytically active species too large to be able to leave the zeolite pores (so-called ship-in-bottle approach) but still leave enough room for the access of substrates and the exit of products. By their nature, SDAs will instead interact closely with the framework and leave little additional space, so there must be different reasons for using metal complexes as SDAs. It can be that their characteristic geometry leads to a new framework type or, more commonly, that their use provides a route to the homogeneous dispersion of metal species (either cationic or metallic) throughout the framework and so to novel catalysts.

Here, we address the inclusion chemistry of metal complexes during the synthesis of different classes of zeolitic solids and establish the extent to which there is structure direction by the complexes. We also discuss the properties of these materials if and when the complexes are thermally decomposed. Any account of the use of metal complexes requires a consideration of their stability under crystallisation conditions, so that we first outline the conditions required to form the most important categories of zeolites and zeotypes. It is then possible to discuss the templating action of the categories of complex that have been reported. The characterisation of these materials is crucial to establishing the role of the complex and their catalytic properties once calcined, so this aspect is also emphasised.

2 Crystallisation Conditions and Charge Balance Considerations

Metal complexes that bear a positive charge can act as SDAs in the same way as alkylammonium cations, so that it is their stability under hydrothermal conditions that is the limiting factor in their applicability. Typically, aluminosilicate zeolites crystallise at temperatures between 100 and 200°C at strongly alkaline pH (10–12), ruling out most organometallics. Under these conditions, the competing ligands are water molecules, hydroxide ions and also aluminate and silicate species in the gel or in the forming framework. Cations such as Ti^{4+}, Cr^{3+}, Fe^{3+} and Zn^{2+} have all been found to substitute for Si in tetrahedral framework sites [4–7]. For other cations, the competition between the complexing ligands and hydroxide species, which may also have low solubility, is the determining factor. Silicas and aluminosilicates also crystallise in fluoride-rich aqueous solutions, where fluoride ions act as mineralisers for silica in place of hydroxide ions [8]. As a consequence, these syntheses take place under near-neutral conditions, and so fluoride ions rather than hydroxide ions are the competing ligands for the metal cations.

Aluminophosphate-based zeotypes and their substituted analogues (MAPOs, M = Mg, Mn, Fe, Co, Zn and SAPOs, silicoaluminophosphates) are also widely studied [1, 9, 10]. These crystallise at solution pH values close to 7, often more rapidly than their zeolite analogues, and the MAPOs crystallise the most quickly of these. As a result of these crystallisation conditions, metal complexes are more likely to remain intact long enough to act as SDAs. The forming aluminophosphate frameworks also compete for the metal cations and show a different selectivity over silicates because of their greater ionicity [11]. The transition metals Mn^{2+}, Fe^{3+}, Co^{2+} and Zn^{2+} all show a strong tendency to adopt framework cation locations in MAPOs, in part due to their ability to adopt tetrahedral coordination geometry [12, 13]. By contrast, Cu^{2+} and Ni^{2+} do not favour tetrahedral sites and are not observed to enter the framework cation sites of MAPO or SAPO zeotypes.

3   Different Classes of Metal Complex SDAs

3.1   Cobaltocenium Ions: Organometallic SDAs

Cationic cobaltocenium (alternatively, cobalticinium) complexes have been found to act as effective SDAs for clathrasils and high-silica zeolites, particularly in fluoride-mediated syntheses. They are by far the most important organometallic complexes reported as templates, which is a consequence of their high thermal and chemical stability (chemically, they can usefully be considered as very large alkali metal cations). Examples of clathrasils (silicas with cage structures) templated by cobaltocenium $[Co(\eta^5 - C_5H_5)_2]^+$, $[Co(Cp)_2]^+$, include the NON, AST and DOH frameworks reported by van de Goor et al. [14], in which examples the complexes were located crystallographically in the cages. The three letter codes are those designated by the International Zeolite Association Structure Commission to denote unique topology types [15]. Part of the cage structure of cobaltocenium nonasil is shown in Fig. 1. Cobaltocenium ions have also been found to give porous zeolites and zeotypes. Behrens reported the crystallisation of the 10R (comprising ten tetrahedral cations and ten oxygen atoms) one-dimensional channel alumino-silicate zeolite ZSM-48 [16] and Valyocsik the formation of a levynite-like zeolite ZSM-45 (LEV) [17].

The most remarkable example of this templating approach was the synthesis by Balkus of the first extra-large pore, high-silica zeolite UTD-1 (University of Texas, Dallas) in the presence of permethylcobaltocenium, $[Co(Cp(CH_3)_5)_2]^+$ [18]. UTD-1 contains one-dimensional channels bounded by 14Rs and is the only new topology type obtained through the use of organometallic SDAs. The cobalt complex can be removed by calcination; however, TEM shows that a few clusters of cobalt oxide remain. Washing the calcined material with hydrochloric acid removes all the detectable cobalt. This leaves, in the case of the aluminosilicate, an extra-large-pore Brønsted acid catalyst.

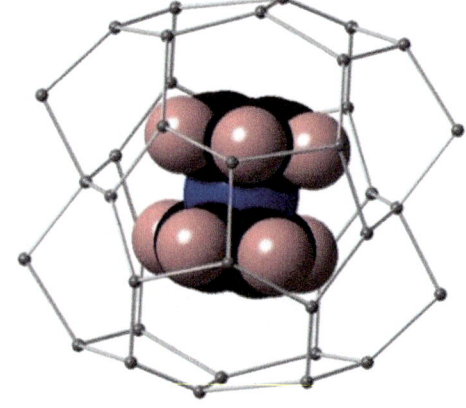

Fig. 1 Large cage of the clathrasil nonasil (NON) containing the cobaltocenium cation $[Co(Cp)_2]^+$. (Framework Si atoms represented by grey spheres, linked, with oxygen atoms of the framework omitted for clarity. Cobalt atom, blue; C, black; H, white.) Fractional atomic coordinates taken from van de Goor et al. [14]

Lobo et al. carried out structural characterisation on the calcined form of UTD-1 [19]. The authors report that the powder X-ray diffraction patterns for UTD-1 indicate that the material is an intergrowth of polymorphs, with TEM and electron diffraction showing that UTD-1 has fault planes parallel to the axis of the 14R channels. With the aim of improving the quality of UTD-1 product, Wessels et al. carried out its synthesis in a fluoride medium. The resulting material was denoted UTD-1F, and no evidence for structural disorder was found, in contrast to the calcined form of conventional UTD-1 [20]. Figure 2 shows the arrangement of the $[Co(Cp(CH_3)_5)_2]^+$ complexes in UTD-1F.

The discovery of UTD-1 opened up the possibility of the preparation of stable, extra-large-pore zeolites capable of catalysing conversions of large hydrocarbons too bulky to gain access to the internal acid sites of 12R zeolites such as beta and Y. Subsequently, other extra-large-pore zeolites have been prepared using more readily available alkylammonium ions [21].

Cobaltocenium cations have also been reported to direct the crystallisation of aluminophosphates AlPO-5 and AlPO-16 from fluoride-based syntheses [22, 23]. In these materials, charge balance is achieved by the presence of fluoride ions coordinated to Al^{3+} cations in otherwise neutral frameworks.

Further, Warrender showed that cobaltocenium ions can template MgAPO-35, which has the LEV topology type and where the positive charge on the cation is balanced by the negative charge on the framework resulting from Mg^{2+}/Al^{3+} substitution [24]. In the as-prepared state, these materials, in which around 80% of the cages are occupied by cobaltocenium ions, display its characteristic yellow colour. Calcination removes the organic cyclopentadienyl groups, leaving a purple solid with absorbance ($\lambda = 490, 580$ nm) characteristic of octahedral Co^{3+} [25]. Attempts to prepare AlPOs using the larger permethylcobaltocenium ion have not yet been successful, however.

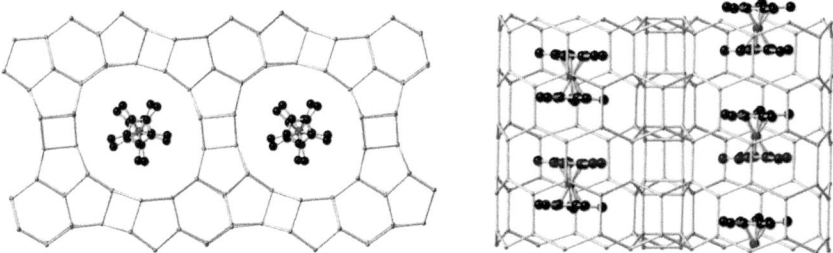

Fig. 2 $[Co(Cp(CH_3)_5)_2]^+$ complexes in the 14-ring channels of UTD-1F. The structure is viewed along (left) and perpendicular to (right) the 14R channels. (Si atoms, grey; Co atoms, blue; C atoms, black; O and H atoms omitted.) Fractional atomic coordinates taken from Wessels et al. [20]

3.2   Crown Ether Complexes of Alkali Metal Cations

The use of crown ethers (macrocycles with oxygens as donor atoms) as SDAs was first reported in 1990 by Delprato et al. [26]. The crystallisation of faujasite-type aluminosilicates with Si/Al ratios greater than 3 was investigated using different macrocycles. The successful macrocycles had to withstand the synthetic conditions used and give a good fit to the faujasite supercage. Crown ethers were found to be suitable as structure-directing agents, and 15-crown-5 (1,4,7,10,13-pentaoxacyclopentadecane) produced cubic zeolite Y or faujasite (FAU) with a Si/Al ratio of approximately 5. Delprato et al. reported that there are eight 15-crown-5 molecules in the cubic unit cell, equating to one 15-crown-5 molecule per faujasite supercage. In the same study, 18-crown-6 (1,4,7,10,16-hexaoxacyclooctadecane) was found to direct the synthesis to a hexagonal polytype of zeolite Y, known as EMC-2 (EMT), again with a Si/Al ratio of approximately 5.

The structure of EMC-2 differs from zeolite Y because its layers of sodalite cages, which are linked by double six-membered rings (D6Rs), are related by a mirror plane rather than an inversion centre and give rise to different types of large cage. Whereas the FAU topology contains a single type of supercage with tetrahedral symmetry, EMT contains two kinds of large cages, one smaller than that of FAU and one larger. Burkett and Davis conducted a Raman spectroscopy investigation to determine the conformation of 18-crown-6 within EMC-2 [27] and found that the Raman spectrum of an aqueous solution of a sodium-crown ether complex closely matches that from 18-crown-6 in EMC-2. The authors concluded that the EMT supercages contain a large concentration of water, and therefore the sodium cations will complex to both 18-crown-6 and water molecules. They also suggested that the structure-directing effect of 18-crown-6 does not arise from space-filling effects alone but is also dependent on electrostatic attraction between the positively charged sodium-crown ether complex and the negatively charged framework, as well as having a modulating effect on the sodium concentration present in the gel. A crystallographic study of EMC-2 confirmed that 18-crown-6 is present as a sodium complex [28]. The crown ether is located in both types of large cages of EMT (shown in Fig. 3). In the smaller of these, there is an 18-crown-6 complex that fits the cage closely, and in the larger one, there is a Na^+ 18-crown-6 complex on one side of the cage, anchored by three sodium ions associated with 6-membered rings of the framework. The crown ether in the large cage is also complexed with water molecules. There is some conformational disorder of 18-crown-6 in the large cage as the Na–O bond distances do not allow simultaneous coordination of oxygen atoms to both types of sodium ions. The authors state that 18-crown-6 is very suitable for the depressions of the sodalite cage layers, and the resulting stacking of the molecules is specific to EMC-2; therefore, the crown ether can be thought of as 'structure directing'. When a mixture of 15-crown-5 and 18-crown-6 are used, FAU/EMT intergrowths can be produced and have been widely studied [29–34].

Zeolites of topology type RHO [35, 36] and KFI [37] have also been synthesised with higher Si/Al ratios when using 18-crown-6 than in its absence. The 18-crown-6 occupies the *lta* cages common to these materials. Chatelain et al. reported that

(a)   (b)

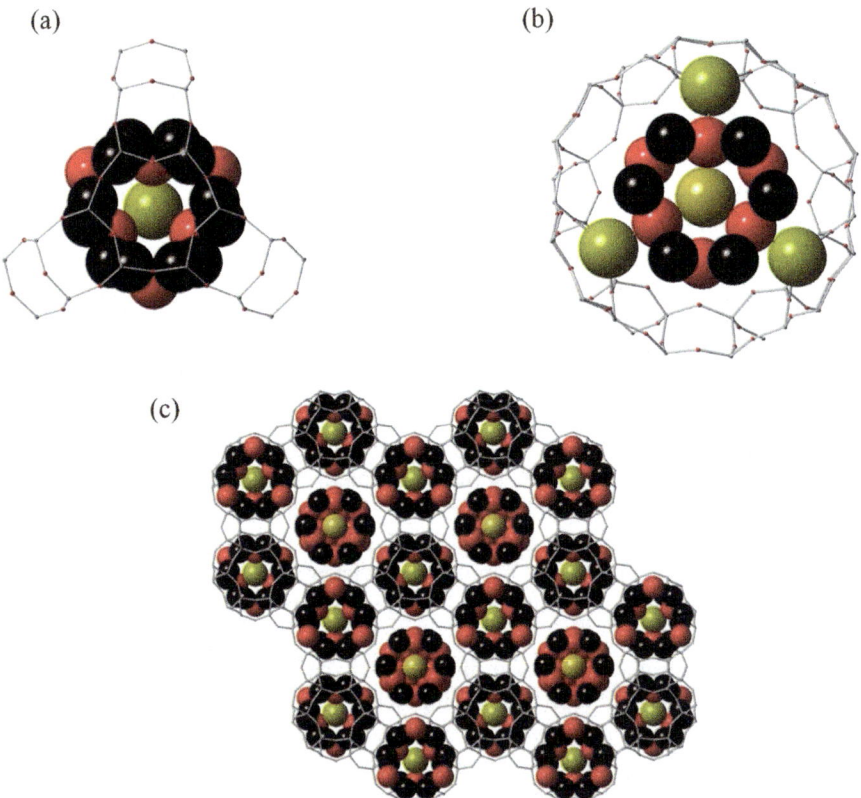

(c)

Fig. 3 Images of the structure of EMC-2 containing Na complexes of 18-crown-6. (**a**) Medium cage, (**b**) large cage and (**c**) projection of layers of crown ether molecules within the framework. (C, O and Na atoms of complex represented by black, red and yellow spheres, while H atoms are omitted, and the framework is represented as linked tetrahedral cations.) Fractional atomic coordinates taken from Baerlocher et al. [28]

Si/Al ratios of as high as 4.5 can be achieved in zeolite Rho when crown ethers are used as templates in addition to sodium and cesium cations [35], whereas zeolite Rho was originally prepared template-free with Si/Al = 3.2 [38]. More recently, Ke et al. report the synthesis of high-silica Rho through the use of high concentrations of crown ether complexes [36]. In this case, Si/Al ratios of up to 16 can be obtained. The authors suggest bulkier SDAs result from crown ether molecules assembling in a metal complex under the conditions of very high SDA concentration and that the inclusion of these reduces the charge density. The aim of this was to enhance the Si content present in Rho – the resultant Rho zeolites with increased Si/Al ratios have higher hydrothermal stability.

Zeolite ZK-5 (KFI) can also be crystallised using 18-crown-6 as an additional SDA, with high crystallinity and with higher Si/Al ratios than with the original organic-free preparation [39]. Chatelain et al. state that there is approximately one 18-crown-6 located within each *lta* cage [37].

3.3 Transition-Metal Azamacrocycle Complexes

Macrocyclic polyamines (Fig. 4) possess many features that make them suitable structure-directing agents for microporous solids, particularly aluminophosphate zeotypes, where their geometry favours the formation of cage structures linked by small 8R windows. The protonated form of tetramethylcyclam present at the pH values of aluminophosphate synthesis acts as a template for MnAPO and SAPO forms of STA-6(SAS) and for the Co- and ZnAPO forms of STA-7(SAV), a polytype of the CHA structure type comprising D6Rs [40]. Further studies of this kind found both tetramethylcyclam and cyclam act as SDAs for STA-7 when used together with tetraethylammonium (TEA$^+$) or other small molecule co-templates [41]. Consideration of the structure of STA-7, which possesses two types of cages of different size (see Fig. 5), indicates that while the larger cage is a suitable site for the cyclam and related molecules, the smaller co-templates are a close fit to the smaller cage. Related studies show that trimethyltriazacyclononane can act to template MAPO-18(AEI) while Kryptofix 222 templates the *lta* cage, giving MAPO-42 (LTA) on its own and SAPO STA-14(KFI) in the presence of TEA$^+$ co-template [41, 42]. Tet-A is a template for a layered aluminophosphate fluoride that can be converted topotactically to the microporous AlPO-41 (AFO) by heating [43].

However, using macrocycles in this way is expensive and moreover does not make full use of their unique complexing behaviour with late first-row transition metals. Table 1 gives the complexation constants for cyclam (the most readily available of these macrocycles), tetramethylcyclam and cyclen [44]. Attempts have therefore been made to use their metal complexes as SDAs, bearing in mind that the transition-metal cations Mn^{2+}, Fe^{3+}, Co^{2+} and Zn^{2+} are well known to be able to substitute for Al in the AlPO framework during hydrothermal synthesis (also in the presence of tetramethylcyclam) [12, 13]. In these cases, the $AlPO_4$ framework provides preferential sites over those available in the macrocycle. The situation is different for Ni^{2+} and Cu^{2+}: firstly, these elements do not favour the environment offered by the zeotype framework, so that in typical amine-templated syntheses NiAPO and CuAPOs are rarely if ever observed; furthermore, their high complexation constants favour them remaining complexed.

A series of examples serves to demonstrate this. [Ni(tmtact)]$^{2+}$ was confirmed as an SDA in MgAPO STA-6 by spectroscopic means, and its inclusion was associated with a tetragonal to orthorhombic symmetry change, while single crystal diffraction unambiguously located the complex in the larger cages of CoAPO STA-7 [45]. By contrast, [Ni(cyclen)]$^{2+}$ gave CoAPO-18(AEI) [45]. Inclusion of both [Ni(tmtact)]$^{2+}$ and [Ni(cyclam)]$^{2+}$ into the SAPO form of STA-6 has also been reported [45, 46]. Upon calcination, Ni^{2+} is released from the macrocycle and can be reduced in hydrogen to Ni^+.

Similar templating behaviour has been observed for the [Cu(cyclam)]$^{2+}$ complex, where the catalytic relevance of Cu-SAPOs gives the studies additional significance. This is due to the catalytic activity of the extra-framework Cu^{2+} for

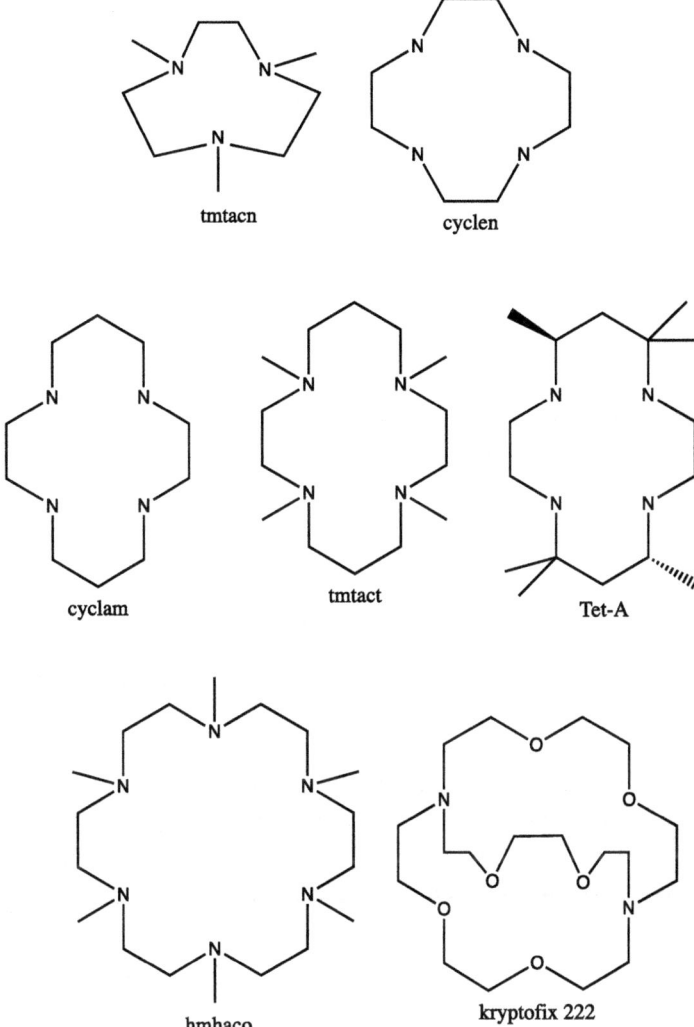

Fig. 4 Structures and acronyms of azamacrocycles used in the synthesis of microporous materials (tmtacn = 1,4,7-trimethy-1,4,7-triazacyclononane; cyclen = 1,4,7,10-tetraaza-cyclododecane; cyclam = 1,4,8,11-tetraazacyclotetradecane; tmtact = 1,4,8,11-tetramethyl-1,4,8,11-tetraaza-cyclotetradecane; Tet-A = 5,7,7,12,14,14-hexamethyl-1,4,8,11-tetra-azacyclotetradecane; hmhaco = 1,4,7,10,13,16-hexamethyl-1,4,7,10,13,16-hexaazacyclooctadecane; Kryptofix 222 = hexaoxa-4,7,13,16,21,24-diaza-1,10-bicyclo[8,8,8]hexacosane)

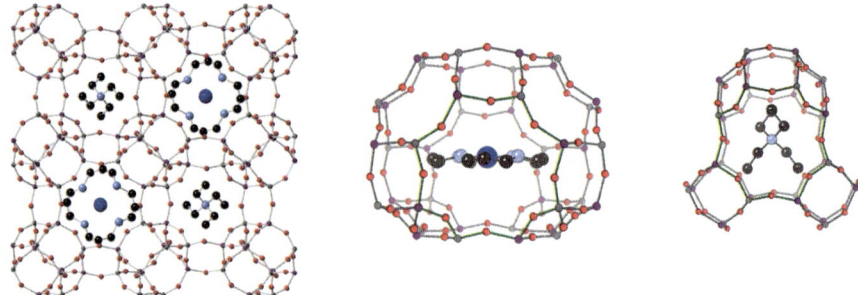

Fig. 5 As-prepared Cu-SAPO STA-7 (P4/n) viewed (left) down the c-axis showing the Cu-cyclam complex in the larger cage and tetraethylammonium in the smaller cage. The larger and smaller cages viewed along the b-axis are also shown (middle and right) showing the positions of the co-templates. (All framework atoms shown; Cu atoms, dark blue; N atoms, light blue; C atoms, black. H atoms omitted.) Fractional atomic coordinates taken from Picone et al. [51]

Table 1 Complexation constants of selected first-row transition metals (as log K) for the azamacrocycles cyclam, tmtact and cyclen [44]

	Co^{2+}	Ni^{2+}	Cu^{2+}	Zn^{2+}
Cyclam	12.7	22.2	26.5	15.5
Tmtact	7.58	8.65	18.3	10.35
Cyclen	13.8	16.4	24.8	16.2

the selective catalytic reduction of NO by ammonia and the hydrothermal stability of the SAPO framework, for example, for SAPO-34 and SAPO-18 [47–50].

The synthesis of Cu(cyclam)-SAPO STA-7 using $[Cu(cyclam)]^{2+}$ as a template is the most important example of the inclusion of a copper complex of an azamacrocycle [51]. The co-templated synthesis of SAPO STA-7 using $[Cu(cyclam)]^{2+}$ and TEA^+ (or N,N-diisopropylethylamine) as SDAs is highly selective, giving STA-7 with a unit cell composition of $Cu_{1.2}(cyclam)_2(TEA^+)_2Al_{24}Si_6P_{18}O_{96}·xH_2O$ (Fig. 5). Diffraction identifies TEA^+ within the smaller cages and the copper complex in the larger cages, although the difference in symmetry between the larger cage (fourfold rotation) and the complex (mirror plane) results in positional disorder of the ethylene and propylene groups of the cyclam. Further, the Cu^{2+} cations in the as-prepared materials give absorptions in the visible at 508 nm typical of $[Cu(cyclam)]^{2+}$ and an ESR signal characteristic of complexed Cu^{2+} ($Cu^{(II)}N_4$).

Calcination of Cu-SAPO STA-7 prepared in this way has been followed in situ by synchrotron microcrystal IR spectroscopy, with and without polarisation, which shows that the complex loses coordinated water below 250°C and breaks down at 400°C [52]. The organics are removed at lower temperature from Cu-SAPO STA-7 than from SAPO STA-7, presumably due to a catalytic effect of the copper. UV-visible and ESR spectroscopies of the hydrated calcined solid indicate dispersed Cu^{2+} cations, and powder X-ray diffraction of the dehydrated calcined solid reveals Cu^{2+} in 6R and 8R sites. IR spectroscopy using NO as a probe molecule for Cu^{2+} gives a multi-peak spectrum at ca. $1{,}900$ cm^{-1}, characteristic of NO chemisorbed on Cu^{2+}. This fine structure was attributed to adsorption on the different sites observed by IR, but recent computational simulation of the structurally related Cu SSZ-13(CHA) suggests that at least part of this fine structure originates from dynamic effects [53].

Calcined Cu-SAPO STA-7 shows good selectivity and activity in the ammonia SCR of NO_x, in line with that expected for Cu^{2+} species distributed within small-pore SAPO zeotypes, and the structure shows reasonable hydrothermal stability under accelerated ageing conditions. The strength of this one-pot synthesis route for generating Cu-SAPO catalysts for SCR is that it removes the requirement for an additional post-synthesis treatment step of Cu^{2+} ion exchange or impregnation, which can result in structural damage and inhomogeneous copper distribution. By contrast, the new route ensures a homogeneous distribution of copper throughout the crystals.

Similar synthetic studies have been performed on Cu-SAPO STA-6, which results from preparations with $[Cu(cyclam)]^{2+}$ but in the absence of TEA$^+$ [24]. Calcination results in the distribution of Cu^{2+} throughout the pores, in 6R and 8R sites, and with similar spectroscopic characteristics. Figure 6 shows the positions of extra framework Cu^{2+} cations in STA-6. No catalytic characterisation has been performed, but its one-dimensional channel system would be expected to possess less favourable transport properties than STA-7.

Additionally, Wheatley and Morris reported the synthesis of [F, Cu-cyclam]-AlPO SAS, a distorted analogue of STA-6 synthesised using complexed cyclam in a

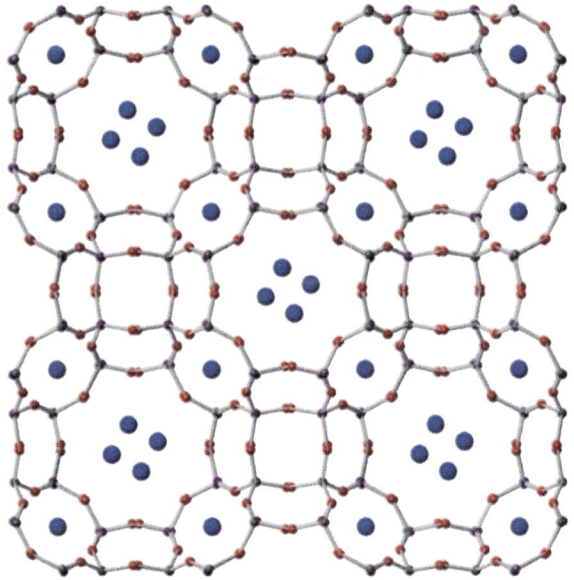

Fig. 6 Refined structure of calcined Cu-SAPO STA-6, showing framework and Cu atoms (only a small fraction of the Cu sites is occupied) (All framework atoms shown; Cu atoms, blue; Al atoms, purple; P/Si atoms, grey; O atoms, red). Fractional atomic coordinates taken from Warrender [24]

fluoride medium [54] with fluoride ions bound to the Al atoms in the SAS framework, as shown in Fig. 7.

The smaller [Cu(cyclen)]$^{2+}$ complex has been reported as an SDA for MgAPO-
18(AEI), which could also be calcined to leave Cu^{2+} cations distributed over 6R and
8R sites [24]. Complexation constants for cyclen are shown in Table 1 [44].

Finally, the approach has been extended to the use of 'bis-cyclam' and 'bis-
cyclen' complexes of Cu^{2+} [24]. The molecular structure of 'bis-cyclen' is given in
Fig. 8. While not successful for the preparation of SAPOs, 'bis-cyclen' has been
used successfully to template the large-pore MgAPO-5 (AFI), giving a bright blue
solid (indicating the complex had been incorporated intact) with unit cell composition (Cu$_2$L)$_{0.5}$Mg$_2$Al$_{10}$P$_{12}$O$_{48}$ (L = 'bis-cyclen').

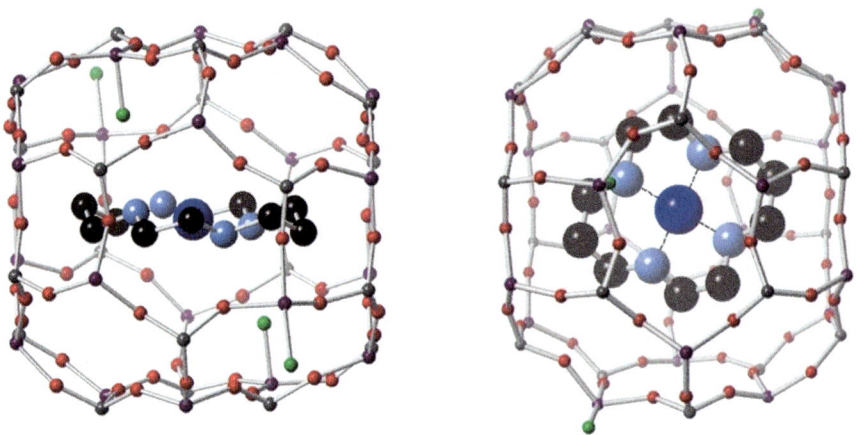

Fig. 7 Structure of [F, Cu-cyclam]-AlPO SAS viewed along (left) the b-axis and (right) the c-axis
(all framework atoms shown, including F atoms bound to Al, plus Cu, N and C of complex, H
atoms omitted). Fraction atomic coordinates taken from Wheatley and Morris [54]

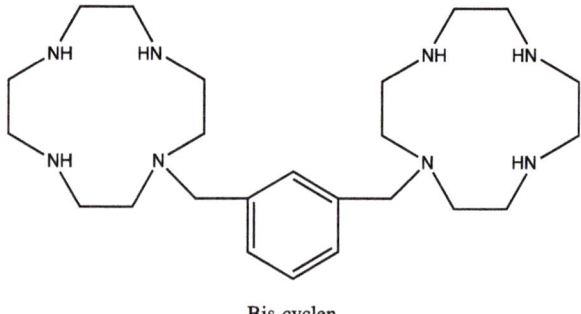

Bis-cyclen

Fig. 8 Molecular structure of 'bis-cyclen' used as an SDA for MgAPO-5

3.4 Transition-Metal Complexes of Linear Polyamines

Linear polyamines offer a cheaper alternative to aza-macrocycles as complexing ligands: examples of those used as complexing ligands in microporous solid synthesis are shown in Fig. 9. The stability of their metal complexes is determined by the number of chelating atoms and by the nature and oxidation state of the transition metal, according to the principles outlined by Irving and Williams [55, 56]. As examples, the stability constants at 25°C, for divalent first-row transition metals of TETA and TEPA, are given in Table 2 [57]. Furthermore, compared to amines and quaternary ammonium salts, polyamines (longer than diethylene-triamine) are less flammable, less volatile and more stable to thermal degradation.

An example of the use of polyamines as complexing agents has been given by Garcia et al. [46], in which $[Ni(DETA)_2]^{2+}$ combined with dipropylamine or

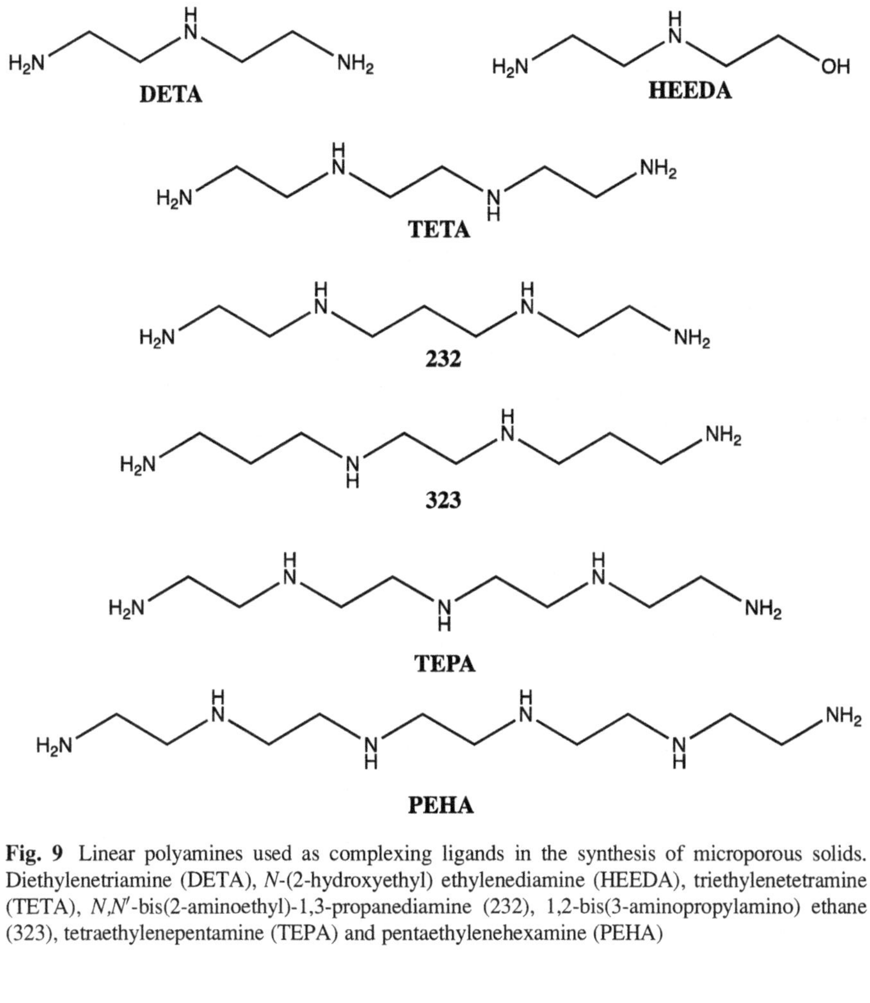

Fig. 9 Linear polyamines used as complexing ligands in the synthesis of microporous solids. Diethylenetriamine (DETA), N-(2-hydroxyethyl) ethylenediamine (HEEDA), triethylenetetramine (TETA), N,N'-bis(2-aminoethyl)-1,3-propanediamine (232), 1,2-bis(3-aminopropylamino) ethane (323), tetraethylenepentamine (TEPA) and pentaethylenehexamine (PEHA)

Table 2 Complexation constants at 25°C of selected first-row transition metals (as log K) for linear polyamines triethylenetetramine (TETA) and tetraethylenepentamine (TEPA) [57]

	Mn^{2+}	Fe^{2+}	Co^{2+}	Ni^{2+}	Cu^{2+}	Zn^{2+}
TETA	4.9	7.8	11.0	13.8	20.4	12.0
TEPA	6.6	9.9	13.3	17.4	22.8	15.1

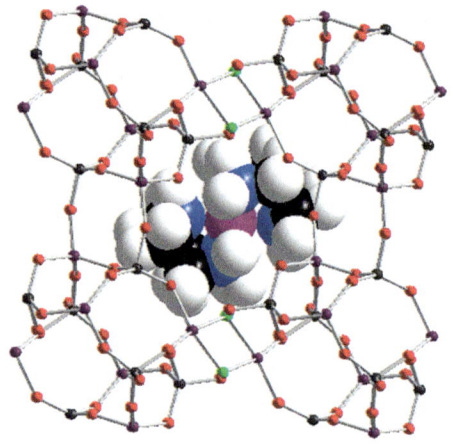

Fig. 10 Part of the structure of Ni(DETA)$_2$-UT-6 showing the energy-minimised position of the μ-fac isomer of the [Ni(DETA)$_2$]$^{2+}$ nickel complex within the cha cage (Al, purple; P, black; O, red; F, green; Ni, large purple; N, blue; C, large black; H, white). Fractional atomic coordinates from Garcia et al. [58]

tripropylamine acts as a structure-directing agent for Ni-AlPO-5 (AFI-type) and MgAPO-34 (CHA-type). The same metal complex added in an aluminophosphate gel containing ammonium fluoride resulted in the crystallisation of UT-6 (triclinic distorted AlPO-34) and orthorhombic F-AlPO-5 [58]. The presence of framework-bound fluoride was found to cause crystallographic distortions of the AlPO$_4$ frameworks in these materials. The integrity of the octahedral nickel complex in both frameworks was confirmed by diffuse reflectance UV-visible spectroscopy ($\lambda_{max} \approx 350$, 550 and 850 nm). Molecular modelling showed that the μ-*fac* isomer of [Ni(DETA)$_2$]$^{2+}$ is energetically favoured over the *mer* isomer in cages in UT-6 (Fig. 10). Calcination leaves Ni-bearing solids that show catalytic activity typical of supported nickel nanoparticles in hydrocarbon transformations [46]. Very recently, the one-pot synthesis of the aluminosilicate zeolite SSZ-13 using the nickel complex of diethylenetriamine has also been reported [59].

Additionally, diethylenetriamine complexed around Co^{2+} has been shown to direct the formation of a cobalt-substituted CoAlPO with the CHA topology [60]. In this case, the metal complex formed in situ (Co^{2+} and the polyamine were added separately to the gel).

Considering copper-polyamine complexes [Cu(TEPA)]$^{2+}$ fulfils the role of template in the formation of SSZ-13 [61, 62] and SAPO-34 [48], the aluminosilicate and silicoaluminophosphate versions of CHA. Furthermore [Cu(TEPA)]$^{2+}$ and [Cu(TETA)]$^{2+}$, together with specific co-templating agents, were used for the synthesis of SAPO-34 [47] and SAPO-18 [49]. In these studies, the group of Corma were able to control the Si/Al and Cu/(Si + Al) ratios in SSZ-13 and change the distribution of Si atoms (which can be either as isolated Si atoms or in aluminosilicate islands) and the Cu/(Si + Al + P) ratio of SAPO-34 and SAPO-18. They then investigated the effect this has on the catalytic activities and hydrothermal stabilities for the selective catalytic reduction of NO_x. Both silicon distribution and Cu loading were controlled by the addition of co-templating molecules (i.e. *N,N,N*-trimethyl-1-adamantammonium for SSZ-13, diethylamine for SAPO-34 and *N,N*-dimethyl-3,5-dimethylpiperidinium for SAPO-18). It was shown that by generating isolated Si atoms (in SAPOs) and optimising Cu loading to charge balance the framework without introducing structural defects, it was possible to enhance the hydrothermal stability of the materials to temperatures of up to 750°C.

Turrina et al. investigated the structure-directing role of a series of linear polyamines (see Fig. 9) in the presence of Cu^{2+} and Ni^{2+} in a SAPO gel by combining spectroscopic techniques and computer simulations [50]. Cu^{2+} complexed by TETA, TEPA, PEHA, DETA and HEEDA templated Cu-SAPO-34 without the addition of further templating agents, while Cu-232 combined with TEA^+ directed the formation of Cu-SAPO-18 over a narrow range of conditions. When Cu^{2+} was replaced with Ni^{2+}, Ni SAPO-34 crystallised with TEPA, 323, DETA and HEEDA while Ni-232 and Ni-TETA formed Ni SAPO-18. A careful analysis of the coordination sphere geometry and composition for the different metal complexes encapsulated in the SAPO frameworks revealed that SAPO-18 was only obtained when the Cu^{2+} or Ni^{2+} is in a square-planar configuration or octahedral with four planar N and one or two axial water molecules. The only exception was the Ni-323 complex which, despite having a coordination geometry similar to Ni-232, favoured crystallisation of SAPO-34. Computational modelling showed that Ni^{2+}-232 with square-planar geometry has a more favourable interaction energy within

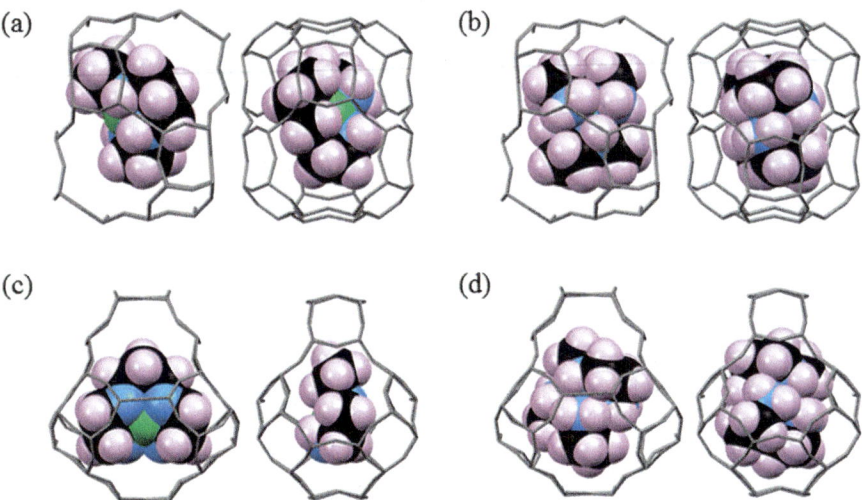

Fig. 11 Energy-minimised positions (two views of each) of $[Ni(232)]^{2+}$ (left) and $[Ni(DETA)_2]^{2+}$ (right) complexes within cha (above, **a**, **b**) and aei (below, **c**, **d**) cages. (Framework grey, C black, H pink, N cyan, Ni^{2+} green.) Reproduced with permission from Turrina et al. [50]

the *aei* cage than in the *cha* cage while the $[Ni(DETA)_2]^{2+}$ complex has a lower energy in the *cha* cage (Fig. 11).

Rietveld analysis of X-ray powder diffraction data of the calcined and dehydrated SAPO-34 materials revealed Cu^{2+} and Ni^{2+} located within the *cha* cage at reasonable distances from the framework oxygen atoms of the 8MR window. A second site has been found for the Ni^{2+} cations at the centre of the six-membered rings of the D6R subunits of the CHA structure. In calcined SAPO-18, Cu^{2+} and Ni^{2+} were located in the centre of the six-membered rings of the D6R subunits of the AEI structure (Fig. 12).

Turrina et al. recently reported the one-pot synthesis of Fe SAPO-34 using Fe^{2+} complexed by DETA, TETA, TEPA and PEHA as SDAs [63]. The inclusion of iron complexes in one-pot synthesis is more difficult than that of copper and nickel complexes because not only are iron cations included more readily into Al framework sites of SAPOs, but the complexation constants of Fe^{2+} are lower than for Ni^{2+} and Cu^{2+} (Table 2). The integrity of the Fe(II)-polyamine octahedral complexes within the *cha* cages and the oxidation state of the iron cations were assessed by a combination of solid-state UV-visible and Mössbauer spectroscopies. Detemplation of the as-prepared solids in oxygen removes the organic molecules included during synthesis, leaving Fe^{3+} cations mainly distributed in extra-framework positions. These materials showed good activity for the selective catalytic reduction of NO by ammonia. Although not as active as Cu^{2+} in SCR at low temperatures, Fe^{3+} materials retain activity at higher temperatures.

3.5   Complexes of Precious Metals

The use of the complexes of precious metals as structure-directing agents for porous solids is less widely studied, despite the catalytic importance of Pt and Pd, for example, for hydrogenations, dehydrogenations and oxidative dehydrogenations. The catalytic performance as bifunctional catalysts of nanoparticles of these metals supported on the acid forms of zeolites is of fundamental importance in hydrocarbon transformations such as hydrocracking and reforming that are integral to petroleum refining [1].

The synthetic challenge to incorporation of these metals during synthesis, rather than by post-synthetic routes, is that their complexes typically decompose under the conditions of hydrothermal synthesis, resulting in the precipitation of reduced metal or metal hydroxide. An early example where this decomposition has been avoided is the work of Mintova and Bein, who were able to use square-planar Pt and Pd ammine complexes ($[M(NH_3)_4]^{2+}$, M = Pt, Pd) below 100°C to direct crystallisation of small-pore, low-silica edingtonite (EDI) zeolite nanoparticles that contain the precious metal complexes (at Pt/Al and Pd/Al ratios of ca. 0.3) [64, 65]. Note that the copper ammine complex is more stable than the Pt and Pd versions and copper ammonia complexes have been included in edingtonite and in ferrierite [64, 66].

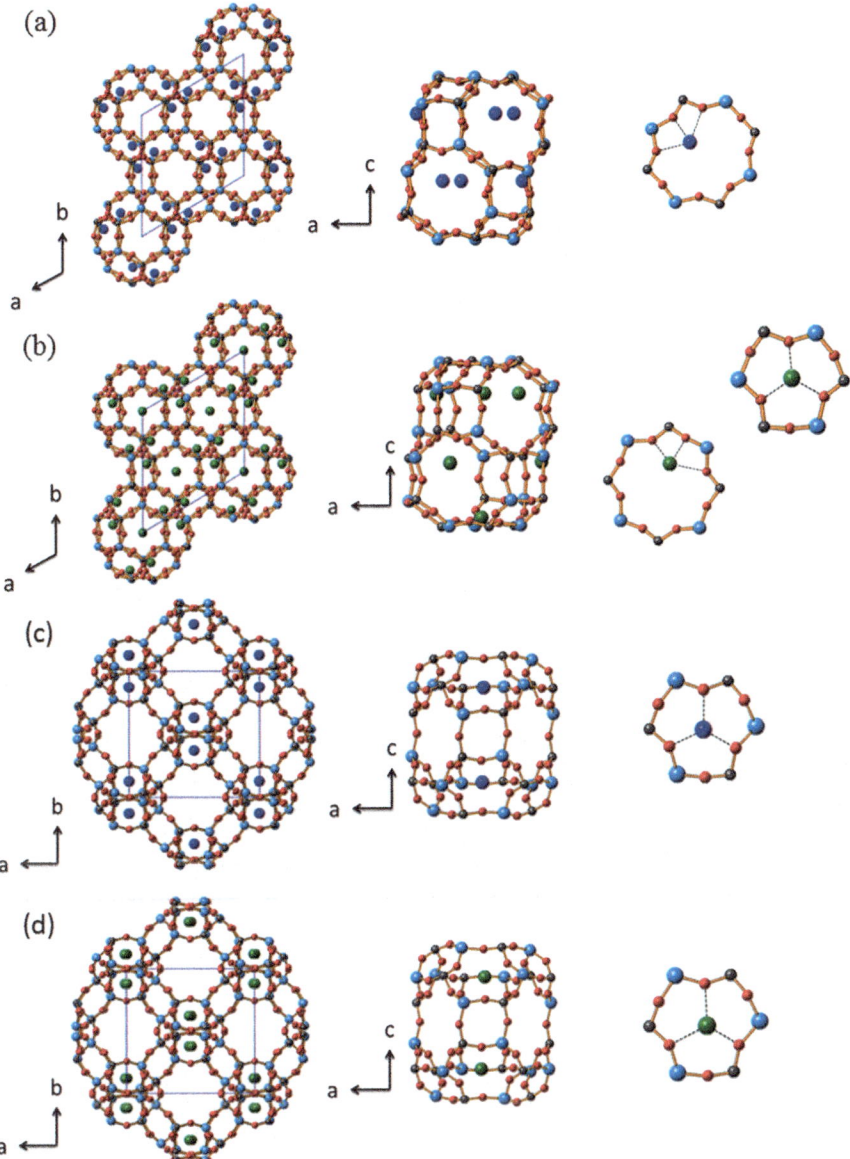

Fig. 12 Structures of calcined Cu-SAPO-34 (**a**), Ni-SAPO-34 (**b**), Cu-SAPO-18 (**c**), Ni-SAPO-18 (**d**) viewed down the *c*-axis showing all symmetry-related positions of Cu^{2+} and Ni^{2+} cations and down the *b*-axis showing the local environments of Cu^{2+} and Ni^{2+} sites. Note that only a small fraction of these sites is occupied (Al light blue, P dark grey, O red, Cu blue, Ni green). Reproduced with permission from Turrina et al. [50]

Returning to precious metals, Garcia et al. showed that it is possible to include Rh(III)-tmtact complexes within the larger cages of the ZnAPO form of STA-7, as demonstrated by both single crystal X-ray diffraction and X-ray absorption spectroscopy, although only a portion of the rhodium ends up in the ZnAPO, while the remainder precipitates [67]. This was not possible for SAPO STA-7, because it crystallises much more slowly than the ZnAPO, by which time all the rhodium had precipitated.

Another direct inclusion route that has successfully been developed to circumvent complex decomposition for these metals is to use thiol complexes rather than amine complexes in the one-pot syntheses. Metal-sulphur adducts are more resistant to decomposition to bulk metal or metal hydroxides and so can be incorporated intact. Choi et al. used Pt, Pd, Ir, Rh and Ag complexes of 3-mercaptopropyl-trimethoxysilane (TMSH) in the synthesis of zeolite A [68]. While not acting as a true structure-directing agent, this does permit quantitative inclusion of the precious metals. For Pt, spectroscopic analysis indicates this is as complexes with a square-planar MS_4 core. It is thought that the siloxane group becomes bonded to the silicate framework. Treatment in O_2 and then H_2 results in highly dispersed precious metal nanoparticles of around 1 nm size, within the *lta* cages, that have been shown to be shape selective in catalysis. More recently, this method has been developed by Moliner et al. to include Pt nanoparticles into a high-silica chabazite via a dual-template method, in which the thiol-stabilised Pt complex of TMSH has been included at the same time as the usual high-silica chabazite (SSZ-13) structure-directing agent, *N,N,N*-trimethyl-1-adamantammonium [69]. The relevance of the high-silica nature of the zeolite host is the high hydrothermal stability that this composition imparts to the framework structure. Remarkably, removal of the organic by heating in O_2 leaves finely dispersed single Pt atoms (in oxide form), while subsequent treatment in H_2 gives Pt nanoparticles ca. 1.5 nm in size. Subsequent O_2/H_2 cycling shows reversible oxidative fragmentation of the nanoparticles is possible, and this opens up a number of shape-selective catalytic possibilities.

4 Concluding Remarks

Metal complexes can act as effective structure-directing agents in the crystallisation of both zeolites and zeotypes such as silicas and SAPOs. Many of the topology types they direct can be considered cage structures (FAU, EMT, RHO, KFI, CHA, AEI, SAS, SAV), but channel types have also been reported (AFI, MFE, DON). In some cases, the complexes can be located crystallographically, but more often, they display disorder within the cages they occupy. Their presence may then be confirmed by a combination of compositional and spectroscopic analyses.

In most cases, the same structure types can more easily and cheaply be prepared using alternative organic or inorganic structure-directing agents, but in some cases, the use of complexes gives both more reproducible syntheses and also frameworks with unique compositional ranges. The major interest in their use as SDAs, however, is in their ability to enable the direct inclusion of transition-metal cations homogeneously throughout the pore space during crystallisation without the need for additional post-synthesis steps. Subsequent calcination liberates the cations from the complexes and distributes them throughout the pore space. The first-row transition-metal cations Fe, Co, Ni and Cu can be included in SAPOs and zeolites as polyamine and cobaltocenium complexes, either alone or in dual-template preparations, while, in a related approach, precious metals such as Pd and Pt can be included as more stable thiol complexes. For small-pore structures, this leads to sinter-resistant and shape-selective catalysts.

References

1. Wright PA (2007) Microporous framework solids. RSC Publishing, Cambridge
2. Herron N (1988) The selective partial oxidation of alkanes using zeolite-based catalysts – phthalocyanine ship-in-bottle species. J Coord Chem 19:25–38
3. De Vos DE, Meinershagen JL, Bein T (1996) Highly selective epoxidation catalysts derived from intrazeolite trimethyltriazacyclononane-manganese complexes. Angew Chem Int Ed 35: 2211–2213
4. Notari B (1996) Microporous crystalline titanium silicates. Adv Catal 41:253–334. (Elly DD, Haag WO, Gates B, eds)
5. Mambrim JST, Pastore HO, Davanzo CU et al (1993) Synthesis and characterization of chromium silicalite. Chem Mater 5:166–173
6. Duke CVA, Latham K, Williams CD (1995) Isomorphous substitution of Fe^{3+} in LTL framework using potassium ferrate (VI). Zeolites 15:213–218
7. Camblor MA, Lobo RF, Koller H et al (1994) Synthesis and characterization of zincosilicates with the SOD topology. Chem Mater 6:2193–2199
8. Camblor MA, Villaescusa LA, Diaz-Cabanas MJ (1999) Synthesis of all-silica and high-silica molecular sieves in fluoride media. Top Catal 9:59–76
9. Wilson ST, Lok BM, Messina CA et al (1982) Aluminophosphate molecular sieves – a new class of microporous crystalline inorganic solids. J Am Chem Soc 104:1146–1147
10. Flanigen EM, Lok BM, Patton RL et al (1986) Aluminophosphate molecular sieves and the periodic table. Pure Appl Chem 58:1351–1358
11. Cora F, Alfredsson M, Barker CM et al (2003) Modeling the framework stability and catalytic activity of pure and transition metal-doped zeotypes. J Solid State Chem 176:496–529
12. Hartmann M, Kevan L (1999) Transition-metal ions in aluminophosphate and silico-aluminophosphate molecular sieves: location, interaction with adsorbates and catalytic properties. Chem Rev 99:635–664
13. Hartmann M, Kevan L (2002) Substitution of transition metal ions into aluminophosphates and silicoaluminophosphates: characterization and relation to catalysis. Res Chem Intermed 28: 625–695
14. van de Goor G, Freyhardt CC, Behrens P (1995) The cobaltocenium cation $[Co^{III}(\eta^5\text{-}C_5H_5)_2]^+$: a metal-organic complex as a novel template for the synthesis of clathrasils. Z Anorg Allg Chem 621:311–322
15. Baerlocher C, McCusker LB, Olson DH (2007) Atlas of zeolite framework types, 6th edn. Elsevier, Amsterdam
16. Behrens P, Panz C, Hufnagel V et al (1997) Structure-directed materials syntheses: metal complexes as structure-directing agents for zeolite-type solids. Solid State Ionics 101:229–234
17. Valyocsik EW (1985) A process for making zeolite ZSM-45. EP0143642 A2

18. Balkus KJ, Biscotto M, Gabrielov AG (1997) The synthesis and characterization of UTD-1: the first large pore zeolite based on a 14 membered ring system. Stud Surf Sci Catal 105: 415–421
19. Lobo RF, Tsapatsis M, Freyhardt CC et al (1997) Characterization of the extra-large-pore zeolite UTD-1. J Am Chem Soc 119:8474–8484
20. Wessels T, Baerlocher C, McCusker LB et al (1999) An ordered form of the extra-large-pore zeolite UTD-1: synthesis and structure analysis from powder diffraction data. J Am Chem Soc 121:6242–6247
21. Jiang JX, Yu JH, Corma A (2010) Extra-large-pore zeolites: bridging the gap between micro and mesoporous structures. Angew Chem Int Ed 49:3120–3145
22. Balkus KJ, Gabrielov AG, Shepelev S (1995) Synthesis of aluminum phosphate molecular sieves using cobalticinium hydroxide. Microporous Mater 3:489–495
23. Schreyeck L, Caullet P, Mougenel JC et al (1997) Synthesis of AlPO$_4$-16 from fluoride-containing media in the presence of various organic templates. Microporous Mater 11:161–169
24. Warrender SJ (2007) Structure direction in the formation of zeolitic materials. PhD thesis, University of St Andrews
25. Cotton FA, Wilkinson G, Murillo CA et al (1999) Advanced inorganic chemistry, 6th edn. Wiley, London
26. Delprato F, Delmotte L, Guth JL et al (1990) Synthesis of new silica-rich cubic and hexagonal faujasites using crown ether-based supramolecules as templates. Zeolites 10:546–552
27. Burkett SL, Davis ME (1993) Structure-directing effects in the crown ether-mediated syntheses of FAU and EMT zeolites. Microporous Mater 1:265–282
28. Baerlocher C, McCusker LB, Chiappetta R (1994) Location of the 18-crown-6 template in EMC-2 (EMT) Rietveld refinement of the calcined and as-synthesized forms. Microporous Mater 2:269–280
29. Dougnier F, Patarin J, Guth J et al (1992) Synthesis, characterization, and catalytic properties of silica-rich faujasite-type zeolite (FAU) and its hexagonal analog (EMT) prepared by using crown-ethers as templates. Zeolites 12:160–166
30. Anderson MW, Agger JR, Hanif N et al (2001) Crystal growth in framework materials. Solid State Sci 3:809–819
31. Alfredsson V, Ohsuna T, Terasaki O et al (1993) Investigation of the surface structure of the zeolites FAU and EMT by high-resolution transmission electron microscopy. Angew Chem Int Ed Engl 32:1210–1213
32. Anderson MW, Pachis KS, Prébin F et al (1991) Intergrowths of cubic and hexagonal polytypes of faujasitic zeolites. J Chem Soc Chem Commun:1660–1664
33. Arhancet JP, Davis ME (1991) Systematic synthesis of zeolites that contain cubic and hexagonal stackings of faujasite sheets. Chem Mater 3:567–569
34. Newsam JM, Treacy MMJ, Vaughan DEW et al (1989) The structure of zeolite ZSM-20: mixed cubic and hexagonal stackings of faujasite sheets. J Chem Soc Chem Commun:493–495
35. Chatelain T, Patarin J, Fousson E et al (1995) Synthesis and characterization of high-silica zeolite RHO prepared in the presence of 18-crown-6 ether as organic template. Microporous Mater 4:231–238
36. Ke Q, Sun T, Cheng H et al (2017) Targeted Synthesis of ultrastable high-silica RHO zeolite through alkali metal–crown ether interaction. Chem Asian J 12:1043–1047
37. Chatelain T, Patarin J, Farré R et al (1996) Synthesis and characterization of 18-crown-6 ether-containing KFI-type zeolite. Zeolites 17:328–333
38. Robson HE, Shoemaker DP, Ogilvie RA et al (1973) Synthesis and crystal structure of zeolite Rho – a new zeolite related to linde type A. Adv Chem Ser 121:106–115
39. Kerr GT (1963) Zeolite ZK-5: a new molecular sieve. Science 140:1412
40. Wright PA, Maple MJ, Slawin AMZ et al (2000) Cation-directed syntheses of novel zeolite-like metalloaluminophosphates STA-6 and STA-7 in the presence of azamacrocycle templates. J Chem Soc Dalton Trans:1243–1248

41. Castro M, Warrender SJ, Wright PA et al (2009) Silicoaluminophosphate molecular sieves STA-7 and STA-14 and their structure-dependent catalytic performance in the conversion of methanol to olefins. J Phys Chem C 113:15731–15741
42. Maple MJ, Philp EF, Slawin AMZ et al (2001) Azamacrocycles and the azaoxacryptand 4,7,13,16,21,24-hexaoxa-1,10-diazabicyclo[8.8.8]hexacosane as structure-directing agents in the synthesis of microporous metalloaluminophosphates. J Mater Chem 11:98–104
43. Wheatley PS, Morris RE (2006) Calcination of a layered aluminofluorophosphate precursor to form the zeolitic AFO framework. J Mater Chem 16:1035–1037
44. Bianchi A, Micheloni M, Paoletti P (1991) Thermodynamic aspects of the polyazacycloalkane complexes with cations and anions. Coord Chem Rev 110:17–113
45. Garcia R, Philp EF, Slawin AMZ et al (2001) Nickel complexed within an azamacrocycle as a structure directing agent in the crystallization of the framework metalloaluminophosphates STA-6 and STA-7. J Mater Chem 11:1421–1427
46. Garcia R, Coombs TD, Shannon IJ et al (2003) Nickel amine complexes as structure-directing agents for aluminophosphate molecular sieves: a new route to supported nickel catalysts. Top Catal 24:115–124
47. Deka U, Lezcano-Gonzalez I, Warrender SJ et al (2013) Changing active sites in Cu–CHA catalysts: deNOx selectivity as a function of the preparation method. Microporous Mesoporous Mater 166:144–152
48. Martínez-Franco R, Moliner M, Franch C et al (2012) Rational direct synthesis methodology of very active and hydrothermally stable Cu-SAPO-34 molecular sieves for the SCR of NOx. Appl Catal B Environ 127:273–280
49. Martínez-Franco R, Moliner M, Corma A (2014) Direct synthesis design of Cu-SAPO-18, a very efficient catalyst for the SCR of NOx. J Catal 319:36–43
50. Turrina A, Eschenroeder ECV, Bode BE et al (2015) Understanding the structure directing action of copper–polyamine complexes in the direct synthesis of Cu-SAPO-34 and Cu-SAPO-18 catalysts for the selective catalytic reduction of NO with NH_3. Microporous Mesoporous Mater 215:154–167
51. Picone AL, Warrender SJ, Slawin AMZ et al (2011) A co-templating route to the synthesis of Cu SAPO STA-7, giving an active catalyst for the selective catalytic reduction of NO. Microporous Mesoporous Mater 146:36–47
52. Eschenroeder ECV, Turrina A, Picone AL et al (2014) Monitoring the activation of copper-containing zeotype catalysts prepared by direct synthesis using in situ Synchrotron infrared microcrystal spectroscopy and complementary techniques. Chem Mater 26:1434–1441
53. Göltl F, Sautet P, Hermans I (2015) Can dynamics be responsible for the complex multipeak Infrared spectra of NO adsorbed to copper(II) sites in zeolites? Angew Chem Int Ed 54:7799–7804
54. Wheatley PS, Morris RE (2002) Cyclam as a structure-directing agent in the crystallization of aluminophosphate open framework materials from fluoride media. J Solid State Chem 167:267–273
55. Irving H, Williams RJP (1953) The stability of transition-metal complexes. J Chem Soc 637:3192–3210
56. Varadwaj PR, Varadwaj A, Jin B-Y (2015) Ligand(s)-to-metal charge transfer as a factor controlling the equilibrium constants of late first-row transition metal complexes: revealing the Irving–Williams thermodynamical series. Phys Chem Chem Phys 17:805–811
57. Smith RM, Martell AE (1975) Critical stability constants Vol. 2: Amines. Plenum Press, New York
58. Garcia R, Shannon IJ, Slawin AMZ et al (2003) Synthesis, structure and thermal transformations of aluminophosphates containing the nickel complex [Ni(diethylenetriamine) 2]$^{2+}$ as a structure directing agent. Microporous Mesoporous Mater 58:91–104; Oliver S, Kuperman A, Lough A et al (1997) Synthesis and characterisation of a fluorinated anionic aluminophosphate UT-6 and its high temperature dehydrofluorination to AlPO4-CHA. J Mater Chem 7:807–812

59. Cui Y, Tong X, Li Y et al (2017) One-pot synthesis of Ni-SSZ-13 zeolite using a nickel amine complex as an efficient organic template. J Mater Sci 52:10156–10162
60. Xu Y-H, Yu Z, Chen X-F et al (1999) Hydrothermal synthesis and characterization of chabazite-type cobaltoaluminophosphate with an encapsulated cobalt complex: $Co_3Al_3(PO_4)_6Co$ $(DETA)_2 \cdot (H_2O)_3$. J Solid State Chem 146:157–162
61. Ren L, Zhu L, Yang C et al (2011) Designed copper–amine complex as an efficient template for one-pot synthesis of Cu-SSZ-13 zeolite with excellent activity for selective catalytic reduction of NOx by NH_3. Chem Commun 47:9789–9791
62. Martínez-Franco R, Moliner M, Thogersen JR et al (2013) Efficient one-pot preparation of Cu-SSZ-13 materials using cooperative OSDAs for their catalytic application in the SCR of NOx. ChemCatChem 5:3316–3323
63. Turrina A, Dugulan AI, Collier JE et al (2017) Synthesis and activation for catalysis of Fe-SAPO-34 prepared using iron polyamine complexes as structure directing agents. Cat Sci Technol 7:4366–4374
64. Kecht J, Mintova S, Bein T (2007) Nanosized zeolites templated by metal-amine complexes. Chem Mater 19:1203–1205
65. Kecht J, Mintova S, Bein T (2008) Exceptionally small colloidal zeolites templated by Pd and Pt amines. Langmuir 24:4310–4315
66. Gomez-Lor B, Iglesias M, Cascales C et al (2001) A diamine copper(I) complex stabilized in situ within the ferrierite framework. Catalytic properties. Chem Mater 13:1364–1368
67. Garcia R (2003) Synthesis and characterisation of aluminophosphate-based materials prepared with nickel complexes as structure directing agents. PhD thesis, University of St Andrews
68. Choi M, Wu Z, Iglesia E (2010) Mercaptosilane-assisted synthesis of metal clusters within zeolites and catalytic consequences of encapsulation. J Am Chem Soc 132:9129–9137
69. Moliner M, Gabay JE, Kliewer CE et al (2016) Reversible transformation of Pt nanoparticles into single atoms inside high-silica chabazite zeolite. J Am Chem Soc 138:15743–15750

Struct Bond (2018) 175: 201–244
DOI: 10.1007/430_2017_9
© Springer International Publishing AG 2017
Published online: 25 October 2017

Chiral Organic Structure-Directing Agents

Luis Gómez-Hortigüela and Beatriz Bernardo-Maestro

Abstract Chirality is crucial for life. The preparation of enantiopure chiral compounds is highly desirable in the chemical industry, especially in the pharmaceutical sector. In this context, the design of chiral solids able to discriminate between
enantiomers of chiral compounds, either during adsorption or asymmetric catalytic
processes, is one of the greatest challenges nowadays in chemical research. Zeolite-
type materials represent ideal candidates to achieve enantioselective chiral solids
since they could combine their high stability, surface area, and shape-selectivity
with a potential enantioselectivity that could be enhanced by the confinement
effect. Despite the occurrence of chiral zeolite frameworks and the strong interest
in preparing these chiral solids, very little success has been met in preparing these in
homochiral form. The main strategy to induce chirality in zeolite materials has been
the use of chiral structure-directing agents, in an attempt to transfer their chiral
feature into the nascent zeolite structure. However, although many chiral organic
species have directed the crystallization of zeolite frameworks, some of them even
being chiral, there is only one unique very recent example of success in transferring
the chirality from the organic structure-directing agent into an enantioenriched
chiral zeolite material. Chiral coordination compounds have been very successful
in transferring their chirality onto inorganic frameworks through the development
of extensive H-bond host–guest interactions, but these chiral materials usually
collapse upon removal of the guest species. In this chapter we report the different
types of chiral molecules, both organic and organometallic compounds, used so far
as structure-directing agents in an attempt to promote the crystallization of
homochiral zeolites; we analyze in detail the possible reasons for the general failure
in transferring their chirality, and we propose approaches to prepare known chiral
zeolite frameworks in homochiral form. Furthermore, we also review a different

L. Gómez-Hortigüela (✉) and B. Bernardo-Maestro
Instituto de Catálisis y Petroleoquímica (ICP-CSIC), C/ Marie Curie 2, Madrid 28049, Spain
e-mail: lhortiguela@icp.csic.es

approach we have followed in our group in order to induce chirality in zeolite materials, consisting in the development of chiral spatial distributions of dopants embedded in otherwise achiral zeolite frameworks.

Keywords Chirality • Enantiomer • Host-guest chemistry • Structure-directing agents • Templates • Zeolites

Contents

1　Introduction: Enantiopure Chiral Zeolitic Frameworks, a Highly Desired Yet Elusive Target in Zeolite Science

Chirality is the property of objects that are not superimposable with their mirror images; the two mirror images are called *enantiomers* [1]. Depending on the dimensionality where the object is referred, there are several kinds of chiral objects. For instance, a spiral is a chiral object in a 2-dimensional space since it does not superimpose with its mirror image (without leaving the 2-dimensional planar space). The 3-dimensional world where we live is plenty of chiral objects. The classical example is our hands (which indeed gave the name of chirality to this property: the term "*chiral*" is derived from "*cheir*" meaning "*hand*" in Greek), which are not superimposable to each other. Other examples of chirality are those objects containing helicoidal structures, such as propellers, springs, or spiral stairs. In nature, chirality is mainly expressed in asymmetric chemical compounds at the molecular level, although asymmetry also manifests sometimes at macroscopic level (see, for instance, the spiral shells of mollusks). The most frequent expression of chirality is when chemical species have a so-called *stereogenic* center, i.e. a tetrahedral atom with four different substituents attached, what gives place (except in particular cases like *meso* compounds) to species that are not superimposable to their mirror images. In this case, the absolute configuration of the two enantiomers of the chiral compound is determined by the spatial arrangement of those substituents, which gives (R) and (S)-enantiomers for each stereogenic center (following the Cahn-Ingold-Prelog rules [2]). Nevertheless, there are also chiral compounds not based on stereogenic centers but on helicoidal structures, with *helicene* as the typical example [3].

Chirality is also ubiquitous in living organisms. In fact, *homochirality*, defined as chirality in its pure enantiomeric form, is crucial for the correct functioning of life. Since its early origin, for some still unclear reason, life in Earth decided to work in an asymmetric fashion [4], although the last origin for this is still a mystery but is a research topic of great interest and current debate [5]. Most of the biochemical building units of living organisms, in particular aminoacids, sugars and nucleotides from which protein and nucleic acid macromolecules are built, are homochiral [6]. As a consequence, chiral molecules generally trigger a different biological response on living organisms as a function of their absolute stereochemical configuration (i.e., of their enantiomeric form); in fact, very often only one of the enantiomers develops the desired therapeutic effect, while the other enantiomer is less efficient or even no efficient at all [7]. In extreme cases, such as that of the sadly famous *thalidomide* drug, which was administered to pregnant women in the 1960s, one of the enantiomers (*R*) had the desired clinical effect, whilst the mirror image (*S*) triggered teratogenic effects, in that case resulting in malformations in the fetus [8].

Most of the pharmacological drugs administered today are chiral; in 2006, 80% of the small-size drugs approved by the *U.S. Food and Drug Administration* (FDA) were chiral, and 75% of them were composed of only one of the enantiomers [9]. Hence, one of the greatest challenges today in the chemical industry, especially in the pharmaceutical sector, is to obtain and/or separate the pure enantiomeric forms of chiral compounds of interest. In this context, the search for materials able to perform operations in an enantioselective way, i.e. being able to discriminate or distinguish between the enantiomeric forms of chiral molecules, both during asymmetric catalytic operations and in chiral separations, represents nowadays a tremendous challenge of great interest for researchers in applied chemistry. The major benefit of developing this type of enantioselective operations with chiral materials is the *chiral multiplication* effect, where a minimum amount of a chiral material can give place to large amounts of an enantiomerically pure (or enantioenriched) chiral compound of interest [10].

Though much less frequent, chirality is also manifested on inorganic systems. However, very rarely examples of homochirality are found among inorganic compounds: quartz (SiO_2) was soon recognized to have a chiral structure, however both enantiomorphic crystalline forms of quartz occur in nature in the same (50:50) proportion [11]. In any case, pioneering experiments showed that this type of chiral inorganic systems could transfer their chiral nature into a physico-chemical process (separation or catalysis), thus resulting in enantioselective operations, as was discovered by using pure enantiomorphic forms of quartz crystals [12]. Indeed, it has been proposed that minerals with chiral features (like the framework topology or particular crystal surfaces) such as quartz, clays, or calcite could have played a role on the origin of homochirality in the molecules of life [13, 14]. In this context, zeolite materials (which are nothing else than nanoporous polymorphs of SiO_2) have been considered as ideal candidates to achieve enantioselective-performing chiral solids since they could potentially combine crucial aspects in chemical processes: their high porosity and large surface area as well as their characteristic *shape-selectivity*, with enantioselectivity. The shape-selectivity of zeolite materials is provided by their particular porous structure based on channels and/or cavities of molecular dimensions, promoting a *confinement effect* on guest species adsorbed within the porous systems which greatly influences the outcome of a catalytic reaction. In the same way, such confinement effect could enhance enantioselectivity of a chemical process by reducing the degrees of motion of the guest species, which can only diffuse in particular directions dictated by the porous system of the particular framework structure.

Zeolite materials are typically prepared by hydrothermal methods from concentrated gels in aqueous media, which are then heated in autoclaves at autogenous pressure at temperatures usually ranging between 100 and 200 °C for a certain period of time. As has been dealt with throughout this Volume, very frequently the synthesis of zeolite materials requires the addition of organic molecules that act as structure-directing agents (SDA) [15]; these molecules organize the tetrahedral inorganic units into a particular geometry from which crystallization of a particular zeolite framework type takes place. In particular occasions, a so-called *template effect* occurs in which a molecule imprints its size and shape on the nanoporous structure that crystallizes, establishing a clear correspondence between the shape and size of the SDA and that of the porous system. It should be noted that the crystallization process in the presence of these SDAs involves that the organic molecules will finally remain occluded within the porous systems, and hence they have to be removed in order to empty the pores prior to the use of these materials in adsorption/catalytic processes.

In this context, the shape-selectivity associated with the presence of regular pores and cavities of well-defined dimensions, in a size-scale similar to that of most of chiral organic molecules of interest in the pharmaceutical industry, with the consequent potential to act as molecular sieves, combined with the presence of well-defined isolated active sites, and together with the possibility of modulating the structural properties of the zeolite frameworks through the use of rationally selected organic SDA molecules, have made zeolite materials one of the most prominent candidates to achieve enantioselective-performing chiral solids. In fact, the search for enantiopure chiral zeolite structures represents one of the greatest challenges in zeolite (and materials) science [16–19], since they could have a great impact on applications such as asymmetric catalysis and enantioselective separations.

The first attempts to induce chirality in zeolite materials consisted in the immobilization of homogeneous chiral catalysts [20] or the anchoring of chiral modifiers [21], which led to notable enantioselective operations. Nonetheless, since the chiral entity is included as an extraframework species in the porous system but is not part of the framework itself, these methodologies often impose problems related to the deactivation or lixiviation of the chiral component during adsorption or catalytic applications. A much more encouraging alternative would be the production of chiral zeolite frameworks where chirality would be intrinsic to the structural network and is not manifested on an extraframework loosely-bound component which would be more weakly retained. In fact, several chiral zeolite frameworks do actually exist [18], like it was long ago recognized with polymorph

A of zeolite Beta (BEA) [22], the chiral zincophosphate (CZP) [23], beryllosilicate OSB-1 (OSO) [24], gallogermanate UCSB-7 (BSV) [25], zeolite mineral goosecreekite (GOO) [26], aluminosilicate Linde Type J (LTJ) [27], cobalt-containing aluminophosphate CoAPO-CJ40 (JRY) [28], the recently discovered ITQ-37 (ITV) [29], or SU-32 [30] and HPM-1 [31], germanosilicate and pure-silica forms of the STW framework type. However, despite the occurrence of chiral zeolite frameworks, their main drawback is that they almost invariably crystallize as 50:50 intergrowths of the two enantiomorphic polymorphs (such as polymorph A of zeolite Beta), or as racemic conglomerates, i.e. 50:50 mixtures of both enantiomorphic homochiral crystals, with each individual crystal being enantiomerically pure (such as HPM-1) [32]. Such occurrence of racemic mixtures of the enantiomorphs of chiral zeolite frameworks of course prevents their use in enantioselective operations. Homochirality in the context of chiral inorganic materials very rarely occurs, although some limited but fascinating examples have been reported where homochirality is expressed in a spontaneous, and generally unpredictable, fashion, through symmetry-breaking phenomena [28, 33]. On the other hand, the chiral nature of some additional zeolite frameworks has only recently been recognized, and calorimetry results suggest that they could have a potential application in processes of enantiomeric discrimination [34]. It should be remarked here also the work of Zhang and coworkers who developed a very interesting strategy to enrich the chiral zincophosphate (CZP) in a particular enantiomer through the use of nucleotides as chiral inductor agents, which will be explained more in detail below [35].

In this line, we cannot continue without mentioning at least the tremendous impact on the topic of chirality in microporous materials triggered by metal-organic frameworks (MOFs). These extraordinary materials are built from metallic centers and organic ligands, giving place to porous coordination polymers with variable pore sizes [36]. These materials have prompted a great revolution in current materials science. The organic nature of the ligands soon promoted the use of chiral ligands to produce chiral MOF materials where the asymmetric entity is part of the framework walls, which indeed led to the production of homochiral solid materials [37, 38]; some of them have even led to notable enantioselectivities during adsorption or catalytic processes. In any case, although the development of enantioselective operations with chiral MOF materials is of great interest in current research, these are out of the scope of the present manuscript.

Zeolite networks are built from TO_4 tetrahedral units (with T being typically Si or Al, although other elements as P, Ge, Ga, etc., can also occur); therefore, these units are symmetric, and hence chirality in zeolite materials cannot come from the presence of stereogenic centers as is the typical case for organic chiral compounds. In zeolite materials, chirality arises from the long-range chiral ordering of these TO_4 units arranged in chiral space groups, usually in the form of helicoidal channels [18], and not from the building units themselves as in some MOF materials. For this reason, the induction of chirality in zeolite materials has to come from non-framework external units, such as solvents or chiral additives [39]. In this context, the synthesis methodology of zeolites which involves the addition of the organic SDAs provides an obvious strategy in order to induce chirality in these materials through the use of chiral organic species as SDAs. Thus, the main objective here would be to use chiral organic molecules with particular features able to transfer and imprint their asymmetric nature into the long-range chiral ordering of the TO_4 units building up a chiral zeolite framework [17].

2 Chiral Organic Structure-Directing Agents: General Aspects

For many years, a large number of chiral molecules have been used as SDAs to direct the crystallization of zeolite frameworks. Many of these have indeed succeeded in directing the formation of zeolite frameworks, some of them even being chiral. Figures 1 and 3 report a list of the organic molecules that have successfully led to zeolite frameworks in silica-based (Fig. 1) [17, 40–53] and metalophosphate (Fig. 3) compositions, respectively; Fig. 2 shows additional chiral SDAs that were unsuccessful to direct the crystallization of zeolite frameworks.

If a pure enantiomorph of a potential chiral zeolite framework is to be crystallized, of course one needs to use an enantiomerically pure chiral organic species as SDA in order to promote the transfer of chirality. Typically large amounts of the organic cations are required for the crystallization of zeolite frameworks, which is performed under highly concentrated conditions (especially in fluoride media). In this context, enantiomerically pure chiral compounds are usually very expensive, especially if a chiral separation is required in order to isolate the different enantiomers. Hence, the most obvious strategy to select enantiomerically pure organic chiral cations as SDAs is to search through the so-called *chiral-pool* provided by nature. Since living organisms work in an asymmetric fashion, as previously mentioned, enantiomerically pure chiral natural products are abundant, and usually commercially available at more reasonable costs. In this line, one of the first chiral species used as SDA was derived from (*S*)-sparteine, which is an alkaloid biosynthesized from *L*-lysine that can be extracted from *scotch broom* [54]. Lobo and coworkers synthesized the N(16)-methyl derivative by quaternization of one of the N groups (cation *A* in Fig. 1), and observed the crystallization of boron-containing SSZ-24 in sodium hydroxide medium, the silica-analogue of AlPO-5, displaying the AFI framework topology [40]. The authors proposed that the incorporation of boron in the framework was probably related to the hydrophobicity of this cation, which had just one quaternary N. The AFI framework is composed of one-dimensional 12-ring achiral channels, hence evidencing that the chirality of N-methyl-sparteinium was not imprinted on the framework, despite ^{13}C NMR results clearly showing that the cation was incorporated intact within the zeolite structure. Wagner and coworkers used the same cation (*A*) as SDA, but this time in the presence of lithium hydroxide, and achieved the crystallization of a new zeolite framework, CIT-5 (CFI framework type) [41, 55]. CIT-5 contains extra-large

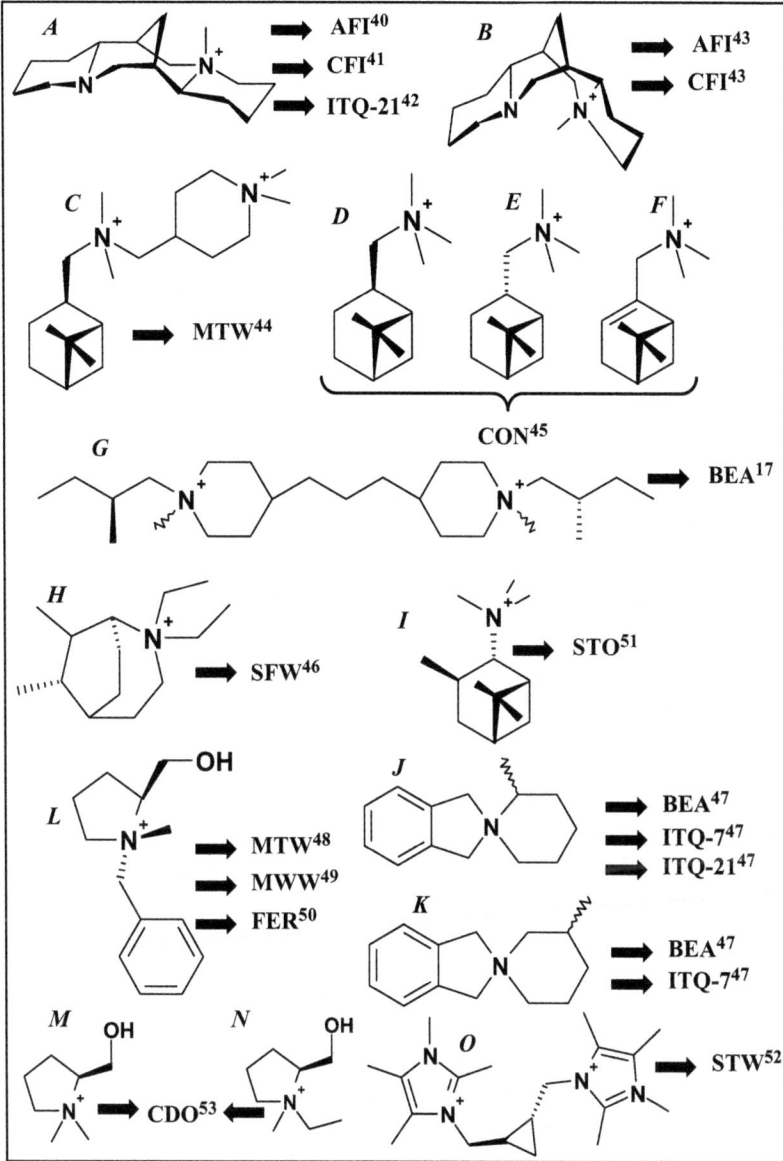

Fig. 1 Chiral organic cations used as structure-directing agents that directed the crystallization of zeolite frameworks (silica-based composition)

14-ring one-dimensional pores, which together with UTD-1 (DON) were the only high-silica extra-large pore molecular sieves at that time. In line with these works, Tsuji and coworkers also used the N(1)-methyl derivative of another diastereoisomer of spartein, N(1)-methyl-α-isosparteinium (cation *B* in Fig. 1), and found that

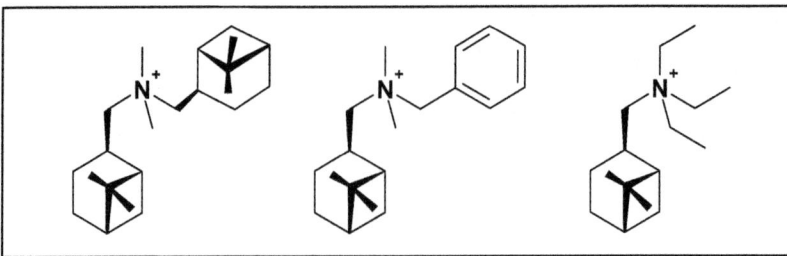

Fig. 2 Additional chiral organic cations derived from the myrtanyl ring used as SDAs that did not lead to crystalline zeolite materials (in silica-based compositions) [44]

this isomer was more efficient for the crystallization of the same CFI framework in the presence of lithium hydroxide than the previous (A) [43]. With the help of molecular simulations, the authors proposed that the better structure-directing ability of B was because of its ability to adopt a configuration when occluded in the CFI channels that developed more Van der Waals interactions with the zeolite framework than diastereoisomer A. However, despite the cation being incorporated intact in the channels, CIT-1 crystallized in the achiral space group *Imma*, and hence no transfer of chirality had occurred. In the presence of boron, the AFI framework was crystallized also with this isospartein derivative (B). In the same work, the authors also prepared another sparteine derivative with two quaternary N ammonium groups (N(1),N'(16)-endo-methyl-sparteinium) as SDA, but this dication was unsuccessful to direct the crystallization of zeolite frameworks, probably because of its high hydrophobicity and rigidity. Corma and coworkers have also successfully used the N(16)-methyl-sparteinium (cation A) for the crystallization of a new zeolite material, ITQ-21, in combination with fluoride anions and germanium in the synthesis gel [42]. ITQ-21 is composed of large cavities and wide pore openings, and exhibited very good catalytic properties. However, again this is not a chiral framework, evidencing once again that the asymmetric nature of the spartein building-block did not play any role during the crystallization process.

Kubota and coworkers designed a series of chiral SDA cations derived from the natural enantiomerically pure (−)-β-pinene monoterpene and its amino-containing derivative (−)-*cis*-myrtanylamine [44]. Quaternization of the latter amine through the addition of three methyl groups yielded trimethyl-*cis*-myrtanylammonium (cation D in Fig. 1), while trimethyl-*trans*-myrtanylammonium (E in Fig. 1) was prepared by a more elaborate procedure from (−)-*trans*-myrtanol; both diastereomeric cations were thus prepared in homochiral form. (−)-*Cis*-myrtanylammonium cation (D) led originally to the production of a new borosilicate zeolite material, CIT-1, with the CON framework type, which was formed by intersecting 10-ring and 12-ring pores [45]; CIT-1 is the ordered polymorph B of SSZ-33. The enantiomer (+)-*Cis*-myrtanylammonium cation was also prepared and led also to the same CIT-1 material (indistinguishable at least from the XRD patterns), suggesting that the homochirality of the cation did not play a role during the crystallization of the ordered polymorph B in CIT-1; indeed, the space group of the CON framework was achiral $(C\ 1\ 2/m\ 1)$. Furthermore, the diastereoisomer trimethyl-*trans*-myrtanylammonium (E in Fig. 1), as well as the unsaturated derivative trimethyl-myrtenylammonium (F in Fig. 1), also led to the crystallization of the same CON framework in the presence of boron [44]. In line with these SDAs, the authors also prepared another chiral cation with the same myrtanyl polycyclic ring, N,N,N-trimethyltricyclo[5.2.1.0^{2,6}]-decaneammonium (cation I in Fig. 1), and in this case they observed the crystallization of SSZ-31 [45], whose structure was later on solved as the interrupted *STO framework type formed as an intergrowth of several polymorphs [51]. Another derivative of $(-)$-*cis*-myrtanylamine [(1,1-dimethyl-4-piperidinium)methyl]-dimethyl-*cis*-myrtanylammonium (cation C in Fig. 1) directed the crystallization of the MTW framework type, formed by 12-ring one-dimensional channels; however, this cation seems too large to be accommodated within the MTW 1D-channels, and indeed ^{13}C NMR results showed that in this case the cation did not resist the hydrothermal treatment and fragments produced upon in-situ decomposition of cation C were incorporated within the zeolite [44]. Several other cations derived from the myrtanyl ring were also prepared in the same work (Fig. 2) in an attempt to produce zeolite Beta enriched in the chiral polymorph A, but these attempts were unsuccessful and gave no crystalline products under the conditions tested [44].

Kubota and Davis prepared cation G with the enantiopure alkyl substituents (although as a mixture of diastereoisomers as a result of new stereogenic N centers upon asymmetric quaternization) (Fig. 1) [17]. This cation led to the crystallization of all-silica zeolite Beta, but details of this work were not reported. In any case, this report is interesting since it represents one of the few examples published in the literature where zeolite Beta is obtained in the presence of a chiral SDA cation (let us remember that one of the polymorphs of zeolite Beta, A, is chiral, and has not yet been crystallized as a pure polymorph, see Sect. 4.1 below).

Xie and coworkers synthesized cation H through a complex organic synthesis sequence (Fig. 1) [46], which was successfully used to prepare the new zeolite SSZ-52. Its framework structure (SFW) was composed of large cavities where pairs of the bulky organic cations were hosted. However, this framework is not chiral (R-$3m$ space group), and indeed no mention to the asymmetric nature of the organic cation was given in the report.

In an investigation through a high-throughput approach to understand the factors governing the crystallization of extra-large pore zeolites, Jiang and coworkers used cations J and K (Fig. 1) as SDAs for the synthesis of germanosilicate materials [47]. Both of these cations (OSDAs 3 and 4 in Ref. [48]) contain one stereogenic center, and hence are chiral compounds. The authors also used two other large SDA cations with stereogenic centers , but these cations had a symmetry plane and hence were achiral *meso*-compounds (OSDAs 5 and 6). In any case, the authors did not mention about the chirality of those SDAS, and hence it is assumable that they used these chiral compounds as a racemic mixture. Results showed that, depending on the synthesis conditions (amount of Ge, Al/B, H_2O, and OSDA), cations J and K drove the crystallization of zeolite Beta, ITQ-7 and ITQ-21. Again the interest here is the crystallization of zeolite Beta, with one of the polymorphs being chiral

(A), by using a chiral cation as SDA. Nevertheless, one would not expect any transfer of chirality to the zeolite since racemic compounds were used.

One of the most typical examples of enantiopure chiral natural products derived from the chiral pool is of course provided by L-aminoacids. As such, these molecules are not proper SDAs for the synthesis of zeolites because of the presence of the acid group. However, we have been able to produce crystalline AlPO-based low-dimensional materials by using L-lysine as organic guest species because of the presence of an additional amino-group in the residue that provides the required basicity [56]; similarly, Dong and coworkers have also produced zinc phosphite/phosphate networks in the presence of L-tryptophan and L-histidine [57]. If one wants to use these molecules as chiral organic SDAs for the synthesis of zeolite materials, one needs to get rid of such acid group, for instance by reducing the acid to an alcohol group. In our group we selected L-proline as chiral precursor because of the rigidity associated with the pyrrolidine ring, which reduces the mobility of the stereogenic center and hence of the asymmetric group. L-proline can be easily reduced to L-prolinol by LiAlH$_4$, and the secondary amine can then be alkylated to give an enantiopure quaternary ammonium compound which retains enantiopurity (such alkylation reactions do not involve the stereogenic center). Moreover, the addition of two different alkyl groups to the N atom generates a new stereogenic center with four different substituents attached. Interestingly, depending on the synthesis route employed, we were able to obtain the different diastereoisomers. Starting from L-prolinol, the addition of benzyl chloride produces (S)-N-benzyl-prolinol; subsequent addition of methyl iodide occurs through only one possible stereochemistry due to the steric constraint provided by the bulky benzyl substituent, giving place exclusively to the (1S,2S)-quaternary ammonium cation ((1S,2S)-2-Hydroxymethyl-1-benzyl-1-methylpyrrolidinium) (cation L in Fig. 1). In contrast, if the substituent attached first to L-prolinol is the smaller methyl group, subsequent addition of the bulky benzyl group can now take place from both sides (due to the less bulky nature of methyl), hence giving place to a (50:50) mixture of both diastereoisomers [48]. Moreover, we were able to obtain a mixture enriched in isomer (R,S) (up to 77%) by successive recrystallizations [50]. When the (S,S) diastereoisomer was used as SDA for the synthesis of pure-silica zeolites in fluoride media [48], this led to the crystallization of ZSM-12 (MTW); this framework is composed by 12-ring one-dimensional channels. Interestingly, the use of the mixture of isomers (S,S and R,S) as SDA prevented the crystallization of this framework, evidencing that the latter diastereoisomer could not direct the crystallization of the framework, and even inhibited the formation of ZSM-12 by the (S,S) isomer (experiments with the mixture of isomers where the total amount of the (S,S) isomer was the same as in the experiment with enantiopure (S,S) that did lead to MTW also resulted in amorphous materials). Molecular simulations showed a worse fitting of the molecular structure of the (R,S) isomer than that of the (S,S)-isomer, thus explaining their different structure-directing efficiency. These diastereomeric compounds also led to the formation of FER-type materials with a certain separation between the FER sheets in the presence of tetramethylammonium as co-SDA [50]. However, the bulky nature of these cations forced the FER sheets to expand a certain distance to accommodate the organic chiral cations between them, a common case in FER-systems, and hence this expanded FER systems could host both diastereomeric cations with no observable differences. In a subsequent work, the (*S,S*) cation directed also the crystallization of the MWW framework in the presence of tetramethylammonium as co-SDA [49]. All this series of works showed diastereoselective structure-directing effects during the synthesis of zeolite materials, which are expected due to the different shape/size of the diastereoisomers. However, still all the zeolite materials produced were not chiral (MTW, FER, and MWW), evidencing once again the difficulty in transferring the chirality from the asymmetric nature of the organic cation into the zeolite framework. In a very recent work, Martínez-Franco et al. have used other quaternary ammonium derivatives of *L*-prolinol (cations *M* and *N* in Fig. 1), which allowed for the synthesis of Al-containing CDO precursors, but no mention to chirality was given in the report [53].

Finally, very recently Davis and coworkers have successfully used cation *O* (Fig. 1) to promote the enantioselective crystallization of the chiral STW framework [52]; this will be explained in detail in Sect. 4.3 below.

Figure 3 shows the organic chiral species that have been used as SDAs for the synthesis of zeolitic materials based on compositions other than SiO_2 [35, 58–66]. The synthesis of aluminophosphate-based (AlPO) microporous frameworks is more versatile than that of silica-based materials since it can take place at neutral and acid pHs, hence allowing the use of (usually protonated) amines (rather than quaternary ammonium compounds) which are more efficient SDAs than for the synthesis of zeolites (which usually occurs at higher pHs where amines remain neutral, though in cases they can also produce zeolites). Following the strategy of searching in the chiral pool, Komura and coworkers prepared the chiral diamine (*S*)-(+)-1-(2-pyrrolidinylmethyl)pyrrolidine (*P* in Fig. 3) from *L*-proline [58]. When *P* was used as SDA for the synthesis of AlPO materials in fluoride medium, a new material denoted GAM-1 was produced. GAM-1 was precursor of the AlPO-CHA framework, and the chiral diamine with two pyrrolidine rings nicely fits within the CHA cages. However, no transfer of chirality from the molecule to the AlPO framework seems to have taken place.

In the same line, in our group we used (S)-N-benzyl-prolinol (*Q* in Fig. 3) as an enantiomerically pure chiral SDA for the synthesis of microporous aluminophosphates [59, 60]. A series of MAPO materials with the AFI framework was obtained in the presence of different dopants (Mg, Zn, Co, Si, V) [59]. This structure is based on one-dimensional 12-ring channels, as previously explained, and hence apparently no transfer of chirality would have occurred. However, we have followed a different approach here in an attempt to induce chirality through chirally ordered spatial distributions of dopants, which will be explained in Sect. 5. When we increased the concentration of the organic SDA and the Zn content, we were able to produce the Zn-containing version of the SAO framework for the first time [60]; a similar Zn-SAO material was obtained with the related molecule N-benzylpyrrolidine, suggesting that the presence of the hydroxymethylene substituent and the stereogenic center did not play an important role in the crystallization of this achiral framework.

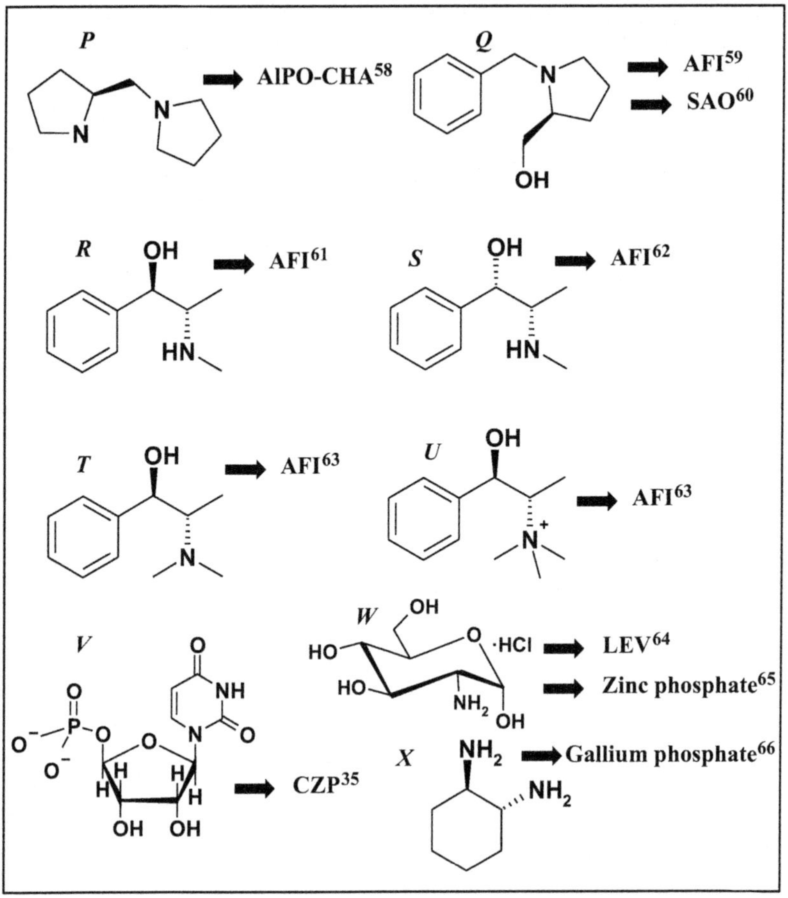

Fig. 3 Chiral organic molecules used as structure-directing agents that directed the crystallization of zeotype frameworks (AlPO, ZnPO, or GaPO-based compositions)

We have also been recently studying the structure-directing efficiency of $(1R,2S)$-$(-)$-ephedrine and its diastereoisomer $(1S,2S)$-$(+)$-pseudoephedrine as chiral precursors for the production of AlPO-based frameworks. Ephedrine and pseudoephedrine are alkaloid natural products which are found in nature in the *Ephedra* species of plants [67], and as such are commercially available in enantiopure form at a reasonable cost. These alkaloids display a biological activity on the sympathetic nervous system, although their pharmacological behavior is different depending on their absolute configuration. We have extensively studied the behavior of these molecules as SDAs for the synthesis of nanoporous aluminophosphates [61, 62]. As usual for these phenyl-containing aromatic amines, these molecules led to the crystallization of the AFI framework in the presence of different dopants (Mg, Zn, or Co); the most stable AFI system was MgAPO-5

[68]. In particular, in order to promote the development of helicoidal arrangements of dopants, as will be explained in Sect. 5 below, we were interested in analyzing the supramolecular behavior of these chiral molecules driven by the development of π-π type interactions between the aromatic rings on the one side, and of H-bond interactions between the OH (as H-bond acceptor) and NH_2^+ (as H-bond donor) groups of consecutive dimers packed within the AFI nanochannels. Indeed, a combination of fluorescence spectroscopy and molecular simulations showed that ephedrine displays a much stronger supramolecular trend than its diastereoisomer pseudoephedrine due to a distinct conformational space of the two isomers, which is in turn driven by the development of different intramolecular H-bonds as a consequence of their different stereochemistry [62] (see Fig. 4): the most stable conformer of ephedrine involves the alkyl chain in an extended-configuration that enables the supramolecular aggregation, whilst the most stable conformation of pseudoephedrine has the alkyl chain in a folded-conformation that inhibits the formation of dimers (these are only formed under high SDA concentrations). Moreover, such distinct stereochemistry also triggers the occurrence of different orientations of the π-π stacked dimers of these chiral molecules embedded within the AFI nanochannels [69] (Fig. 4). On the other hand, we also added one ((1R,2S)-methyl-ephedrine) and two ((1R,2S)-dimethyl-ephedrinium) methyl groups to ephedrine, and observed again the crystallization of the AFI framework when using these new chiral derivatives as SDAs [63]; results showed that the hydrothermal resistance of the dimethyl-quaternary ammonium cation was limited. In this case, we observed that the addition of methyl groups involved a notable reduction of the incorporation of supramolecular dimers within the channels. We ascribed this to the more difficult (for methyl-ephedrine) or impeded (for dimethyl-ephedrinium) formation of H-bonds between consecutive dimers when embedded within the AFI frameworks that reduce the repulsive electrostatic interactions established between the closely located positive charges of the N atoms under the dimer configuration (see Fig. 4, bottom-left). In summary, throughout our series of studies analyzing the structure-directing behavior of these chiral derivatives, we have observed interesting packing effects governing the supramolecular chemistry of these molecules during the crystallization of nanoporous frameworks as a consequence of their particular stereochemistry. However, as previously stated, the AFI framework is not chiral itself, which denotes the inability of these molecules (under these conditions) to promote the formation of chiral frameworks. Nonetheless, these molecules are suitable candidates to promote instead the development of chiral spatial distributions of dopants, as will be explained in Sect. 5.

Although their role has not been that of an organic chiral SDA to direct the crystallization of zeolite frameworks, worth is mentioning the use of two chiral biomolecular additives in the preparation of zeolite materials. The first example is the use of chiral nucleotides as chiral auxiliaries for the enantioenrichment of crystals of the chiral zincophosphate (CZP) [35]. CZP has an intrinsically chiral topology, which crystallizes as a racemic mixture of crystals (50:50 in $P6_122$ or $P6_522$ space groups). In their work, the authors proposed the use of a ribonucleotide (uridine-5′-monophosphate, ump, V in Fig. 3) as a chiral additive in order to create

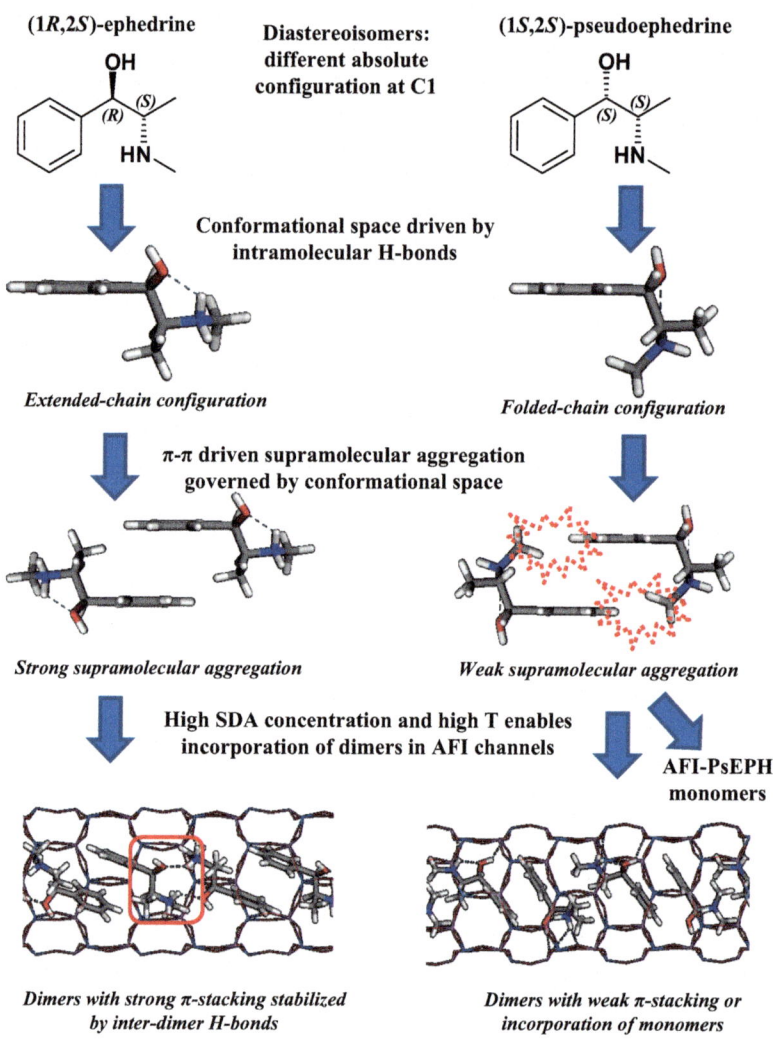

Fig. 4 Schematic picture with the structure-directing effect of (1R,2S)-ephedrine (*left*) and (1S,2S)-pseudoephedrine (*right*) and their associated supramolecular chemistry during occlusion within the AFI channels. Adapted with permission from [62], Copyright (2015) American Chemical Society

homochiral solids or at least enantioenrich the CZP crystals, under the hypothesis that these chiral compounds might develop enantioselective (diastereomeric) interactions with the surface of the CZP enantiomorphic crystals through their phosphate groups. Such interactions of the enantiopure nucleotide should be different with the two enantiomers of CZP, and hence this could potentially lead to a differentiation between the enantiomorphic crystals. Indeed, the authors observed by Circular Dichroism Spectroscopy and single-crystal structural refinements that the addition of a small amount of ump to the crystallization mixture did result in an enantioenrichment of the CZP crystals in one of the enantiomers (the one with $P6_122$ space group, giving an approximate enantiomeric excess of 85%). The authors showed the crucial importance of the binding mode of the nucleotide to the growing $ZnPO_4$-based zeotype surface which had to occur through the phosphate group (inosine-5'-monophosphate nucleotide that usually binds to metals through the hypoxanthine ring was not able to promote a similar asymmetric crystallization). This work represented a milestone in the crystallization of enantiopure or enantioenriched chiral zeolite frameworks, although to our knowledge this strategy has not been extended to other systems so far. Nonetheless, in their work Zhang and coworkers have used ump as a chiral auxiliary, but not as a chiral SDA. Of course one rapidly makes the connection between phosphate-containing chiral nucleotides with their potential as SDAs for the synthesis of zeolitic materials, especially those based on PO_4^{3-} networks (such as microporous aluminophosphates). These should be ideal candidates to promote the crystallization of chiral microporous AlPO-based frameworks since the asymmetric unit and one of the building units of the zeolitic network (PO_4^{3-}) are directly linked and an impression of the asymmetric nature of the chiral organic precursor into the zeolite framework should be more easily achievable. Indeed, a direct connection by covalent bonds of L-proline aminoacid and phosphate groups has enabled the transfer of chirality from the organic component into a homochiral 3-dimensional zinc phosphonate with helical channels [70]. However, the role of these biomolecular nucleotide-based species as SDAs might be severely hindered by the low hydrothermal stability of these compounds (Zhang and coworkers showed that at an intermediate crystallization temperature of 120 °C the chirality induction effect was lost, probably due to the hydrolysis of the nucleotide into phosphate groups and uridine [35]). In this context, in our group we have used adenosine-monophosphate as SDA for the synthesis of AlPO materials, but these experiments have been unsuccessful so far [71], partly because of its low hydrothermal stability. In any case, diastereomeric interactions between biomolecules essential for life and inorganic crystals have a very strong interest in the scientific community because of their particular relevance to the unsolved problem of the origin of homochirality in the molecules of life [5].

Another important class of chiral biomolecules is provided by D-sugars which are part of biopolymers. In this context, there is one chiral sugar derived from D-glucose which contains an amino-group that provides the required basicity for SDA molecules, D-glucosamine (W in Fig. 3). This molecule has been used for the crystallization of two novel zinc phosphate phases, a mesoporous phase and a lamellar phase [65], but no transfer of chirality was apparent. In another work, we used D-glucosamine as additive for the synthesis of SAPO-35 (LEV structure type) in an attempt to promote the formation of mesopores. However, we observed an unexpected behavior: the addition of this amino-containing sugar led to a dramatic reduction of the crystal size [64]. This surprising effect was ascribed to a binding of D-glucosamine molecules on the growing surface of the SAPO-35 crystals which hinders further growth, favoring instead the nucleation process and hence the occurrence of small crystals. However, in none of the presented examples the intrinsic chirality of D-glucosamine has been transferred to the resulting porous material, but worth is further exploring the ability of these chiral species to transfer their chirality in zeolite frameworks, especially those based on AlPO composition since the crystallization of these is more favorable in the presence of amines (instead of quaternary ammonium cations). In any case, we note again the low hydrothermal stability of D-glucosamine under the crystallization conditions that will force the use of low crystallization temperatures, thus reducing the chances of getting 3D-zeolite frameworks.

Although it did not lead to the production of zeolite-type frameworks, for the sake of completeness we also include the use of enantiopure $(1R,2R)$-$(-)$-1,2-diaminocyclohexane (X in Fig. 3) as organic SDA [66]. In this case, the organic chiral amine led to the production of a new low-dimensional gallium phosphate material composed of infinite chains of GaO_6 octahedra bridged by phosphate groups, which make the material unstable upon calcination. Interestingly, when the racemic amine is used, another material is obtained, with the same inorganic chains but a different arrangement of the organic ammonium cations.

In summary, we have presented in this section a list of chiral organic species that have been used as SDAs for the crystallization of zeolite frameworks. Except for one case that will be explained below [52], all these chiral SDAs share the fact that they have not been able to imprint their asymmetric nature onto the nascent zeolite frameworks during the structure-directing phenomenon.

3   Chiral Organometallic Structure-Directing Agents: Transfer of Chirality from Transition-Metal Complexes Through H-Bonds

Another important class of chiral compounds is provided by organometallic compounds, i.e. coordination complexes with transition metals. In these complexes, the ligands are organized around the transition metal in a particular spatial configuration. Their unique chemical properties involve that they can adopt particular conformations that cannot be obtained with usual organic molecules. Interestingly, their geometrical configuration brings the possibility of having chiral coordination complexes. In this case, chirality is not given by the presence of stereogenic centers, as typically in organic molecules, but by the spatial configuration of the ligands. Certain octahedral complexes with a six coordination number in which there are several bi- (or tri)-dentate ligands possess 3-dimensional structures which are not superimposable with their mirror images. Chirality in these compounds is manifested through the spatial configuration whereby the bidentate ligands bind to a given transition metal in an octahedral environment (see Fig. 5). This type of chirality is often referred to as *propeller chirality*: a right-handed twist of a propeller is not superimposable to its left-handed version (Fig. 5-bottom).

Fig. 5 Schematic picture of the *propeller* chirality of octahedral transition-metal coordination compounds with bidentate ligands

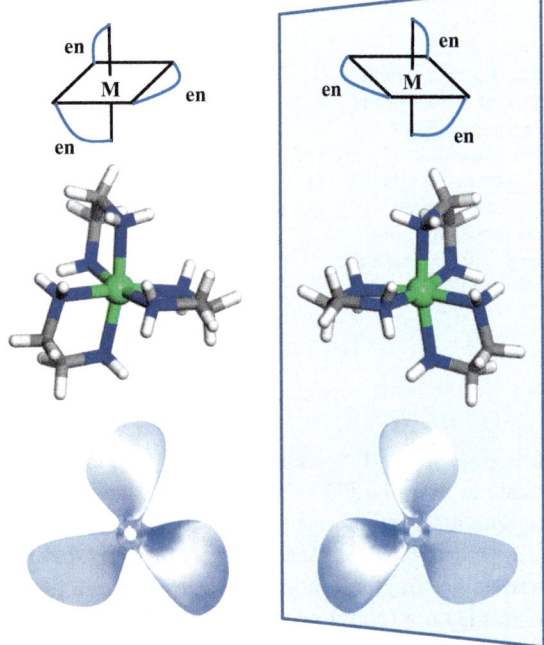

This type of organometallic compounds has been extensively used for the synthesis of zeolitic materials usually based on compositions other than silica (mainly aluminophosphates and gallogermanates) [72–81]. Of course for these metal complexes to act as SDA species, they must be stable under the hydrothermal synthesis conditions, what precludes the use of a large number of complexes where an exchange of ligands would readily occur. Figure 6 reports the most commonly used chiral octahedral complexes, which are stable enough to resist the hydrothermal synthesis conditions.

Balkus et al. introduced the use of metallophthalocyanines coordination complexes to produce zeolites with the FAU structure, although not all the cavities were occupied with these complexes [82]. Since then, many studies have analyzed the use of chiral coordination complexes as chiral SDAs. The essential requirement for these complexes is that they must resist the crystallization process without racemizing. In this regard, several chiral octahedral complexes with Co^{3+} cations are kinetically stable under hydrothermal conditions, and hence have been successfully used for the preparation of low-dimensional aluminophosphate, gallophosphate, and zincophosphate-based materials (see Fig. 6). Morgan and coworkers used the Co(III) tris(ethylendiamine) [Co(en)$_3^{3+}$] complex (cation *A* in Fig. 6) as SDA and produced a low-dimensional chiral AlPO framework [Co (en)$_3$Al$_3$P$_4$O$_{16}$·3H$_2$O] [72], in which the chirality of the organometallic complex was directly imprinted onto the aluminophosphate layers. In this type of low-dimensional AlPO frameworks, Al is usually linked to 4 P atoms (through O

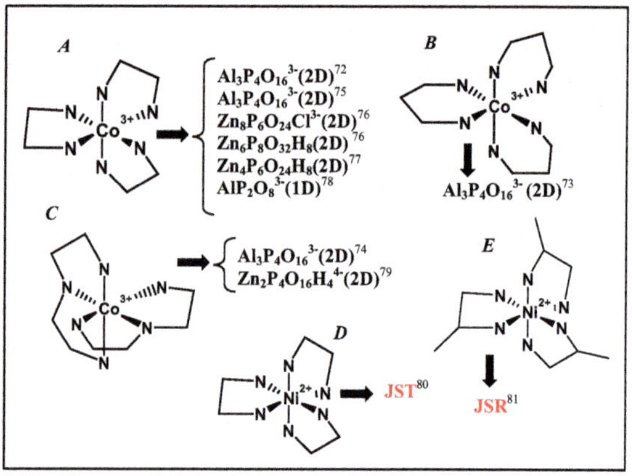

Fig. 6 Some of the most common chiral organometallic coordination complexes used as SDAs for the crystallization of low-dimensional or zeotype frameworks (AlPO, ZnPO, or GaGeO-based compositions); chiral zeolite frameworks are shown in *red*

bridges), whilst P is usually bonded to less than 4 Al (through O bridges), giving place to terminal PO^- groups. In this material, the $Al_3P_4O_{16}$ framework formed a unique macroanionic sheet, and the chiral $Co(en)_3^{3+}$ cations and water molecules were located in the space between the layers. In the $Al_3P_4O_{16}$ sheets, the 4-rings had a tricyclic structural motif which resembles a [3.3.3] propellane, hence providing a chiral feature (chiral pockets) where the chiral $Co(en)_3^{3+}$ cations sit. However, the use of racemic DL-$Co(en)_3^{3+}$ involved that both enantiomers were included in the material, and hence overall the crystal structure of this layered AlPO framework was not chiral: it contained both enantiomers of the complex in the same amount, and hence the same number of chiral pockets of opposite handedness. Nonetheless, the discovery of this framework provided evidence that chiral organometallic complexes were able to induce chiral features in an inorganic lattice. In fact, the authors found that such a transfer of chirality must be directly related to the extensive H-bond network established between the complex (through the NH_2 groups) and the terminal PO^- groups. Bruce and coworkers used in the same time Co(III) tris-(1,3diaminopropane) $[Co(tn)_3^{3+}]$ (cation *B* in Fig. 6) to produce a new $Co(tn)_3 \cdot Al_3P_4O_{16} \cdot 2H_2O$ low-dimensional framework [73], where again the $Co(tn)_3^{3+}$ cations were located in the interlayer space between the AlPO anionic layers. Remarkably, in this new material each single crystal contained only one enantiomer of the complex despite the use of a racemic mixture in the synthesis, providing an example of spontaneous resolution to give a racemic conglomerate (a mixture of crystals each containing a single enantiomer). This material was a chiral solid, its chirality being due to the presence of such single enantiomer of the complex between the layers, but the chirality was not transferred to the inorganic framework. In a subsequent work, Bruce and coworkers used *trans*-Co-bis-(2-aminoethyl)amine) $[Co(dien)_2^{3+}]$ complex (cation *C* in Fig. 6) to produce a new chiral low-dimensional material, $Co(dien)_2 \cdot Al_3P_4O_{16} \cdot 3H_2O$ [74] (space group $P6_522$). This material consisted of chiral aluminophosphate macroanionic layers stacked in a helical fashion occluding a single enantiomer of $[Co(dien)_2^{3+}]$ between the layers (see Fig. 7-top), although again the use of a racemic mixture involved the production of a crystalline racemic conglomerate (a 50:50 mixture of homochiral crystals with opposite handedness). Nonetheless, the incorporation of a single enantiomer of the complex on each crystal involved a strong chiral molecular recognition phenomenon during the crystal growth. Again H-bond interactions (see dashed circles in Fig. 7) were believed to play a fundamental role during the transfer of chirality from the complex to the chiral layer.

In a step further, Gray and coworkers produced a new chiral aluminophosphate framework based on a stacking of chiral macroanionic $Al_3P_4O_{16}^{3-}$ sheets similar to those found previously, but in this case they used enantiomerically pure d-Co(en)$_3^{3+}$ as the SDA (cation A in Fig. 6) [75]. The main interest of this new material was that there was only one enantiomer of the chiral AlPO sheet in any crystal of the material, with only enantiomer d-Co(en)$_3^{3+}$ intercalated between the chiral AlPO sheets (see Fig. 7-bottom). Interestingly, the chiral $Al_3P_4O_{16}^{3-}$ sheets here were different than those of $[Co(en)_3Al_3P_4O_{16}·3H_2O]$ [72] mentioned previously, prepared with the same cation but in a racemic form (remember that in the latter a mixture of both

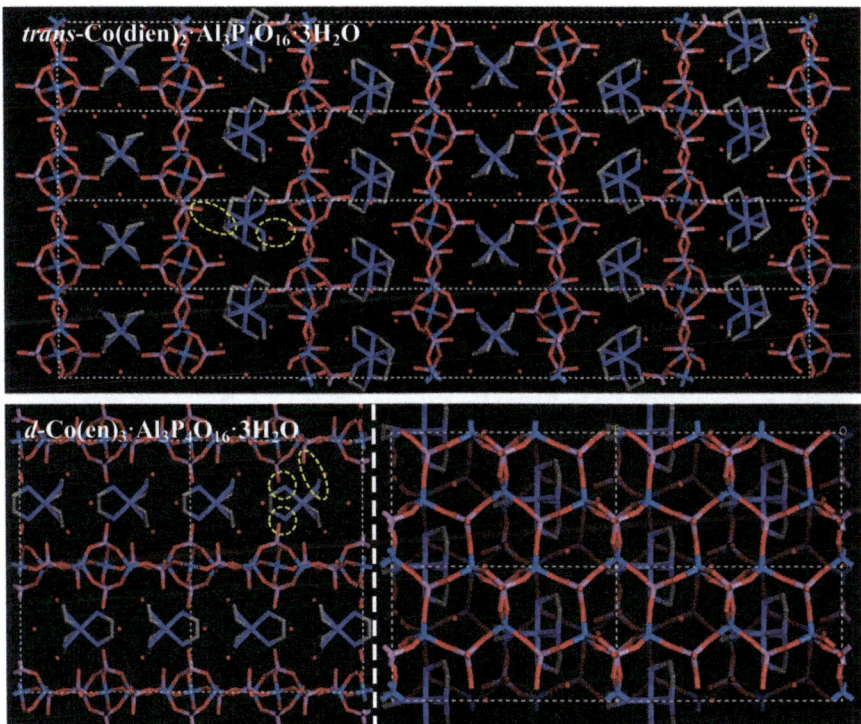

Fig. 7 Structure of *trans*-Co(dien)$_2$·Al$_3$P$_4$O$_{16}$·3H$_2$O (*top*) [74] and of d-Co(en)$_3$·Al$_3$P$_4$O$_{16}$·3H$_2$O (*bottom*) [75] (H atoms are not shown). O atoms are shown in *red*, Al in *blue*, P in *pink*, N in *dark blue*, Co in *light blue,* and C in *grey*

enantiomers was present in the framework). Instead the AlPO sheets in the present case were similar to those prepared with [Co(dien)$_2$$^{3+}$] (where only one enantiomer was present on each single crystal) [74]. This work provided evidence that homochiral inorganic frameworks can be built using homochiral structure-directing guest species. As usual, an extensive H-bond network seemed to be responsible for the transfer of chirality to the chiral inorganic sheets (see dashed circles in Fig. 7-bottom).

Optically pure d-Co(en)$_3$$^{3+}$ (A in Fig. 6) has also led to the production of a chiral 1-dimensional aluminophosphate chain compound, d-Co(en)$_3$[AlP$_2$O$_8$]·6.5H$_2$O (AlPO-CJ22) [78] (Fig. 8). The framework structure of AlPO-CJ22 was similar to

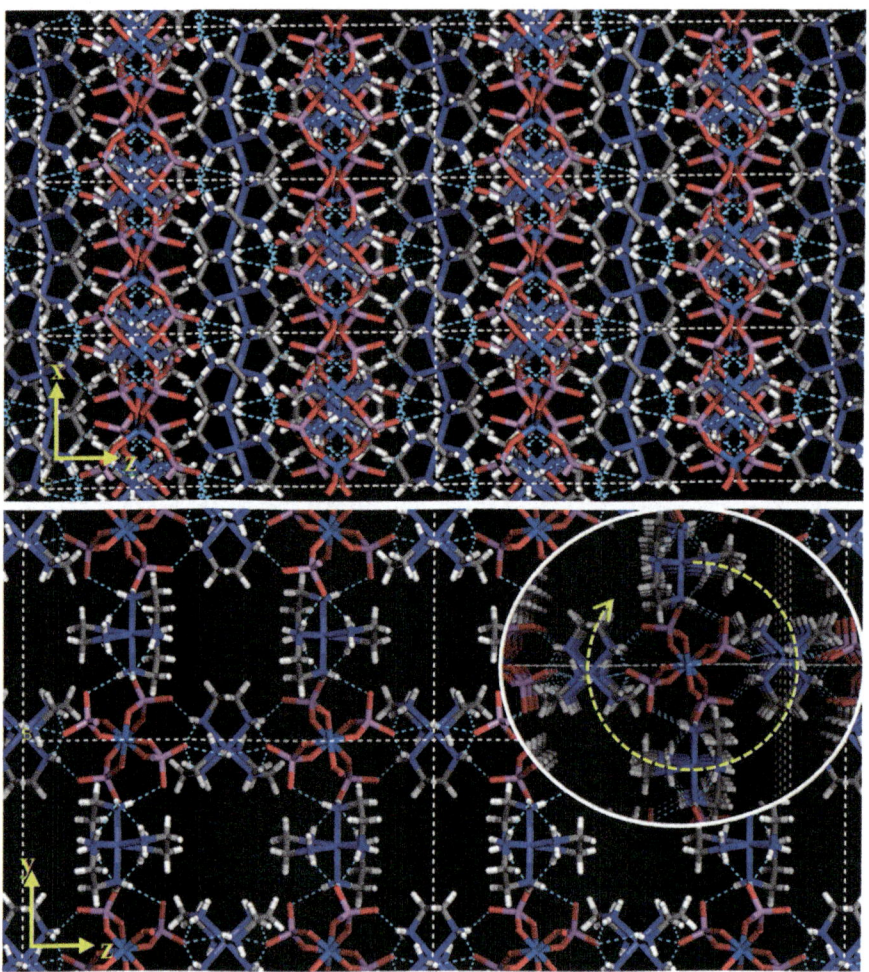

Fig. 8 Two views (*top* and *bottom*) of the structure of d-Co(en)$_3$·AlP$_2$O$_8$·6.5H$_2$O, highlighting the H-bond helix around the AlPO chain [78]. O atoms are shown in *red*, Al in *blue*, P in *pink*, N in *dark blue*, Co in *light blue*, C in *grey*, and H in *white*. *Dashed blue line* indicates H-bonds

that of the previously reported $Co(en)_3[AlP_2O_8] \cdot xH_2O$ [83], in which racemic Co $(en)_3$ was disordered between the chains. In contrast, the d-$Co(en)_3^{3+}$ cations in AlPO-CJ22 displayed a chiral order in which they follow a 2_1 screw axis along the [84] direction, interacting with the AlPO chains through extensive H-bonds in a helical fashion between the N atoms and the terminal O atoms of the inorganic chain (see Fig. 8, inset). Indeed, measurement of the optical rotation clearly revealed that the Co complex cations have predominantly the Λ-configuration. Reversibly, optical rotation measurement of the corresponding material produced with l-$Co(en)_3^{3+}$ showed the predominance of the Co complexes in the Δ-configuration [78].

Chiral framework materials with other compositions such as gallophosphates [85, 86], borophosphates [87], or zincophosphates [76, 77, 79] have also been prepared in the presence of chiral Co complexes. Racemic $Co(en)_3^{3+}$ (A in Fig. 6) led to the discovery of two zincophosphate materials, $[Co^{II}(en)_3]_2[Zn_6P_8O_{32}H_8]$ and $[Co^{III}(en)_3][Zn_8P_6O_{24}Cl] \cdot 2H_2O$ [76]; both frameworks contained chiral structural motifs, whose chiral features were transferred from the metal complex, thus resulting in a spontaneous separation of the enantiomers of $Co(en)_3^{3+}$ as Λ- and Δ-configurations in the structure in alternate interlayer regions. In fact, the orderly separation of the enantiomers of the Co complex clearly involved that the inorganic sheets had chiral molecular recognition ability for the asymmetric nature of the Co complex. Such recognition ability was again promoted by the extensive host–guest H-bond network established. Indeed, Wang and coworkers [77] clearly demonstrated that such chiral enantiospecific molecular recognition ability was a consequence of the development of extensive H-bonds between the inorganic chiral motif and the amino-groups of the chiral complexes: such stereospecific correspondence was due to the fact that the number of H-bonds is maximized in a particular diastereomeric interaction (between one enantiomer of the chiral inorganic motif and one enantiomer of the complex), and this number of H-bonds is reduced if the alternative diastereomeric interaction (between one enantiomer of the chiral inorganic motif and the opposite enantiomer of the complex) is established, thus decreasing the stability of the system; this had as a consequence a transfer of chirality of the complex to the inorganic framework. Interestingly, Wang and coworkers also managed to prepare a new zincophosphate framework using a racemic mixture of $Co(dien)_2^{3+}$ (cation C in Fig. 6) where the main interest was that it was not a low-dimensional framework material as those mentioned before, but consisted instead of a 3-dimensional open-framework ($[Zn_2(HPO_4)_4][Co(dien)_2] \cdot H_3O$) with multidirectional helical 12-ring channels [79] (see Fig. 9). In this structure, each 12-ring helical channel accommodated one $Co(dien)_2^{3+}$ cation, indeed a pair of enantiomers in alternating rows. The most attractive structural feature of this open-framework was that the 12-ring channels are enclosed by two intertwined helices of the same handedness (yellow and green polyhedra in Fig. 9-right), connected through them by Zn-O-P linkages. Although the presence of both enantiomers of the chiral Co complex implies having the two handedness of the helicoidal channel within the same framework, this work represented an important milestone since it evidenced that the chirality of organometallic compounds can be

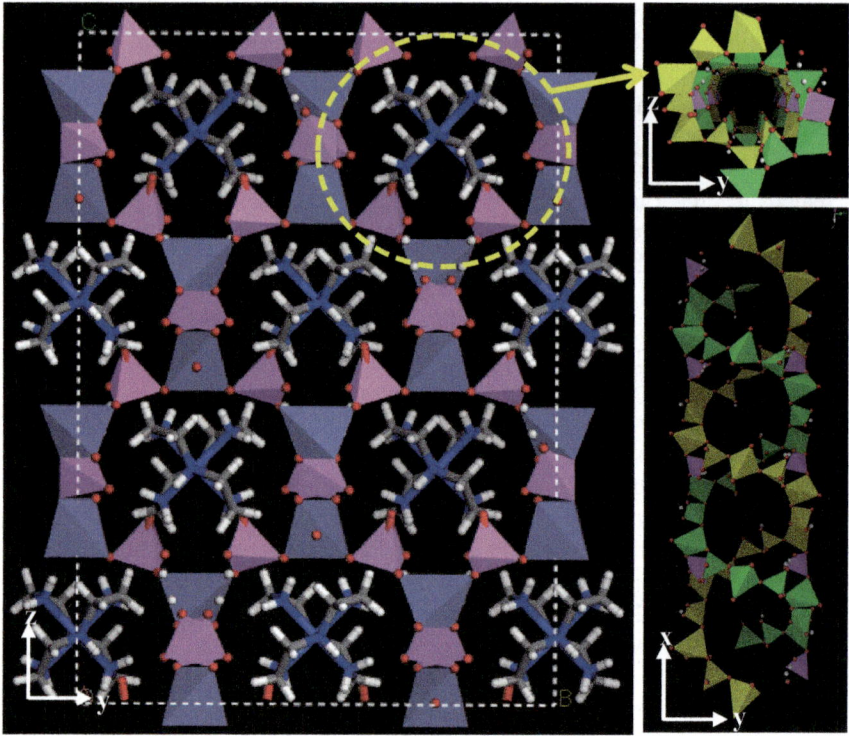

Fig. 9 Structure of ($[Zn_2(HPO_4)_4][Co(dien)_2]\cdot H_2O$ (*left*), highlighting the helicoidal 12-ring channel with the two strands in different colors (*right*) [79]. O atoms are shown in *red*, Zn and P as *grey* and *pink* polyhedra, N in *dark blue*, Co in *light blue*, C in *grey*, and H in *white*. The two intertwined helices are highlighted as *yellow* and *green* polyhedra (*right*). Adapted with permission from [79], Copyright (2003) John Wiley and Sons

transferred not only to chiral pockets or sheets in low-dimensional frameworks, as previously shown, but also to 3-dimensional open-framework materials based on helicoidal (chiral) channels. Once again, the analysis of the H-bond pattern showed this to be the origin for the transfer of chirality. However, despite their 3D open-framework structure, the presence of PO_4 tetrahedra with two O linked to Zn and the other two in terminal positions probably makes this material unstable upon removal of the guest species, hence preventing its use for enantioselective operations.

The persistent effort in using chiral complexes as SDAs carried out by the group of Yu and coworkers finally led to the discovery of two new 4-connected chiral zeolite gallogermanate open-frameworks (JST and JSR), in this case using Ni^{2+} chiral complexes (cations E and D in Fig. 6) [80, 81]. GaGeO-CJ63 (JST) is a zeolitic structure with 3D intersecting 10-ring channels, built exclusively upon 3-rings, and was synthesized using $[Ni(en)_3]^{2+}$ in racemic form as the SDA (Fig. 10-top) [80]. The structure is composed of cages with C_3 symmetry, which is the same as

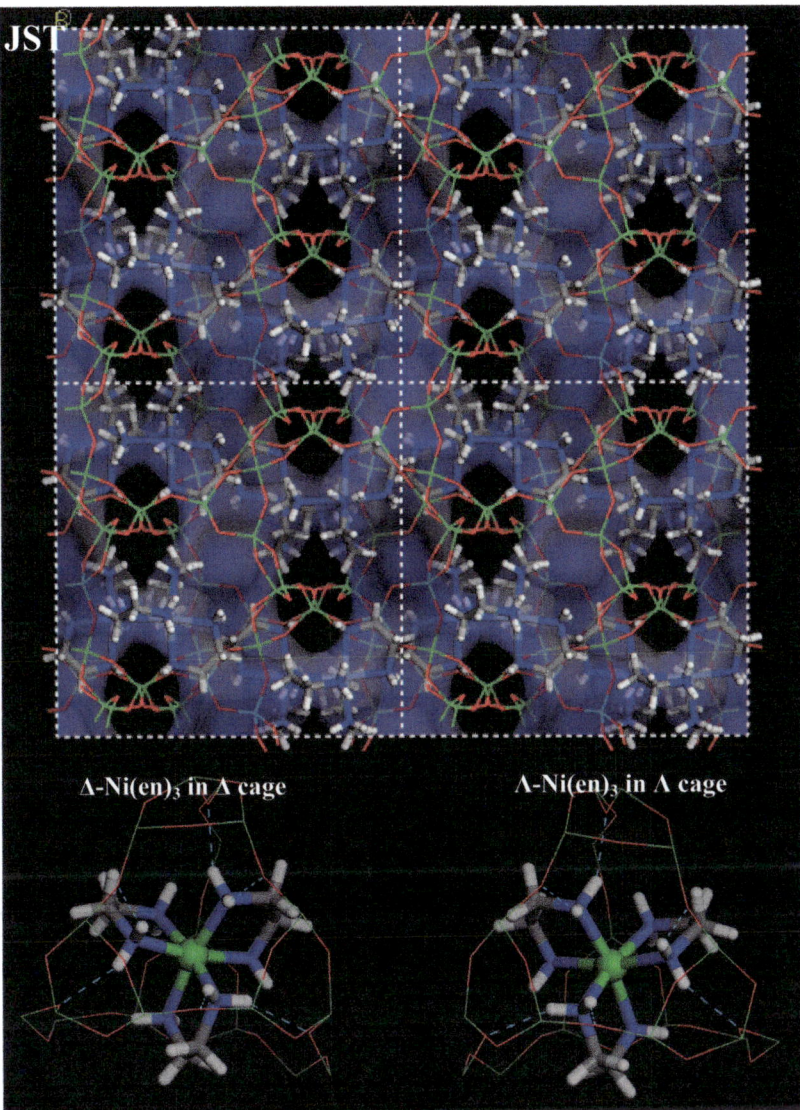

Fig. 10 Structure of JST structure type (GaGeO-CJ63) (*top*) and stereochemical relationship between the chirality of the complex cations and that of the cages (H-bond interactions are indicated by *dashed blue lines*) [80]. The free volume is shown as a *blue* surface (*top*). O atoms are displayed in *red*, Ga/Ge in *dark green*, N in *blue*, C in *grey*, H in *white,* and Ni in *light green*. Adapted with permission from [80], Copyright (2011) John Wiley and Sons

that of $[Ni(en)_3]^{2+}$. As a consequence, such cages are also chiral, and indeed their chirality is related to that of the complex cations, as illustrated in Fig. 10 (bottom). Molecular simulations showed that the diastereomeric complex–cage interactions found in the experimental host–guest systems (Λ-complex$\cdots\Lambda$-cage and Δ-complex$\cdots\Delta$-cage) are stronger than those of the alternative diastereomeric pairs (Λ-complex$\cdots\Delta$-cage and Δ-complex$\cdots\Lambda$-cage), clearly demonstrating a new example of chiral molecular host–guest recognition phenomenon, induced once again by the development of H-bond interactions (see dashed blue lines in Fig. 10-bottom). However, the use of racemic $[Ni(en)_3]^{2+}$ in the synthesis involves that both enantiomers were incorporated within the framework, each of them confined within the corresponding chiral cage, and hence the overall structure was not homochiral. In a subsequent work, Yu and coworkers also prepared a new chiral zeotype gallogermanate framework (JSR) in the presence of another chiral Ni^{2+} complex cation, Ni-tris(1,2-diaminopropane) $[Ni(1,2\text{-PDA})_3]^{2+}$ (cation E in Fig. 6) [81]. This structure contained novel chiral cavities that occurred as pairs of enantiomers containing or surrounded by the corresponding chiral enantiomers of the Ni complex, providing another example of molecular chiral recognition between the coordination complex and particular asymmetric structural features of zeolite frameworks through the development of H-bond interactions.

In summary, this section provides a list of examples in which the chirality of organometallic coordination compounds has been successfully transferred into oxide-based frameworks, either through chiral pockets, sheets, cages, or channels. All these materials share the common feature that the impression of the chirality of the complex takes place through the development of extensive H-bond interactions between the amino-groups (H-bond donor) of the ligands and O atoms (H-bond acceptors) of the inorganic networks. Indeed, in the case of low-dimensional frameworks, these H-bond interactions are mainly established with the negatively charged $P\text{-}O^-$ groups that interrupt the 3-dimensionality of the framework, maximizing in this way the electrostatic interactions. In contrast, the main drawback associated with the ability of these chiral complexes of imprinting their chirality is that they establish very strong interactions with the frameworks, and as a consequence these host–guest materials are usually not stable upon removal of the chiral guest species in order to generate porosity, what has prevented so far any type of enantioselective application.

4 Targeting Known Chiral Zeolite Frameworks: Cases of Study

So far in this chapter we have dealt with the use of chiral organic or organometallic compounds in an attempt to transfer their chirality into inorganic frameworks. In this section, we change the approach to tackle the problem of searching for chiral zeolites, and we will deal with the chirality of three of the most common zeolite frameworks that are known, and make some considerations of how chiral organic SDA species could potentially transfer their asymmetric nature into them in order to produce enantiopure chiral zeolites.

4.1   The Beta Zeolite Case: The Quest for Chiral Polymorph A

Zeolite Beta (*BEA) was first synthesized by Mobil researchers in 1967 using tetraethylammonium hydroxide as the organic SDA [88]. However, it was not until many years later that its structure was solved independently by Newsam et al. [89] and by Higgins et al. [90] in 1988. Zeolite Beta is a heavy intergrowth of two distinct but closely related polymorphs, the so-called polymorph A and polymorph B, typically in a ratio of 44:56. Both polymorphs have fully three-dimensional pore systems based on 12-rings built upon the same layers, but they differ in the way these layers are stacked on top of each other. In polymorph A, the stacking of the layers takes place through 90° rotations always in the same direction, in either a right-handed 4_1 sequence (RRRR...), giving place to one enantiomer (with space group $P4_122$) or in a left-handed 4_3 sequence (LLLL...), giving place to the other enantiomer (with space group $P4_322$). As a consequence, the cages in polymorph A are arranged in a helical fashion around a four-fold screw axis along the "c" direction, resulting in a helical channel which can correspondingly fold in a right- or left-handed manner, giving place to a chiral framework which can occur as the two possible enantiomers [18] (see Fig. 11-top). In contrast, polymorph B results of a recurrent alternation of the same sheets stacked through 90° rotations of the layers in right- and left-handed fashions (with a sheet stacking sequence of RLRLRL...) giving place to an achiral framework (space group $C2/c$) (see Fig. 11-middle). The underlying problem here is that both polymorphs are in principle equally stable, and hence they occur with almost equal probability in zeolite beta, thus giving place to a highly faulted intergrowth of the two polymorphs. Several other stacking sequences generating other polymorphs have been proposed and some of them synthesized [91].

The discovery of the structure of zeolite Beta triggered a quest for enriching this framework in the chiral polymorph A, if possible in a homochiral (or enantioenriched form). Davis and Lobo were the first to report a zeolite beta with enantioselective catalytic properties (in the epoxide aperture of *trans*-stilbene oxide) by using a chiral SDA in the synthesis, although details were not given [92]; this demonstrated the possibility of the chiral polymorph A of zeolite beta to transfer its chiral nature into a particular chemical process of enantioselective adsorption or catalysis. Molecular simulations showed that a good match of the chiral pore geometry and that of the adsorbed molecule is required for enantioselectivity to manifest [93]. Manning and coworkers claimed enantioselective adsorption properties of zeolite Beta using chiral hydrobenzoin as a probe [94], but in this case the enantioselective adsorption

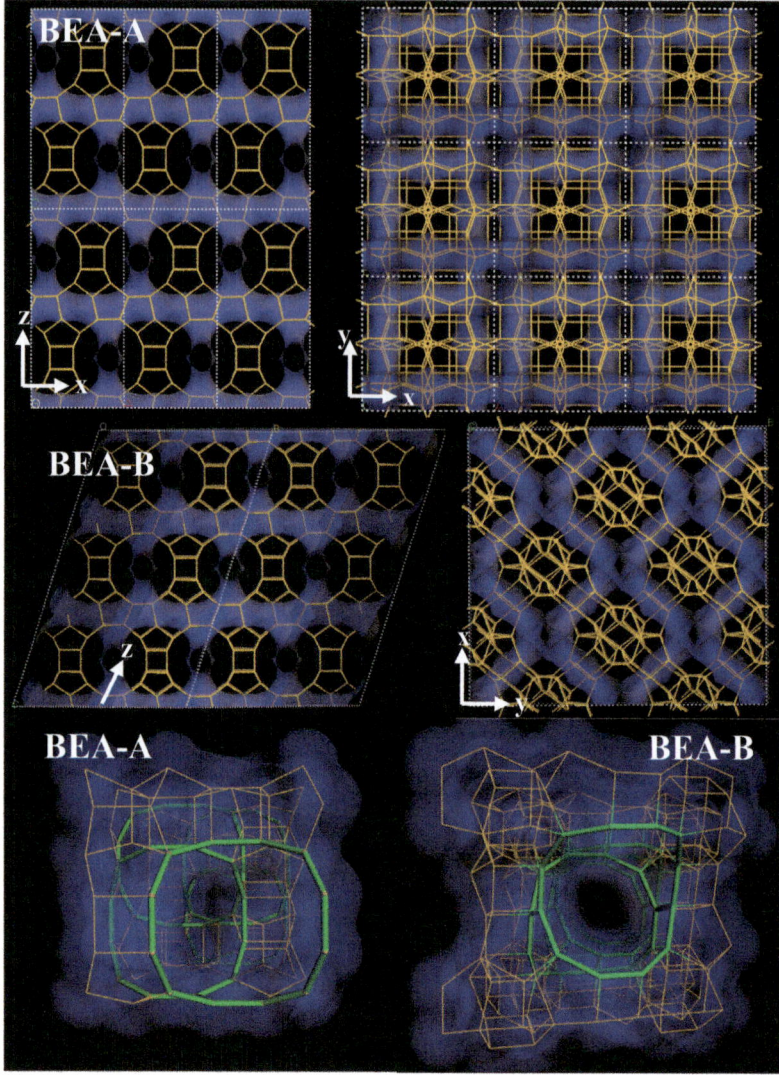

Fig. 11 Different stacking of the Beta sheets giving place to polymorph A (*top*) or B (*middle*), and generation of the distinct channels in the "*c*" direction (*bottom*)

probably came from an enantiomeric equilibrium between the pure enantiomers previously adsorbed on zeolite Beta and the racemic (adsorption) solution until the enantiomeric composition of both became equal.

Despite the great interest in this material, there are not many reports in the literature dealing explicitly with the use of chiral SDA cations to direct the crystallization of this framework towards the chiral polymorph A in homochiral form (see Fig. 1 above). Takagi and coworkers developed a new synthetic methodology of zeolite beta in acidic medium at low temperature in order to enable the presence of chiral amines or chiral rhodium complexes (as secondary SDAs) in an attempt to transfer their chirality to the beta zeolite. In fact the authors reported a slight enrichment in polymorph A, and they claimed that this material was able to enantioselectively adsorb a chiral amine (bis(α-methyl)benzylamine) [95]; however, the authors compared the adsorption from pure enantiomeric solutions but not from a racemic mixture. On the other hand, several reports exist about a partial enrichment of polymorph A in zeolite beta [96]. Xia and coworkers reported that the addition of W, Pd, or Pt to the synthesis of zeolite beta resulted in a partial enrichment in polymorph A, which indeed showed a certain enantioselectivity (with enantiomeric excesses, ee, up to 10%) in the hydrogenation of tiglic acid. However, the origin for this enantioselectivity (in the absence of chiral inductors during the synthesis of the zeolite) is difficult to understand [97]. Recently two groups have claimed new synthetic protocols of zeolite beta that lead to a partial enrichment of polymorph A. On the one hand, Taborda and coworkers used an aging-drying method of the starting synthesis gels (for long times) prior to the conventional hydrothermal treatment which led to enrichments in polymorph A up to 68–70% [98, 99]. On the other hand, Tong and coworkers developed a similar synthesis strategy but this time using an extremely high concentration of the synthesis gels through a dramatic reduction of the water content by heating the gels at 80 °C for several days until reaching a H_2O/SiO_2 ratio of 0.3; this also caused a partial degradation of the tetraethylammonium cations (through Hofmann elimination) to give triethylamine [84, 100]. In fact, it was later demonstrated that the presence of the tertiary amine played a crucial role for the enrichment in polymorph A [101]. These polymorph A-enriched zeolite beta samples doped with Ti showed a surprising enantioselective catalytic activity in the asymmetric epoxidation of bulky alkenes (β-methylstyrene and 1-phenyl-1-cyclohexene), giving enantiomeric excesses up to 11%, even despite the absence of using any chiral precursor in the synthesis of the catalytic materials.

These special stereoselective catalytic properties of zeolite beta further prompt the interest in having a polymorph A-enriched zeolite beta enriched in turn in one of the enantiomorphic forms. If we compare the channel systems of both polymorphs A and B, it can be noticed that the difference between them is on the geometry of the channels along the c axis, which are helicoidal in the case of polymorph A and straight in polymorph B (see Fig. 11-bottom), while channels in the other two dimensions are similar in both polymorphs. Therefore, if one wants to favor the stacking of the beta sheets in the homochiral RRRR... (or LLLL...) fashion of polymorph A, the SDA cation should be rationally designed as to (1) be chiral, (2) be large enough as to interact with more than one sheet along the c axis, (3) align preferably along the c axis, which is the stacking direction, while preventing its alignment with the a and b channels, and (4) contain a particular geometric shape with a chiral feature that prompts folding in a helicoidal fashion (in one direction and preventing the opposite) so that it adsorbs on the growing surface with such a chiral folding which favors the stacking of the subsequent sheet always in one direction while avoiding the opposite. We are currently working on rationally designing this type of chiral SDA cations.

4.2 The ITV Case: Searching for Chiral Structure-Directing Agents

A new chiral germanosilicate zeolite framework, ITQ-37 (ITV), has been recently reported by Corma and coworkers [29], which was in turn the first zeolite with a channel system in the range of mesopores. ITQ-37 crystallizes in the chiral space group *P4₁32* (or its enantiomorph *P4₃32*), and is built upon a single tertiary building unit $(T_{44}O_{145}(OH)_7)$ comprising one D4R with one OH group, three laumontite cages, and three D4R with two OH groups. The 3D arrangement of these units creates a channel system with a single gyroidal surface, with these large cavities connected to others through 30-rings, giving place to a chiral framework. The organic SDA originally used for the synthesis of ITQ-37 is achiral (even despite having four stereogenic centers, but the symmetry plane in the center of the molecule makes it an achiral *meso* compound, cation *A* in Fig. 12-right), and consequently the polycrystalline ITQ-37 product is expected to be a racemic mixture of the two enantiomorphs of the chiral structure (*P4₁32* or *P4₃32*).

In a systematic study about the production of extra-large-pore zeolites with increasingly larger SDA molecules, Yu and coworkers were able to produce also ITQ-37 but using a simpler ammonium achiral cation (*B* in Fig. 12) [47, 102]. On

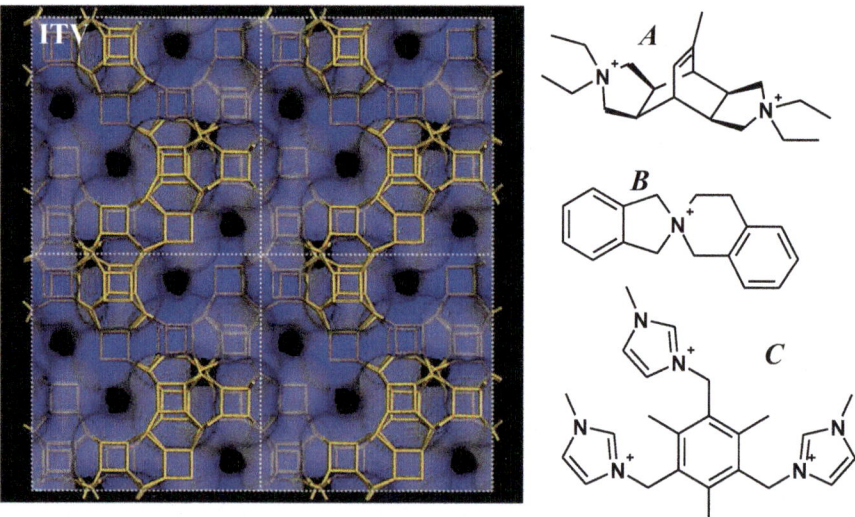

Fig. 12 The ITQ-37 framework structure (ITV) (*left*), and the organic cations used as SDAs that have directed its crystallization so far (*right*)

the other hand, Chen and coworkers have recently reported a new SDA trication based on imidazolium rings for the synthesis of the same structure (cation *C* in Fig. 12) [103]. Interestingly, the authors found by fluorescence spectroscopy that these organic species arranged as supramolecular dimers within the ITV framework forming a sandwich-type structure with the phenyl rings stacked through π-π type interactions. However, in both cases again the achiral nature of the cations used as SDAs should lead to a mixture of the ITV enantiomorphs. Therefore, the challenge here is to find a large-enough enantiopure chiral SDA able to direct the formation of this extra-large pore zeolite and stabilize its intrinsic low framework stability (due to the very low framework density) while being able to develop strong interactions with the framework so that it can imprint its asymmetric nature on the zeolite structure. In this context, the use of supramolecular chemistry, not only through hydrophobic π-π type interactions like cation *C* (Fig. 12) but also through H-bonds, which should extend the molecular chirality into a supramolecular level, could in principle facilitate the achievement of this goal.

4.3 The STW Case: Breaking the Racemic Conglomerate

In 2008 the group of Zou reported the synthesis of a new chiral zeolite based on a germanosilicate network with abundance of D4R units, SU-32 (STW structure type) [30]. The structure of this framework was fascinating as it comprised a 10-ring helicoidal channel along the *c* axis (Fig. 13; the helical channel is highlighted by dashed yellow lines). STW contains a large number of D4R units, which are stabilized by the presence of Ge and fluoride, both species known to stabilize this secondary building unit. The main structural feature of this chiral framework is that it does not crystallize as intergrown polymorphs as zeolite beta, but in contrast each single crystal grows as one unique enantiomorphic form (in $P6_122$ or $P6_322$ space groups). Despite the chirality of this framework, the organic molecule used originally as SDA (diisopropylamine, *A* in Fig. 13) was nonchiral, and consequently SU-32 was a racemic conglomerate, with each crystal being chiral itself but the overall polycrystalline powder being a 50:50 mixture of the two enantiomorphic crystals. Later on, Zhang and coworkers managed to produce STW as pure germanosilicate (and also in the presence of Co and Cu), but using *N,N*-diethylethylendiamine as SDA (*B* in Fig. 13), again an achiral species [104].

Rojas and Camblor have recently managed to produce the Ge-free all-silica version of the STW framework type by using an effective imidazolium-based SDA cation (2-ethyl-1,3,4-trimethyl-imidazolium, cation *C* in Fig. 13) which provided a great stability to this framework that compensated the absence of Ge to stabilize the D4R units (stabilized instead by the presence of fluoride) [31]. Interestingly, the authors were able to unravel the location of the organic cations. Cation *C* has one ethyl group which can rotate (around the CH_3–CH_2 bond axis), giving different conformers. The authors proposed that the conformers which are hosted in the STW

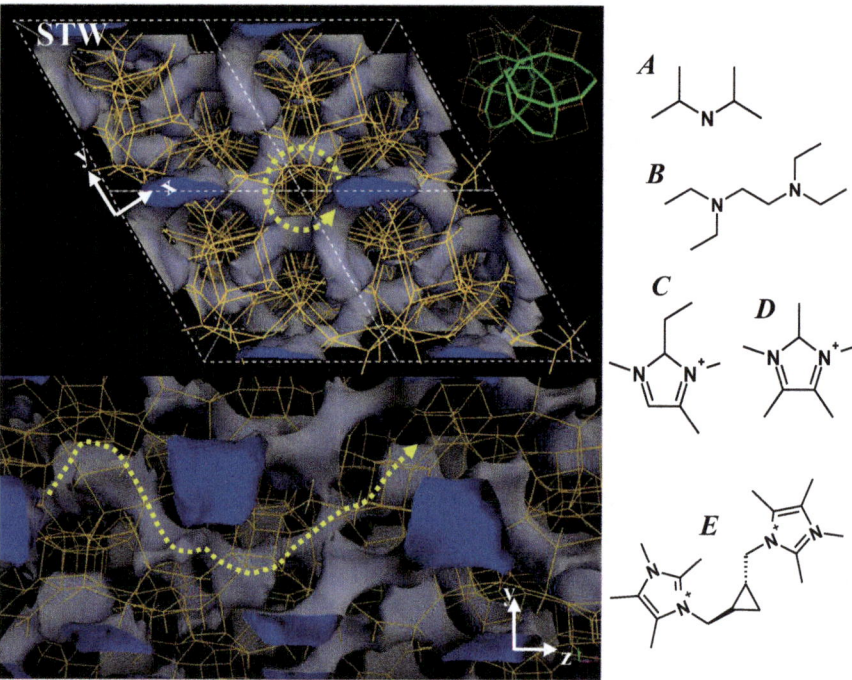

Fig. 13 The STW framework structure (*left*), with details of the helical channel (highlighted by *yellow dashed lines*; a detail of the channel is highlighted by *green* sticks at top-right), and the organic species used as SDAs that have directed its crystallization so far (*right*)

cavities are indeed chiral as long as the confinement within the cavities prevents a free-rotation of the ethyl substituent so as to lock the conformations; they could therefore be considered as pseudoatropoisomers [32]. However, an interconversion of the two mirror-image conformers of cation C is expected in the absence of such confinement, i.e. in the synthesis gel, so thermodynamics (both mirror-image conformers will have the same stability in solution) will ensure the presence of the two conformers in an equal proportion and hence the crystallization of each diastereomeric pair (one conformer will drive the $P6_122$ enantiomorph and the other mirror-image conformer will drive the $P6_322$ enantiomorph). Hence, at the end the overall HPM-1 sample is again a racemic conglomerate (with each crystal being enantiopure), as the authors evidenced by Mueller matrix microscopy [32].

Following the successful use of substituted imidazolium-based organic cations for the synthesis of the STW structure, Schmidt et al. designed computationally new imidazolium-based organic cations that optimized the interaction with the STW framework [105]. Indeed, their computational protocol predicted the pentamethyl-imidazolium derivative (cation D in Fig. 13) as the most efficient SDA for this framework, as was indeed confirmed experimentally.

Based on the strong structure-directing ability of such pentamethyl-imidazolium units for the cavities of the STW framework, Davis et al. have very recently designed a new chiral bis-imidazolium SDA dication in an attempt to enrich the chiral STW framework in one of the enantiomorphs [52]. Chirality in the STW framework is imposed by how two consecutives cavities connect, rotated by + or −60°, defining the two enantiomorphic crystals; therefore, in order to impose a particular STW handedness through the use of a chiral SDA, this has to interact with two consecutive cavities. Following this reasoning, the authors designed again with the aid of molecular simulations an organic SDA with two pentamethyl-imidazolium rings linked by an asymmetric rigid chain based on a *trans*-disubsti-tuted cyclopropane unit (cation *E* in Fig. 13). Simulations showed that each enantiomer should fit much more favorably within just one of the enantiomers of the STW framework. Interestingly, the use of this chiral SDA did enable an enrichment of the STW structure in one of the enantiomers. The success in enantioselectively directing the crystallization of one enantiomorph of STW is linked to the particular molecular structure of cation *E*, with two imidazolium rings that will site in two consecutive STW cavities along the helical channel, with the handedness of the channel being imposed by the chirality of the rigid *trans*-cyclopropane linker which forces the dication to fold in one direction, following one particular handedness of the helicoidal pattern in the STW framework. More-over, the authors were also able to produce the mirror-image of the chiral SDA, which as expected led to an enrichment of the other enantiomorph of the STW framework. Remarkably, the use of these enantiomerically enriched STW zeolites in asymmetric catalysis did lead to certain enantiomeric excesses (up to 10%) in the asymmetric ring-opening of large epoxides (in particular 1,2-epoxyoctane) as well as to an enantioselective adsorption of small chiral alcohols (2-butanol). In order to contrast the validity of the statement that the chirality of the STW framework did promote the enantioselective behavior, the authors used also the mirror-image enantiomerically enriched STW material which led to mirror-image catalytic and adsorption behaviors, as expected. Therefore, this very recent fascinating work represents the first clear example of a production of an enantiomerically enriched chiral zeolite framework through the use of rationally designed chiral organic SDAs, which indeed behaves enantioselectively in asymmetric catalytic and adsorption processes. This successful case, after a so long quest for homochiral zeolite frameworks, will surely prompt an intensive effort to search for new enantiomerically enriched chiral zeolite frameworks, possibly with larger pores, to enable the processing of larger chiral molecules of interest, in particular in the pharmaceutical industry.

5 Chiral Spatial Distributions of Dopants: A New Concept of Chirality in Zeolites

Except for the very recent case just reported about the enantiomerically enriched STW zeolite [52], there has been very limited success in transferring the chirality from enantiopure organic species into homochiral open-framework zeolite structures. In this context, it is interesting to mention the work of Castillo and coworkers [106] which used molecular simulations in an attempt to understand the factors governing the transfer of chirality from chiral zeolite frameworks onto chiral sorbates (similar conclusions could be transferred in the opposite direction, applied to the imprinting of chirality from the organic SDAs to the nascent zeolite framework). In their work, the authors studied the performance of SOF, STW, and ITN frameworks to discriminate between enantiomers of CHBrClF and 4-ethyl-4-methyloctane. The authors concluded that a proper match between the pore geometry and the packing of the chiral sorbate molecules within the channel system is required for enantioselectivity to occur: a tight fitting of the chiral sorbate in the zeolite channels will favor enantiodifferentiation. Indeed, they proposed that the main reason for the observation of enantiodiscrimination is the different long-range asymmetric packing interactions between the enantiomers of the chiral sorbate. Hence, when transferred to the imprinting of the asymmetric nature of chiral organic SDAs onto zeolite frameworks, a proper match of the molecular shape and that of the zeolite framework is as important as the occurrence of lateral asymmetric long-range packing interactions of the SDA species embedded in the zeolite channels.

On the other hand, another possibility for the lack of transfer of chirality is related to the high mobility that is usually associated with the SDAs (which are the chiral carriers) confined within the microporous systems. This is caused by the typical lack of a tight fitting of the SDAs and the surrounding pores/cavities [17]. Such high mobility involves that the SDAs can freely rotate during the crystallization process, what would dissipate their asymmetric feature. The failure in transferring the chirality is also explained by the typically weak and non-directional interactions established between the terminal alkyl groups of the organic SDA species and the oxide zeolite networks. Section 3 has clearly evidenced that the transfer of chirality is notably enhanced with the use of chiral coordination complexes based on polyamine ligands, which have abundant primary amino-groups which develop strong H-bond interactions with the O atoms of the framework, and especially with the negatively charged O atoms (in terminating PO^- groups). However, the main drawback for the use of these chiral complexes as SDAs is that so far they have led to frameworks which are not stable upon removal of the guest species, possibly a consequence of those strong H-bond interactions that stabilize the systems.

Another likely cause for the failed transfer of chirality might come from the different dimension of chirality between the host and guest species, in which chirality from the organic species is typically at an atomic (or molecular) level, thus generating an asymmetric environment at a local (short-ranged) level, compared to the chirality of zeolite structures which is associated with a long-range chiral ordering (commonly in the form of helical channels) of the achiral TO_4 units. In this line, if we want to optimize the transfer of chirality from the organic molecules, we need to expand their chirality into a larger level so that the dimensions of both chiral features are adapted to each other. In fact, this was the main molecular feature of cation E (Fig. 13) that enabled the transfer of chirality onto the STW framework: the use of large dications with a chiral spacer in the middle which spanned two consecutive cavities of the chiral framework, with the asymmetric nature of the spacer favoring only one of the enantiomorphic frameworks.

Based on all these considerations, we have proposed in our group a new strategy to induce chirality in zeolite frameworks. In order to transfer their chirality onto zeolite frameworks, chiral SDAs should (1) expand their asymmetric nature into the long-range, possibly through supramolecular asymmetric packing interactions, in order to adapt the host/guest chiral dimensions, (2) their motion should be restricted in order to enhance the manifestation of the asymmetric environment, and (3) they should establish strong H-bond interactions with the zeolite framework.

In an attempt to expand their asymmetric nature into a long-range level, it becomes crucial to change from the traditional concept of isolated organic SDA species, whose structure-directing effect takes place through single molecular units, thus generating chiral microenvironments at a local level, into self-assembling entities which can develop supramolecular chiral entities stabilized by intermolecular interactions, thus extending the chiral feature into the supramolecular level. An easy way of achieving this self-assembly in aqueous systems is by using chiral molecules with aromatic rings, since these tend to form supramolecular aggregates in water stabilized by π-π interactions between the aromatic rings (Fig. 14-top), as explained in detail in [107]. The formation of these supramolecular aggregates involves a first extension of the chiral environment to the supramolecular level. In turn, the formation of these dimers will notably reduce the motion of the SDAs when confined within zeolite pore systems. Moreover, in the case of being confined in zeolite systems based on channels, the molecular chirality also involves that the lateral or packing interactions between consecutive dimers will also be asymmetric, with one particular fitting (in turn a particular rotation between consecutive dimers around the channel axis) being favored, thus developing a supramolecular helicoidal arrangement of the dimers (Fig. 14-top), with one rotation being favored with respect to the opposite (mirror-image) one, and hence developing homochiral supramolecular entities with a long-range chiral dimension that could be potentially transferred into the zeolite framework.

Apart from this strategy, in order to maximize our chances of transferring chirality, we still need to enhance the interactions between the zeolite frameworks and a localized area of the organic SDAs, preferably associated with the stereogenic center so that the chiral feature is more strongly manifested. As previously mentioned, the most successful chiral SDAs used so far are chiral coordination complexes which imprint their chirality by the establishment of extensive H-bond interactions through primary amine groups. Therefore, it would be convenient to use amines which can develop H-bond interactions (rather than quaternary

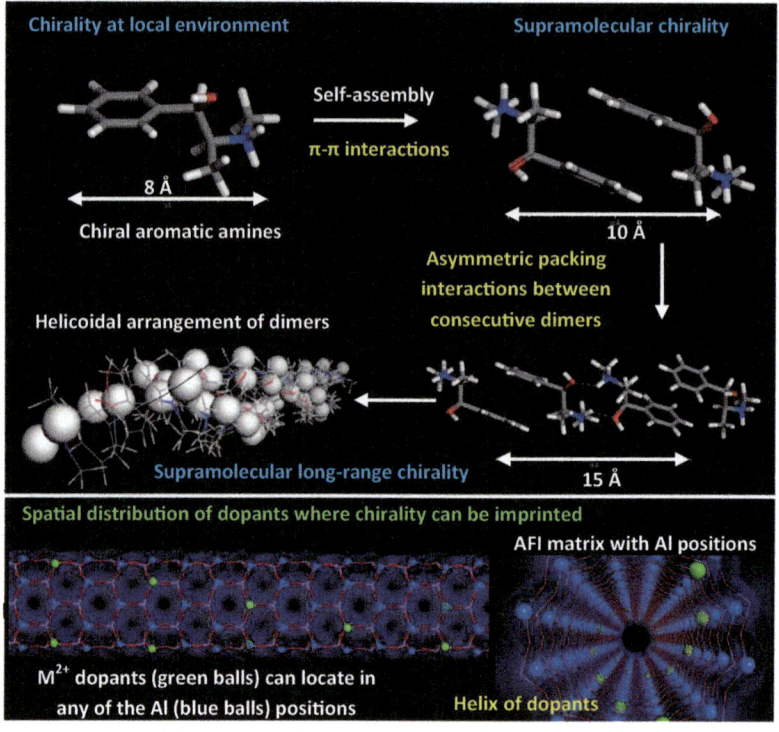

Fig. 14 Schematic illustration of the concepts of asymmetric self-assembly of chiral aromatic amines in order to extend the chiral environment into the chiral dimension typical of zeolite structures (*top*), and AFI matrix with the different positions where dopants can be incorporated (*blue* balls, Al atoms), and a possible helicoidal distribution of dopants (*green* balls) (*bottom*)

ammonium cations). In this context, amines are better SDAs for the synthesis of AlPO-based (rather than silica-based) zeolitic frameworks due to the neutral-to-acidic pHs at which their synthesis is performed, which enables a protonation of the amines and hence a development of strong interactions with the network. On the other hand, the strongest and more localized interactions during a zeolite crystallization process are established between the positive charge associated with the protonated ammonium groups of the organic molecules and the negative charge associated with defects in the zeolite network, commonly in the form of low-valence isomorphic substitutions of Si^{4+} by Al^{3+} in zeolites, or Al^{3+} by divalent metals like Mg^{2+}, Zn^{2+} or Co^{2+} in AlPO networks. In this context, the dopants in the zeolitic networks can adopt a large number of different spatial configurations depending on the available T positions where the dopant is to be incorporated (Fig. 14-bottom, blue balls). This spatial distribution of dopants arises a new degree of freedom and consequently provides a matrix where a chiral feature could be potentially imprinted, in the form of asymmetric distributions of dopants embedded in otherwise achiral frameworks (Fig. 14-bottom, green balls), which represents a new concept of chirality in zeolite materials that we have recently proposed [108–110]. In fact, the strong interactions established between the positive charge of protonated ammonium groups containing H-bond donor groups and the negative charge associated with the incorporation of dopants make possible that these organic species can direct the spatial incorporation of dopants, as we have demonstrated in other FER systems [111, 112]. Therefore, it would in principle be feasible that if we manage to promote such supramolecular helicoidal arrangements of the organic aromatic SDA molecules through asymmetric packing interactions, and these aromatic amines (or rather protonated ammonium cations) develop strong electrostatic (and H-bond) interactions with the dopants, such helicoidal organic arrangement could be imprinted on the zeolite network through a helicoidal spatial distribution of dopants, hence inducing chirality in the zeolite framework. Interestingly, Song and coworkers have discovered a new zeolite framework (JRY) where a helix of cobalt atoms occurs within the framework, leading to a chiral material [28]. In fact, despite using achiral ethylamine as SDA, circular dichroism experiments did surprisingly reveal a certain enantiomeric excess in the solid material, which was ascribed to a symmetry-breaking event amplified by secondary nucleation.

Following this reasoning, we designed a chiral (enantiopure) aromatic amine as SDA derived from L-proline, (S)-$(-)$-N-benzyl-prolinol (BPM) [59, 106] (molecule Q in Fig. 3), which directed the crystallization of the AFI AlPO-based framework in the presence of different dopants. Interestingly, molecular mechanics simulations showed that the most stable packing of dimers of BPM along the AFI channels involved an asymmetric supramolecular arrangement where consecutive dimers were always rotated by an angle of $-90°$ (this represented the most stable rotation angle, see Fig. 15-top-left, blue line), thus leading to a helicoidal supramolecular arrangement of the organic SDA species, which was our initial target (Fig. 15-top-right) [106]. Of course the occurrence of just one particular rotation angle being more stable than the rest (and especially than the mirror-image rotation angle in the opposite direction) was a direct consequence of the introduction of the hydroxymethyl substituent in the pyrrolidine ring that made the molecule chiral. This was evidenced by the fact that rotations between consecutive dimers in one or the opposite direction for the achiral equivalent molecule, benzylpyrrolidine, were equally stable (see Fig. 15-top-left, red line). However, fluorescence spectroscopy results showed that the actual incorporation of these BPM molecules confined within the AFI 1-dimensional channels involved both monomer and dimer species, with relative amounts of each depending on the type of dopant and the protonation state of the SDA: protonated dimers seem to cause an electrostatic repulsion between consecutive dimers that disfavored the occlusion of dimers [114, 115]. The non-exclusive incorporation of dimers would of course disrupt the supramolecular helicoidal arrangement of the dimers. Moreover, we observed that the interaction between the ammonium groups of BPM dimers and the divalent dopants located in the different positions (replacing Al in the channel walls) was weak since these ammonium groups sited close to the center of the channels and hence away from the channel walls (with distances larger than 5 Å), and hence they could not interact with the dopants. Therefore, it did not seem likely that these BPM

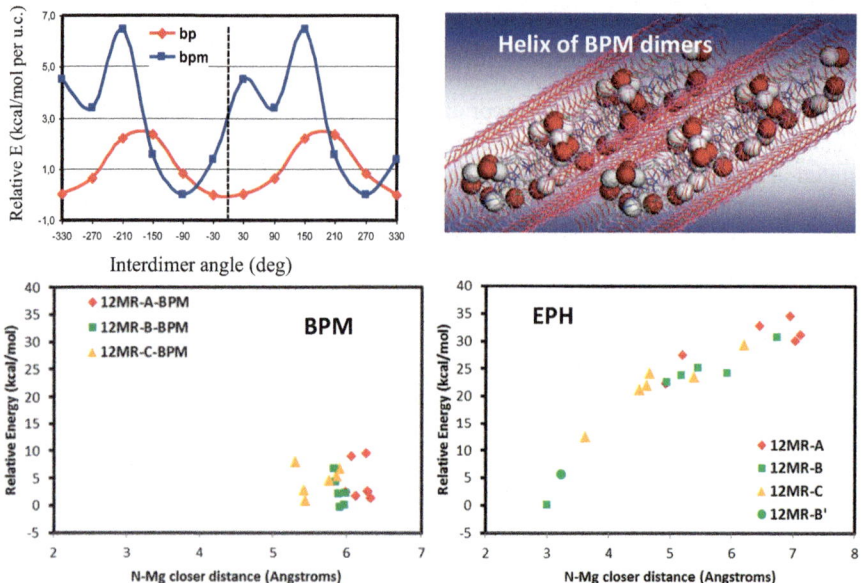

Fig. 15 Relative energy diagram for the rotation angle between consecutive dimers for (S)-N-benzyl-prolinol (BPM, *blue*) and benzylpyrrolidine (BP, *red*) (*dashed line* indicates 0° rotation) (*top-left*), and helicoidal arrangement of BPM (*top-right*). *Bottom*: stability for different replacement positions of Al by Mg as a function of the Mg···N interatomic distance for BPM (*left*) or ephedrine (EPH, *right*) (different colors indicate location of Mg in consecutive 12-rings along the AFI channels). Reproduced (adapted) from [108] and [113] by permission of the PCCP Owner Societies

dimers would direct the spatial incorporation of dopants, as evidenced by the fact that similar stabilities were found for AFI/BPM systems with Mg sited in different Al positions (see Fig. 15-bottom-left). As a consequence, a transfer of chirality from the supramolecular dimer arrangement into a helicoidal distribution of dopants did not seem likely to occur for this particular chiral SDA molecule.

We then selected another molecule, (1R,2S)-ephedrine (EPH, molecule *R* in Fig. 3), that satisfied the same requirements previously commented, while being potentially able to interact more strongly with the framework walls through H-bond interactions. The selection of this chiral precursor was based on the simultaneous occurrence of (1) an aromatic ring which promotes the formation of supramolecular dimers through π-π type interactions, (2) two stereogenic centers that impart a strong asymmetric molecular structure, (3) H-bond donor and acceptor groups that trigger an asymmetric packing (lateral) interaction between consecutive dimers, potentially leading to long-range chiral orderings, and (4) a secondary amine which will be protonated in acidic medium to form R-NH_2^+-R' positively charged groups that will develop strong electrostatic interactions with the negative charge associated with divalent dopants. As previously mentioned, the use of this molecule as SDA for the synthesis of nanoporous aluminophosphates led to the crystallization of the AFI framework [61, 62]. Molecular mechanics simulations then showed the occurrence of one particular rotation between consecutive dimers which was more stable than the others (+30°, see Fig. 16-top-left), and hence consecutive dimers will tend to be rotated by this angle, developing again a helicoidal supramolecular arrangement of EPH dimers (Fig. 16-top-right) [113]. This particular rotation represented a more stable packing configuration since it enabled the development of a stronger double H-bond interaction between the NH_2^+ and O(H) groups of consecutive dimers (Fig. 16-bottom-left). Interestingly, in this case the NH_2^+ did locate close to the channel walls and did interact strongly with them. In fact, there was a clear relationship between the stability of the different locations of the divalent dopants with their distance to the ammonium group (see Fig. 15-bottom-right): the dopants prefer to locate close to the ammonium positive charge. This is an indication that the ammonium groups will direct the spatial distribution of dopants in the AFI channels, and hence suggests that the helicoidal supramolecular arrangement of the EPH dimers could potentially be transferred into a helicoidal spatial distribution of dopants (Fig. 16-bottom-right).

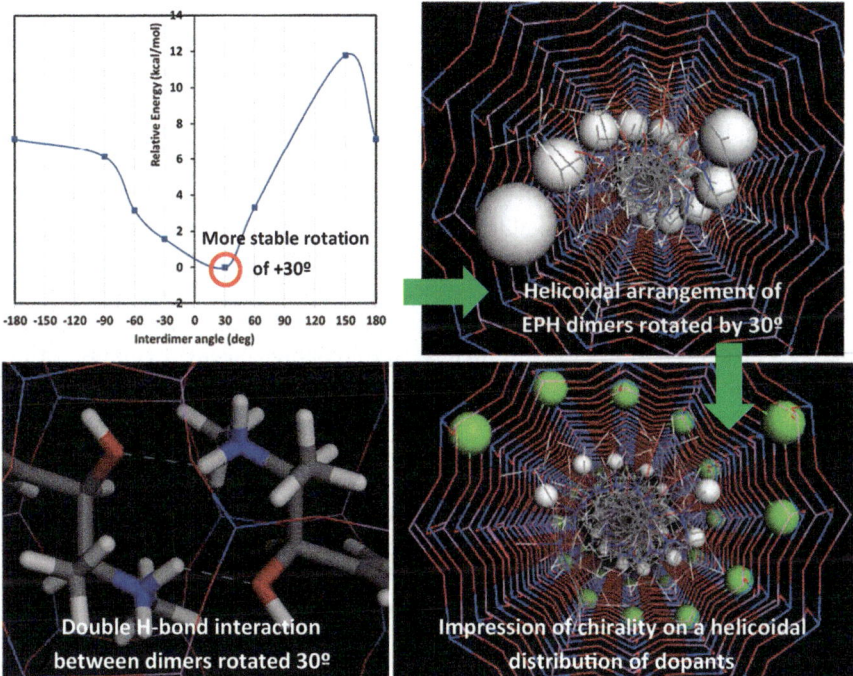

Fig. 16 Relative energy diagram for the rotation angle between consecutive dimers for (1R,2S)-ephedrine (*top-left*), with the rotation of 30° being more stable because of a stronger double H-bond interaction between dimers (*bottom-left*), helicoidal arrangement of EPH dimers rotated by +30° (*top-right*), and impression on a chiral distribution of dopants (*bottom-right*). Reproduced (adapted) from [113] by permission of the PCCP Owner Societies

Nonetheless, although we have clearly demonstrated the exclusive incorporation of ephedrine dimers within the AFI framework (depending on the synthesis conditions) [68, 69], so far we have not been able to experimentally demonstrate the occurrence of the chiral distribution of dopants. Confirmation of such special configuration for the dopants is extremely difficult (if not impossible) by diffraction techniques (either electron or X-ray diffraction methods) due to the similar scattering power of Mg dopants and Al and by the subtle changes that would be induced in the diffraction pattern as a consequence of these particular dopant distributions. Therefore, in an attempt to demonstrate the occurrence of this concept of chirality and its potential use in enantioselective operations, we are currently working on the adsorption of different chiral probe molecules within these materials.

6 Summary

In this chapter we have analyzed in detail the potential of transferring chirality into zeolite frameworks through the use of chiral structure-directing agents. First attempts by using chiral organic species led invariably to a failure in imprinting their chirality onto zeolite networks, possibly because of a lack of strong and specific host–guest interactions, high mobility of the chiral precursors within the channels and/or cavity systems, and the different dimension of the chiral feature in the organic unit (at a local environment) and in the zeolite framework (at a long-range level).

The use of chiral organometallic complexes has been more successful in transferring their chirality onto inorganic frameworks through the establishment of extensive H-bond interactions; however they usually promoted the crystallization of low-dimensional frameworks, or 3D zeolitic open-frameworks but where the guest species are removed with difficulty and probably causing a collapse of the structure. Nevertheless, these works clearly demonstrate the fundamental role of H-bond interactions when imprinting chirality on an inorganic lattice.

There has been a very recent example where a computationally guided rational design of a chiral SDA enabled the production of an enantiomerically enriched chiral zeolite framework, which represents a fundamental milestone in the quest for a homochiral zeolite structure, a highly desired yet elusive target in zeolite science. This work not only demonstrates the feasibility of producing homochiral zeolite materials where chirality is imprinted through the use of rationally selected chiral SDAs, where the aid of molecular simulation techniques has proven fundamental, but also confirms that this enantioenriched chiral zeolite materials are able to perform enantioselective operations during both asymmetric catalysis and adsorption (although so far the enantiomeric excesses found are not very high). Hence, this fundamental work will probably prompt a more intensive search for other chiral zeolite frameworks in order to produce active, robust, and efficient enantioselective catalysts, which will potentially be of enormous relevance in the chemical industry, especially in the pharmaceutical sector.

Acknowledgements Funding from the Spanish Ministry of Science and Innovation (MICINN) through projects MAT2012-31127 and MAT2015-65767-P is acknowledged. BBM acknowledges the Spanish Ministry of Economy and Competitivity for a predoctoral (BES-2013-064605) contract.

References

1. Lough WJ, Wainer IW (eds) (2002) Chirality in natural and applied science. CRC Press. ISBN 0-632-05435-2
2. Chan RS, Ingold CK, Prelog V (1966) Specification of molecular chirality. Angew Chem Int Ed 5(4):385–415
3. Martin RH (1974) The helicenes. Angew Chem Int Ed Engl 13(10):649–660
4. Gardner M (1964) The ambidextrous universe. Symmetry and asymmetry from mirror reflections to superstrings, 3rd revised edn. Penguin Books
5. Guijarro A, Yus M (2009) The origin of chirality in the molecules of life: a revision from awareness to the current theories and perspectives for this unsolved problem. RSC Publishing. ISBN 978-0-85404-156-5
6. Mason SF (1984) Origins of biomolecular handedness. Nature 311:19–23
7. Patil PN (2002) Chirality in medicinal chemistry. In: Lough WJ, Wainer IW (eds) Chapter 6 in chirality in natural and applied science. CRC Press, pp 139–178. ISBN 0-632-05435-2
8. Rouhi M (2005) Top pharmaceuticals: thalidomide. Chem Eng News 83:122–123
9. Thayer AM (2007) Centering on chirality. Chem Eng News 85:11–19
10. Haris KDM, Thomas SJM (2009) Selected thoughts on chiral crystals, chiral surfaces, and asymmetric heterogeneous catalysis. ChemCatChem 1:223–231
11. Hazen RM, Sholl DS (2003) Chiral selection on inorganic crystalline surfaces. Nat Mater 2:367–374
12. Kavasmaneck PR, Bonner WA (1977) Adsorption of amino acid derivatives by d- and l-quartz. J Am Chem Soc 99:44–50
13. Bonner WA (1991) The origin and amplification of biomolecular chirality. Orig Life Evol Biosph 21:59–111
14. Hazen RM, Filley TR, Goodfriend GA (2001) Selective adsorption of L- and D-amino acids on calcite: implication for biochemical homochirality. Proc Natl Acad Sci U S A 98 (10):5487–5490
15. Moliner M, Rey F, Corma A (2013) Towards the rational design of efficient organic structure-directing agents for zeolite synthesis. Angew Chem Int Ed 52:13880–13889
16. Davis ME (2014) Zeolites from a materials chemistry perspective. Chem Mater 26:239–245
17. Davis ME (2003) Reflections on routes to enantioselective solid catalysts. Top Catal 25:3–7
18. Yu J, Xu R (2008) Chiral zeolitic materials: structural insights and synthetic challenges. J Mater Chem 18:4021–4030
19. Dubbeldam D, Calero S, Vlugt TJH (2014) Exploring new methods and materials for enantioselective separations and catalysis. Mol Simul 40:585–598
20. Mc Morn P, Hutchings GJ (2004) Heterogeneous enantioselective catalysts: strategies for the immobilisation of homogeneous catalysts. Chem Soc Rev 33:108–122
21. Davis ME (1998) Zeolite-based catalysts for chemicals synthesis. Microporous Mesoporous Mater 21:173–182
22. Treacy MMJ, Newsam JM (1988) 2 New 3-dimensional 12-ring zeolite frameworks of which zeolite beta is a disordered intergrowth. Nature 332:249–251
23. Rajic N, Logar NZ, Kaucic V (1995) A novel open framework zincophosphate – synthesis and characterization. Zeolites 15:672–678

24. Cheetham AK, Fjellvag H, Gier TE et al (2001) Very open microporous materials: from concept to reality. Stud Surf Sci Catal 135:158–158
25. Bu X, Feng P, Gier TE et al (1998) Hydrothermal synthesis and structural characterization of zeolite-like structures based on gallium and aluminum germanates. J Am Chem Soc 120:13389–13397
26. Rouse RC, Peacor DR (1986) Crystal structure of the zeolite mineral goosecreekite, $CaAl_2Si_6O_{16}\cdot5H_2O$. Am Mineral 71:1494–1501
27. Broach RW, Kirchner RM (2011) Structures of the K^+ and NH_4^+ forms of Linde J. Microporous Mesoporous Mat 143:398–400
28. Song XW, Li Y, Gan L et al (2009) Heteroatom-stabilized chiral framework of aluminophosphate molecular sieves. Angew Chem Int Ed 48:314–317
29. Sun J, Bonneau C, Cantín A et al (2009) The ITQ-37 mesoporous chiral zeolite. Nature 458:1154–1157
30. Tang LQ, Shi L, Bonneau C et al (2008) A zeolite family with chiral and achiral structures built from the same building layer. Nat Mater 7:381–385
31. Rojas A, Camblor MA (2012) A pure silica chiral polymorph with helical pores. Angew Chem Int Ed 51:3854–3856
32. Rojas A, Arteaga O, Kahr B et al (2013) Synthesis, structure and optical activity of HPM-1, a pure silica chiral zeolite. J Am Chem Soc 135:11975–11984
33. Liu X, Xing Y, Wang X et al (2010) Chirality and magnetism of an open-framework cobalt phosphite containing helical channels from achiral materials. Chem Commun 46:2614–2616
34. Dryzun C, Mastai Y, Shvalb A et al (2009) Chiral silicate zeolites. J Mater Chem 19:2062–2069
35. Zhang J, Chen SM, XH B (2009) Nucleotide-catalyzed conversion of racemic zeolite-type zincophosphate into enantioenriched crystals. Angew Chem Int Ed 49:6049–6051
36. Issue S (2014) Metal-organic frameworks. Chem Soc Rev 43:5403–6176
37. Ma L, Abney C, Lin W (2009) Enantioselective catalysis with homochiral metal-organic frameworks. Chem Soc Rev 38:1248–1256
38. Liu Y, Xuan W, Cui Y (2010) Engineering homochiral metal-organic frameworks for heterogeneous asymmetric catalysis and enantioselective separation. Adv Mater 22:4112–4135
39. Morris RE, XH B (2010) Induction of chiral porous solids containing only achiral building blocks. Nat Chem 2:353–361
40. Lobo RF, Davis ME (1994) Synthesis and characterization of pure-silica and boron-substituted SSZ-24 using N(16) methylsparteinium bromide as structure-directing agent. Microporous Mater 3:61–69
41. Yoshikawa M, Wagner P, Lovallo M et al (1998) Synthesis, characterization, and structure solution of CIT-5, a new, high-silica, extra-large-pore molecular sieve. J Phys Chem B 102:7139–7147
42. Corma A, Díaz-Cabañas MJ, Martínez-Triguero J et al (2002) A large-cavity zeolite with wide pore windows and potential as an oil refining catalyst. Nature 418:514–517
43. Tsuji K, Wagner P, Davis ME (1999) High-silica molecular sieve syntheses using the sparteine related compounds as structure-directing agents. Microporous Mesoporous Mater 28:461–469
44. Kubota Y, Helmkamp MM, Zones SI et al (1996) Properties of organic cations that lead to the structure-direction of high-silica molecular sieves. Microporous Mater 6:213–229
45. Lobo RF, Davis ME (1995) CIT-1: a new molecular sieve with intersecting pores bounded by 10- and 12-rings. J Am Chem Soc 117:3766–3779
46. Xie D, McCusker LB, Baerlocher C et al (2013) SSZ-52, a zeolite with an 18-layer aluminosilicate framework structure related to that of the DeNOx catalyst Cu-SSZ-13. J Am Chem Soc 135:10519–10524
47. Jiang J, Xu Y, Cheng P et al (2011) Investigation of extra-large pore zeolite synthesis by a high-throughput approach. Chem Mater 23:4709–4715

48. García R, Gómez-Hortigüela L, Sánchez F et al (2010) Diastereoselective structure directing effect of (1S,2S)-2-hydroxymethyl-1-benzyl-1-methylpyrrolidinium in the synthesis of ZSM-12. Chem Mater 22:2276–2286

49. García R, Gómez-Hortigüela L, Pérez-Pariente J (2012) Study of the structure directing effect of the chiral cation (1S,2S)-2-hydroxymethyl-1-benzyl-1-methylpyrrolidinium in alumino-silicate preparations in the presence of co-structure directing agents. Catal Today 179:16–26

50. García R, Gómez-Hortigüela L, Sánchez F et al (2011) Structure-direction of chiral 2-hydroxymethyl-1-benzyl-1-methylpyrrolidinium in the cotemplated synthesis of ferrierite: fundaments of diastereo-recognition from non-chiral microporous structures. Microporous Mesoporous Mater 146:57–68

51. Lobo RF, Tsapatsis M, Freyhardt CC et al (1997) A model for the structure of the large-pore zeolite SSZ-31. J Am Chem Soc 119:3732–3744

52. Brand SK, Schmidt JE, Deem MW et al (2017) Enantiomerically enriched, polycrystalline molecular sieves. Proc Natl Acad Sci U S A 114(20):5101–5106

53. Martínez-Franco R, Paris C, Martínez-Triguero J et al (2017) Direct synthesis of the alumi-nosilicate form of the small pore CDO zeolite with novel OSDAs and the expanded poly-morphs. Microporous Mesoporous Mater 246:147–157

54. Golebiewski WM, Spenser ID (1988) Biosynthesis of the lupine alkaloids. II. Sparteine and lupanine. Can J Chem 66(7):1734–1748

55. Wagner P, Yoshikawa M, Lovallo M et al (1997) CIT-5: a high-silica zeolite with 14-ring pores. Chem Commun 2179–2180

56. Gómez-Hortigüela L, Álvaro-Muñoz T. Unpublished results

57. Dong Z, Zhao L, Liang Z et al (2010) [Zn(HPO$_3$)(C$_{11}$N$_2$O$_2$H$_{12}$)] and [Zn$_3$(H$_2$O)(PO$_4$)(HPO$_4$)(C$_6$H$_9$N$_3$O$_2$)$_2$ (C$_6$H$_8$N$_3$O$_2$)]: homochiral zinc phosphite/phosphate networks with biofunctional amino acids. Dalton Trans 39:5439–5445

58. Komura K, Horibe Y, Yajima H et al (2016) Synthesis, crystal structure and characterization of novel open framework CHA-type aluminophosphate involving a chiral diamine. Dalton Trans 45:15193–15202

59. Gómez-Hortigüela L, Pérez-Pariente J, Blasco T (2007) (S)-(−)-N-benzylpyrrolidine-2-methanol: a new and efficient structure directing agent for the synthesis of crystalline microporous aluminophosphates with AFI-type structure. Microporous Mesoporous Mater 100:55–62

60. Pinar AB, Gómez-Hortigüela L, McCusker LB et al (2011) Synthesis of Zn-containing microporous aluminophosphate with the STA-1 structure. Dalton Trans 40:8125–8131

61. Álvaro-Muñoz T, López-Arbeloa F, Pérez-Pariente J et al (2014) (1R,2S)-Ephedrine: a new self-assembling chiral template for the synthesis of aluminophosphate frameworks. J Phys Chem C 118:3069–3077

62. Bernardo-Maestro B, López-Arbeloa F, Pérez-Pariente J et al (2015) Supramolecular chem-istry controlled by conformational space during structure-direction of nanoporous materials: self-assembly of ephedrine and pseudoephedrine. J Phys Chem C 119:28214–28225

63. Bernardo-Maestro B, Vos E, López-Arbeloa F et al (2017) Supramolecular chemistry con-trolled by packing interactions during structure-direction of nanoporous materials: effect of the addition of methyl groups on ephedrine derivatives. Microporous Mesoporous Mater 239:432–443

64. Pinilla-Herrero I, Gómez-Hortigüela L, Márquez-Álvarez C et al (2016) Unexpected crystal growth modifier effect of glucosamine as additive in the synthesis of SAPO-35. Microporous Mesoporous Mater 219:322–326

65. Nenoff TM, Thoma SG, Provencio P et al (1998) Novel zinc phosphate phases formed with chiral d-glucosamine molecules. Chem Mater 10:3077–3080

66. Lin HM, Lii KH (1998) Synthesis and structure of [(1R,2R)-C$_6$H$_{10}$(NH$_3$)$_2$][Ga(OH)(HPO$_4$)$_2$]·H$_2$O, the first metal phosphate containing a chiral amine. Inorg Chem 37:4220–4222

67. Abourashed EA, El-Alfy AT, Khan IA et al (2003) Ephedra in perspective – a current review. Phytother Res 17:703–712
68. Bernardo-Maestro B, Roca-Moreno MD, López-Arbeloa F et al (2016) Supramolecular chemistry of chiral (1R,2S)-ephedrine confined within the AFI framework as a function of the synthesis conditions. Catal Today 277:9–20
69. Bernardo-Maestro B, López-Arbeloa F, Pérez-Pariente J et al (2017) Comparison of the structure-directing effect of ephedrine and pseudoephedrine during crystallization of nanoporous aluminophosphates. Microporous Mesoporous Mater. In Press. https://doi.org/10.1016/j.micromeso.2017.04.008
70. Shi X, Zhu G, Qiu S et al (2004) $Zn_2[(S)-O_3PCH_2NHC_4H_7CO_2]_2$: a homochiral 3D zinc phosphonate with helical channels. Angew Chem Int Ed 43:6482–6485
71. Bernardo-Maestro B, Gómez-Hortigüela L. Unpublished results
72. Morgan K, Gainsford G, Milestone N (1995) A novel layered aluminophosphate [Co(en)$_3$Al$_3$P$_4$O$_{16}$·3H$_2$O] assembled about a chiral metal complex. J Chem Soc Chem Commun 425–426
73. Bruce DA, Wilkinson AP, White MG, Bertrand JA (1995) The synthesis and structure of a chiral layered aluminophosphate containing the template Co(tn)$_3{}^{3+}$. J Chem Soc Chem Commun 2059–2060
74. Bruce DA, Wilkinson AP, White MG et al (1996) The synthesis and characterization of an aluminophosphate with chiral layers; trans-Co(dien)$_2$·Al$_3$P$_4$O$_{16}$·3H$_2$O. J Solid State Chem 125:228–233
75. Gray MJ, Jasper JD, Wilkinson AP et al (1997) Synthesis and synchrotron microcrystal structure of an aluminophosphate with chiral layers containing Λ tris(ethylendiamine)cobalt (III). Chem Mater 9:976–980
76. Yu J, Wang Y, Shi Z et al (2001) Hydrothermal synthesis and characterization of two new zinc phosphates assembled about a chiral metal complex: [Co^{II}(en)$_3$]$_2$[Zn$_6$P$_8$O$_{32}$H$_8$] and [Co^{III}(en)$_3$][Zn$_8$P$_6$O$_{24}$Cl]·2H$_2$O. Chem Mater 13:2972–2978
77. Wang Y, Yu J, Li Y et al (2003) Chirality transfer from guest chiral metal complexes to inorganic framework: the role of hydrogen bonding. Chem A Eur J 9:5048–5055
78. Chen P, Li J, Yu J et al (2005) The synthesis and structure of a chiral 1D aluminophosphate chain compound: d-Co(en)$_3$[AlP$_2$O$_8$]·6.5H$_2$O. J Solid State Chem 178:1929–1934
79. Wang Y, Yu J, Guo M et al (2003) [Zn$_2$(HPO$_4$)$_4$][Co(dien)$_2$]·H$_3$O: a zinc phosphate with multidirectional intersecting helical channels. Angew Chem Int Ed 42:4089–4092
80. Han Y, Li Y, Yu J et al (2011) A gallogermanate zeolite constructed exclusively by three-ring building units. Angew Chem Int Ed 50:3003–3005
81. Xu Y, Li Y, Han Y et al (2013) A gallogermanate zeolite with eleven-membered-ring channels. Angew Chem Int Ed 52:5501–5503
82. Balkus KJ, Hargis CD, Kowalak S (1992) Synthesis of NaX zeolites with metallophthalocyanines. ACS Symp Ser 499:347–354
83. Jasper JD, Wilkinson AP (1998) Synthesis of low-dimensional aluminophosphates from higher dimensional precursors: conversion of Λ,Δ-Co(en)$_3$[Al$_3$P$_4$O$_{16}$]·xH$_2$O to the chain compound Λ,Δ-Co(en)$_3$[AlP$_2$O$_8$]·xH$_2$O. Chem Mater 10:1664–1667
84. Tong M, Zhang D, Zhu L et al (2016) An elaborate structure investigation of the chiral polymorph A-enriched zeolite beta. CrstEngComm 18:1782–1789
85. Stalder SM, Wilkinson AP (1997) Synthesis and characterization of a chiral 3D-framework material: d-Co(en)$_3$[H$_3$Ga$_2$P$_4$O$_{16}$]. Chem Mater 9:2168–2173
86. Wang Y, Yu J, Li Y et al (2003) Synthesis and characterization of a new layered gallium phosphate [Co(en)$_3$][Ga$_3$(H$_2$PO$_4$)$_6$(HPO$_4$)$_3$] templated by cobalt complex. J Solid State Chem 170:176–181
87. Yang G, Sevov SC (2001) [Co(en)$_3$][B$_2$P$_3$O$_{11}$(OH)$_2$]: a novel borophosphate templated by a transition-metal complex. Inorg Chem 40:2214–2215
88. Wadlinger RL, Kerr GT, Rosinski EJ (1967) Catalytic composition of a crystalline zeolite. US Patent 3,308,069

89. Newsam JM, Treacy MMJ, Koetsier WT et al (1988) Structural characterization of zeolite-beta. Proc R Soc London Ser A 420:375–405
90. Higgins JB, LaPierre RB, Schlenker JL et al (1988) The framework topology of zeolite-beta. Zeolites 8:446–452
91. Conradsson T, Dadachov MS, Zou XD (2000) Synthesis and structure of $(Me_3N)_6[Ge_{32}O_{64}]$ $(H_2O)_{4.5}$, a thermally stable novel zeotype with 3D interconnected 12-ring channels. Microporous Mesoporous Mat 41:183–191
92. Davis ME, Lobo RF (1992) Zeolite and molecular sieve synthesis. Chem Mater 4:756–768
93. Clark LA, Chempath S, Snurr RQ (2005) Simulated adsorption properties and synthesis prospects of homochiral porous solids based on their heterochiral analogs. Langmuir 21:2267–2272
94. Manning MP, Warzywoda J, Karahan O et al (2004) Enantioselective adsorption of hydrobenzoin on zeolite beta. Stud Surf Sci Catal 154:1957–1960
95. Takagi Y, Komatsu T, Kitabata Y (2008) Crystallization of zeolite beta in the presence of chiral amine or rhodium complex. Microporous Mesoporous Mater 109:567–576
96. Camblor MA, Corma A, Valencia S (1996) Spontaneous nucleation and growth of pure silica zeolite-beta free of connectivity defects. Chem Commun 20:2365–2366
97. Xia QH, Shen SC, Song J et al (2003) Structure, morphology, and catalytic activity of β zeolite synthesized in a fluoride medium for asymmetric hydrogenation. J Catal 219:74–84
98. Taborda F, Willhammar T, Wang Z et al (2011) Synthesis and characterization of pure silica zeolite beta obtained by an aging-drying method. Microporous Mesoporous Mater 143:196–205
99. Taborda F, Wang Z, Willhammar T et al (2012) Synthesis of Al-Si-beta and Ti-Si-beta by the aging-drying method. Microporous Mesoporous Mater 150:38–46
100. Tong M, Zhang D, Fan W et al (2015) Synthesis of chiral polymorph A-enriched zeolite beta with an extremely concentrated fluoride route. Sci Rep 5:11521
101. Lu T, Xu R, Yan W (2016) Co-templated synthesis of polymorph A-enriched zeolite beta. Microporous Mesoporous Mater 226:19–24
102. Qian K, Li J, Jiang J et al (2012) Synthesis and characterization of chiral zeolite ITQ-37 by using achiral organic structure-directing agent. Microporous Mesoporous Mater 164:88–92
103. Chen FJ, Gao ZH, Liang LL et al (2016) Facile preparation of extra-large pore zeolite ITQ-37 based on supramolecular assemblies as structure-directing agents. CrstEngComm 18:2735–2741
104. Zhang N, Shi L, Yu T et al (2015) Synthesis and characterization of pure STW-zeotype germanosilicate, Cu- and Co-substituted STW-zeotype materials. J Solid State Chem 225:271–277
105. Schmidt JE, Deem MW, Davis ME (2014) Synthesis of a specified, silica molecular sieve using computationally predicted organic structure-directing agents. Angew Chem Int Ed 53:8372–8374
106. Castillo JM, Vlugt TJH, Dubbeldam D et al (2010) Performance of chiral zeolites for enantiomeric separation revealed by molecular simulation. J Phys Chem C 114:22207–22213
107. Paris C, Moliner M (2017) Role of supramolecular chemistry during templating phenomenon in zeolite synthesis. Struct Bond. https://doi.org/10.1007/430_2017_11. (in this volume)
108. Gómez-Hortigüela L, Corà F, Catlow CRA et al (2006) Computational study of a chiral supramolecular arrangement of organic structure directing molecules for the AFI structure. Phys Chem Chem Phys 8:486–493
109. Gómez-Hortigüela L, Corà F, Pérez-Pariente J (2012) Chiral distributions of dopants in microporous materials: a new concept of chirality. Microporous Mesoporous Mater 155:14–15
110. Comyns AE (2011) Chiral zeolites, a novel approach. Microporous Mesoporous Mater 138:243

111. Gómez-Hortigüela L, Pinar AB, Corà F et al (2010) Dopant-siting selectivity in nanoporous catalysts: control of proton accessibility in zeolite catalysts through the rational use of templates. Chem Commun 46:2073–2075
112. Pinar AB, Gómez-Hortigüela L, McCusker LB et al (2013) Controlling the aluminum distribution in the zeolite ferrierite via the organic structure directing agent. Chem Mater 25:3654–3661
113. Gómez-Hortigüela L, Álvaro-Muñoz T, Bernardo-Maestro B et al (2015) Towards chiral distributions of dopants in microporous frameworks: helicoidal supramolecular arrangement of (1R,2S)-ephedrine and transfer of chirality. Phys Chem Chem Phys 17:348–357
114. Gómez-Hortigüela L, Pérez-Pariente J, López-Arbeloa F (2009) Aggregation behavior of (S)-(−)-N-benzylpyrrolidine-2-methanol in the synthesis of the AFI structure in the presence of dopants. Microporous Mesoporous Mater 119:299–305
115. Gómez-Hortigüela L, Lopez Arbeloa F, Márquez-Álvarez C et al (2013) Effect of fluorine and molecular charge-state on the aggregation behavior of (S)-(−)-N-benzylpyrrolidine-2-methanol confined within the AFI nanoporous structure. J Phys Chem C 117:8832–8839

Index

Printed by Printforce, the Netherlands